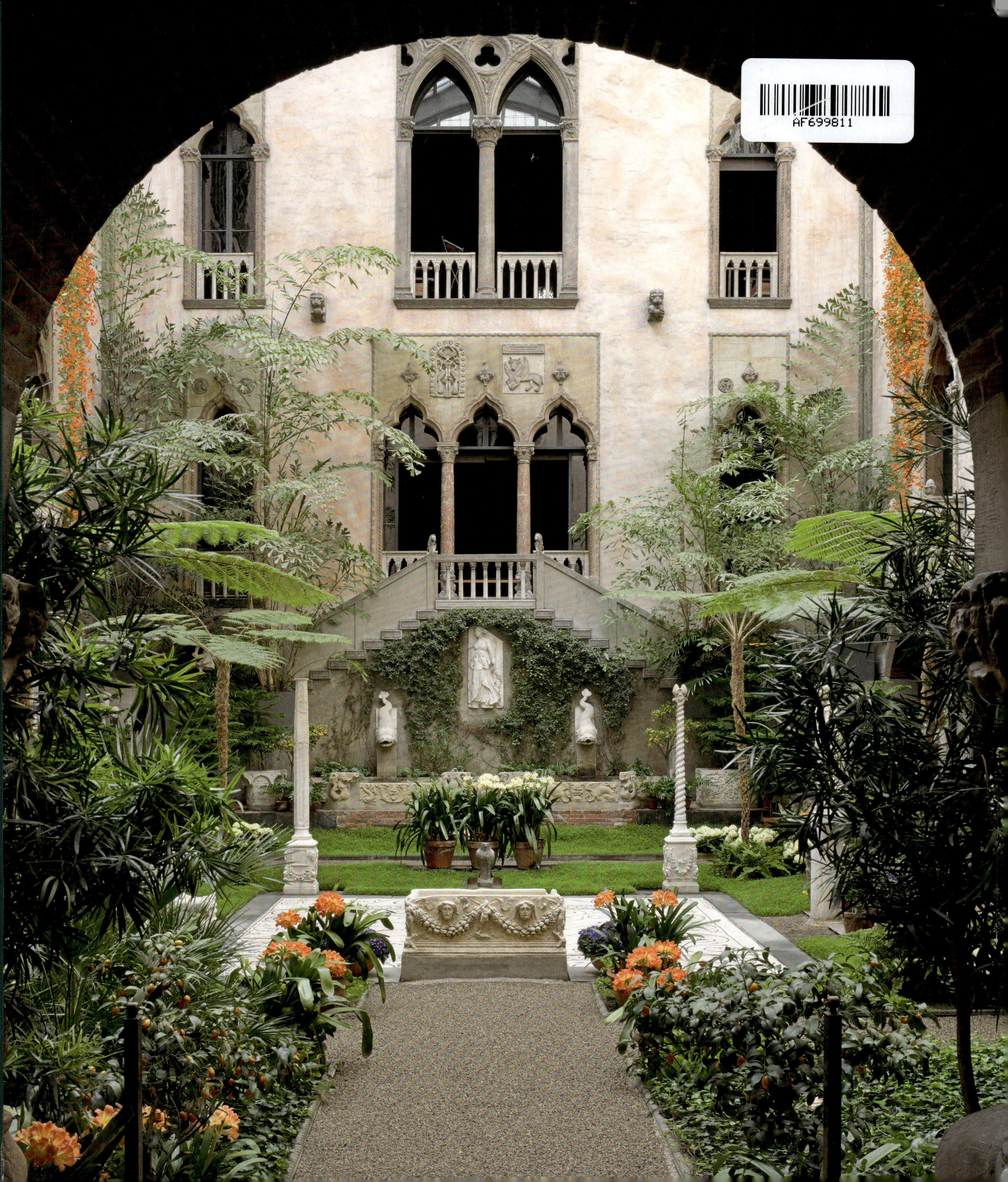

AF699811

Betye Saar: Heart of a Wanderer and *Fellow Wanderer: Isabella's Travel Albums* are supported by the Abrams Foundation, the Ford Foundation, the Wagner Foundation, the Henry Luce Foundation, the Wyeth Foundation for American Art and the Getty Foundation through The Paper Project initiative.

Additional support is provided by an endowment grant from the Mellon Foundation and the National Endowment for the Humanities.

The Museum receives operating support from the Massachusetts Cultural Council, which is supported by the state of Massachusetts and the National Endowment for the Arts.

FELLOW WANDERER

ISABELLA STEWART GARDNER'S TRAVEL ALBUMS

Edited by
Diana Seave Greenwald
Casey Riley

With contributions by Pujan Gandhi, Diana Seave Greenwald, Madeleine Haddon, David Odo, Casey Riley, Jaclyn M. Roessel, and Stephanie Tung

Isabella Stewart Gardner Museum, Boston

Distributed by Princeton University Press, Princeton and Oxford

Contents

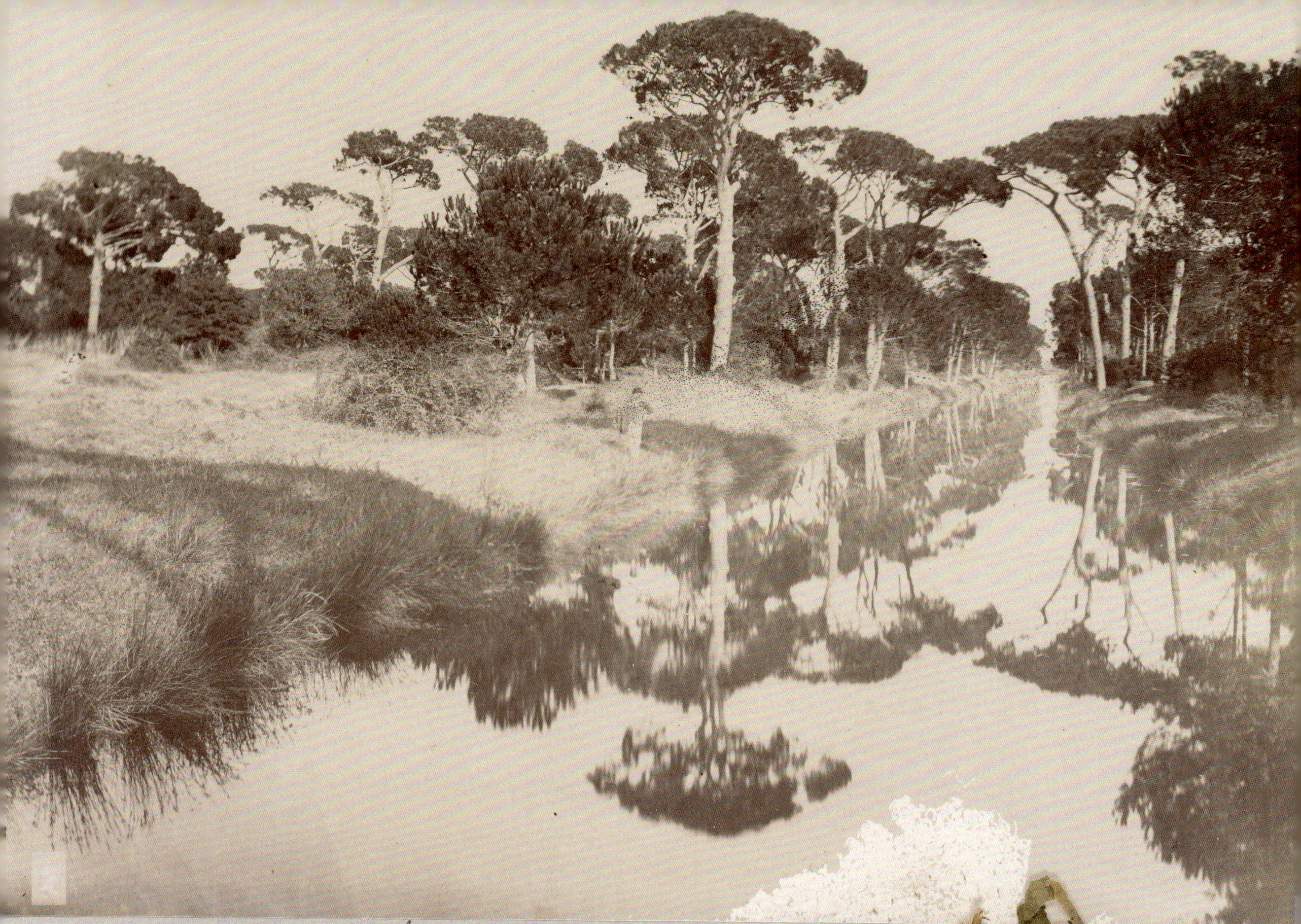

Il Pinetto -

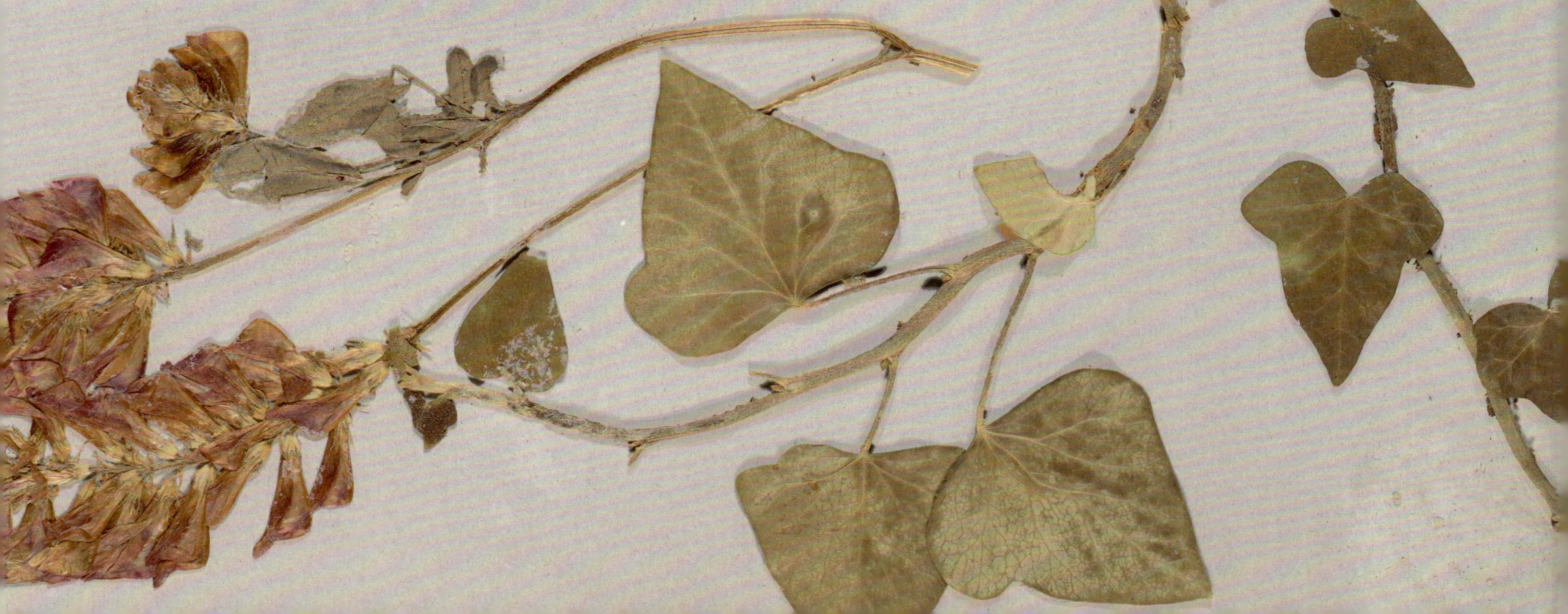

Director's Foreword

Nearly a century after her death, Isabella Stewart Gardner and her museum remain filled with mystery. For all the press coverage on Gardner during her lifetime and the abundance of archival materials she carefully preserved, we still have many lingering questions: why did she create such a specific kind of museum? What inspired her? Isabella left few first-person accounts of her life's experiences and opinions. Her collaged travel albums of found papers, photographs, her own watercolors, and other ephemera are some of our richest sources. Created on her trips to thirty-nine countries between 1867 and the late 1890s, they document where she went, what she saw, how she reacted to new places and cultures, and how her extensive world travels inspired her museum. They are, in short, a visual treasure trove and the focus of *Fellow Wanderer: Isabella Stewart Gardner's Travel Albums.*

This volume's careful analyses of Gardner's travel albums—many of which have never been exhibited or extensively studied—make clear that our founder's sources of inspiration were myriad. Though the museum is often described as a Venetian palazzo transplanted to Boston, this description simplifies Gardner's art historical interests. She had a voracious appetite for visiting and learning about cultures on four continents. Her collection is global in scope and inspired by these wide-ranging experiences.

The travel albums hold significance for us today in several ways. They are personal documents of one American woman's journeys and the people and places she explored; by virtue of that fact, they raise questions of encounter between East and West, colonial power and Indigenous cultures, wealth and scarcity in the late nineteenth century and beyond to the present day. They are also works of art in their own right, manifesting the collage aesthetic that presaged Isabella's installation of her collections in the museum she envisioned. The achronological combinations of European, American, and non-Western objects; visual puns between paintings, sculpture, furniture, and decorative arts; and insertion of personal mementos among the greatest Renaissance masterpieces all speak to the experimentation visible in the travel albums. In that sense, these albums reveal the artistic thought process behind Isabella's distinctive aesthetic leanings, holding lessons for all of us who love museums and understand that their influence on what and how we see is never neutral.

I am enormously thankful to Diana Seave Greenwald, Assistant Curator of the Collection, who recognized the importance of the travel albums and their resonance with contemporary artists and their practice. This publication grew out of a paired exhibition project, connecting two women for whom travel provided essential inspiration:

Ravena, June 1892 (detail) from Isabella Stewart Gardner's
Travel Album: France, Great Britain, Italy, Volume I, page 53

Betye Saar, a twenty-first-century African American feminist artist who has lived and worked in Southern California her entire life; and Isabella, the founder of the first American museum built and led by a woman. For this publication, Diana and coeditor Casey Riley have assembled a series of essays by scholars who draw on the albums to reveal new facts about our founder and the collection. They have fundamentally and permanently shifted our understanding of her engagements with regions worldwide; this will provide a foundation for interpretation projects for years to come.

We are grateful to the funders of *Fellow Wanderer: Isabella Stewart Gardner's Travel Albums* and *Betye Saar: Heart of a Wanderer.* They are the Abrams Foundation, the Ford Foundation, the Wagner Foundation, the Henry Luce Foundation, and the Getty Foundation through The Paper Project Initiative. The Wyeth Foundation for American Art also provided generous funding for this catalogue.

Peggy Fogelman
Norma Jean Calderwood Director

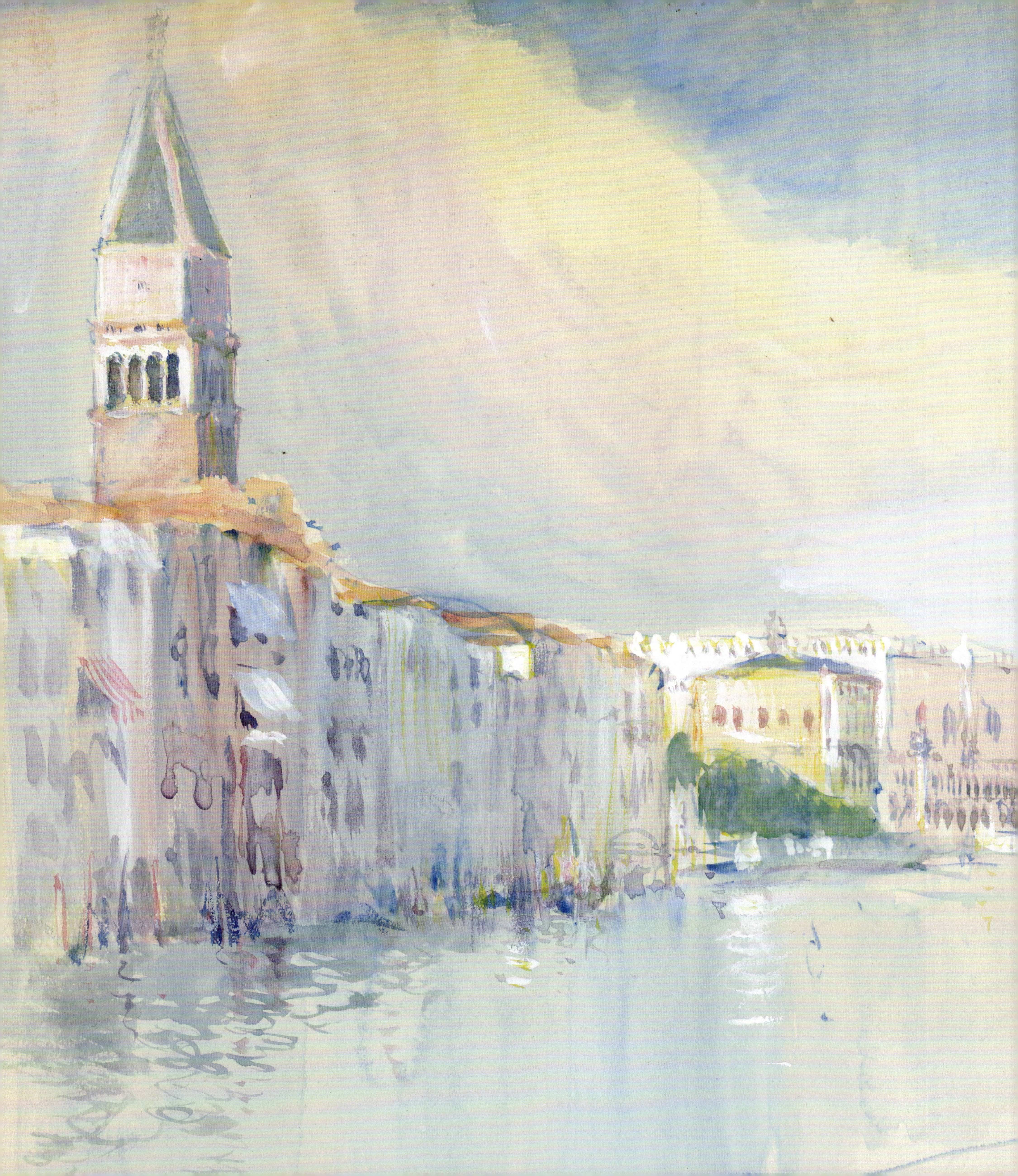

Acknowledgments

This volume and the exhibition it accompanies are the culmination of many years of research, insight, and mutual support. While the origins for this project may be traced to one of our doctoral dissertations, the resulting publication reflects a conversation that has evolved and grown stronger through collaboration. It has been a privilege and a joy to work as coeditors on such complex and fascinating material, and to shape a conversation around works all too often overlooked in the collection of the Isabella Stewart Gardner Museum.

Many staff members have helped to bring this exhibition catalogue to fruition. We wish to thank Peggy Fogelman, the Gardner Museum's Norma Jean Calderwood Director, and Nathaniel Silver, former William and Lia Poorvu Curator of the Collection and Division Head, for their primary roles in supporting this project to its completion. Each offered essential encouragement and advice in framing this exhibition and publication. We are equally grateful to Elizabeth Reluga, Head of Collections Access, and Adrienne Chaparro, Curatorial Assistant, for their extraordinary efforts in managing every facet of this volume. From sourcing crucial images to keeping everyone on deadline to scrutinizing transcriptions of the travel albums, both were invaluable to the success of this project. We also wish to thank Shana McKenna, Archivist, for her longstanding faith in and support of researchers interested in Gardner's photographic materials. Her abundant knowledge of the museum's history has been essential to this project. Ben Blacklock painstakingly went through the travel albums and recorded the locations where Isabella traveled over time, underpinning the maps in this volume.

In this collective endeavor, we are deeply grateful to our scholarly and creative partners, and for the ways in which their work illuminates the history of the Isabella Stewart Gardner Museum. Above all else, we wish to thank our authors for brilliantly expanding the narrative surrounding Gardner's albums: Pujan Gandhi, Madeleine Haddon, David Odo, Jaclyn Roessel, and Stephanie Tung. Their expertise and insights have enlarged our understanding of Gardner, her travels, and the contents of her renowned museum. We particularly appreciate their generosity of time and knowledge-sharing in a virtual academic roundtable hosted in the spring of 2021, during which we considered the artist Betye Saar's travel sketchbooks and artworks alongside Gardner's albums and installations. Heartfelt thanks to Elizabeth Gardner, former Director of Interpretation at the ISGM, for seamlessly organizing and facilitating such a generative gathering. We offer particular gratitude to those collaborators who have brought this publication to life: Michelle Komie, of Princeton University Press,

Detail of watercolor of the Grand Canal in Venice by Joseph Lindon Smith (American 1863–1950) from Isabella Stewart Gardner's *Travel Album: France, Great Britain, Germany, Italy, Austria, and Belgium, Volume I*, 1894, page 12

who championed a partnership with the Gardner Museum, and Frank J. Mondragón, Ria Roberts, Alejandro Stein, from world studio and Nouhtrang Thao, who created a beautiful design for the exhibition this catalog accompanies.

On a personal note, we wish to thank the many friends, mentors, and family members who have supported us throughout this endeavor. Our deepest gratitude to the following, in no particular order: Ned Cooke, Kim Sichel, Miko McGinty, Deneen Howell, Mary Kuhn, Erin Nolan, Emily Voelker, Anjuli Lebowitz, Christina Nielsen, Kate Herlihy, Kaywin Feldman, Matthew Welch, Katie Luber, Sarah Meister, Sophie Hackett, Sarah Kennel, Liz Siegel, Corey Keller, Susan Riley, Finnian Riley, Aloura Charles, Rob Bertrand, Ronan Bertrand, Ava Seave, Hugues Le Bras, and Eloïse Le Bras.

Diana Seave Greenwald
Assistant Curator of the Collection
Isabella Stewart Gardner Musuem

Casey Riley
Chair of Global Contemporary
Curator of Photography and New Media
Minneapolis Institute of Art

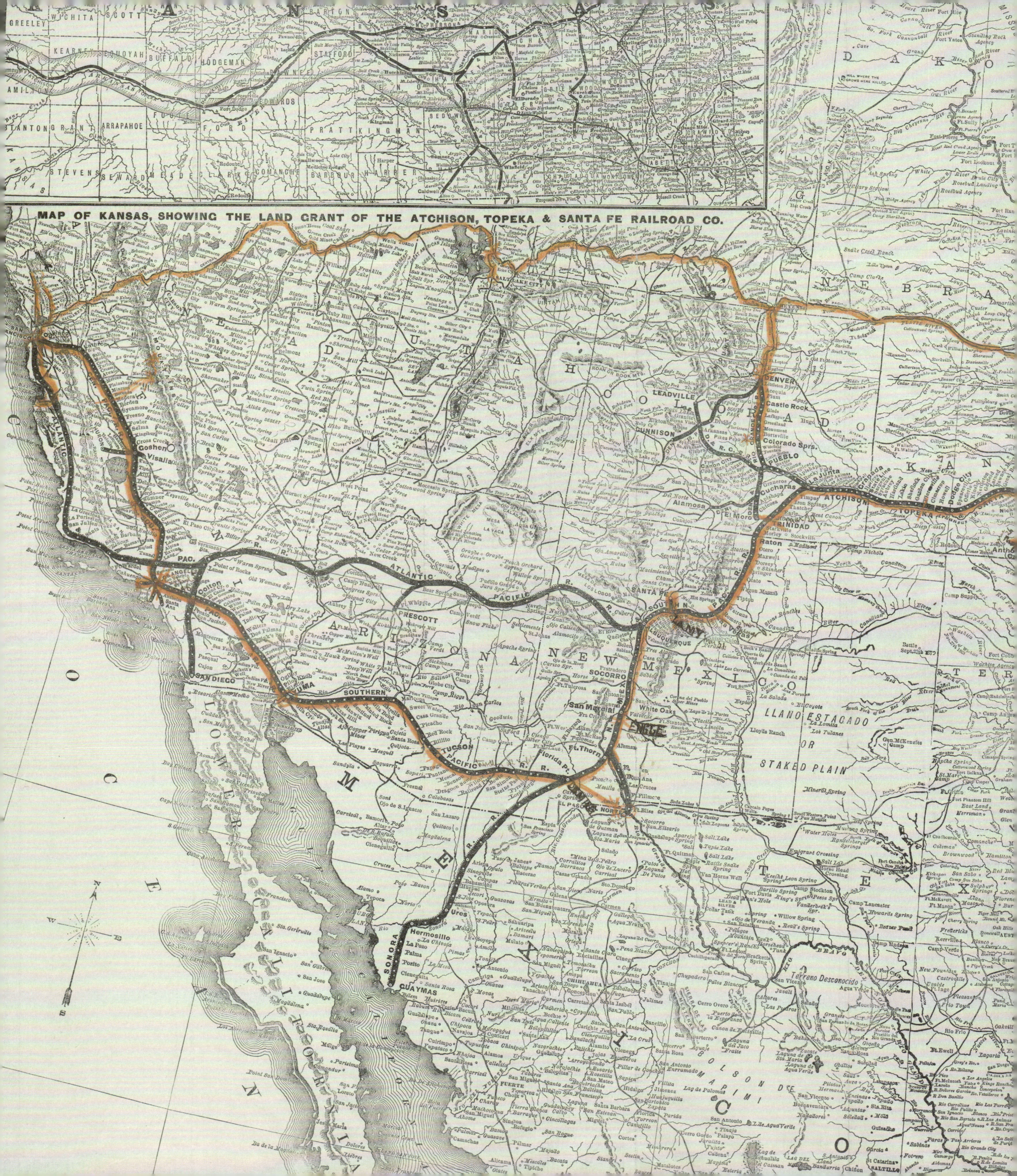
MAP OF KANSAS, SHOWING THE LAND GRANT OF THE ATCHISON, TOPEKA & SANTA FE RAILROAD CO.
ATLANTIC
PACIFIC
SOUTHERN
PACIFIC
ATCHISON
TOPEKA
DENVER
LEADVILLE
GUNNISON
PUEBLO
Colorado Sprs.
Castle Rock
SANTA FE
PRESCOTT
TUCSON
SAN DIEGO
YUMA
SOCORRO
EL PASO
GUAYMAS
Hermosillo
SONORA
CHIHUAHUA
LLANO ESTACADO
OR
STAKED PLAIN
Goshen
Visalia
Colton

Mapping Isabella Stewart Gardner's World Travels

On the following pages and throughout this volume, we have created maps showing the places that Isabella Stewart Gardner visited. Mapping Gardner's travels is complicated for several reasons—from occasionally incomplete archival records to the changing names and borders applied to define places. Traveling at the height of imperialism and during a century where there were wars and shifting territorial control, maps were not stable during Gardner's lifetime. They have since shifted again and again in the century since her death. In addition, the concept of what constitutes a sovereign nation has changed in the intervening hundred years.

We have used current twenty-first-century borders to define which countries Gardner visited. While not a perfect solution, we think it will help readers grasp the scope of her journeys.

Map of the Atchison Topeka Santa Fe Railroad showing the Gardners' route through the western United States and Mexico, enclosed in Isabella Stewart Gardner's *Travel Album: Northern and Eastern Europe, Russia, the United States, Mexico, and Cuba*, 3 March–6 May 1881

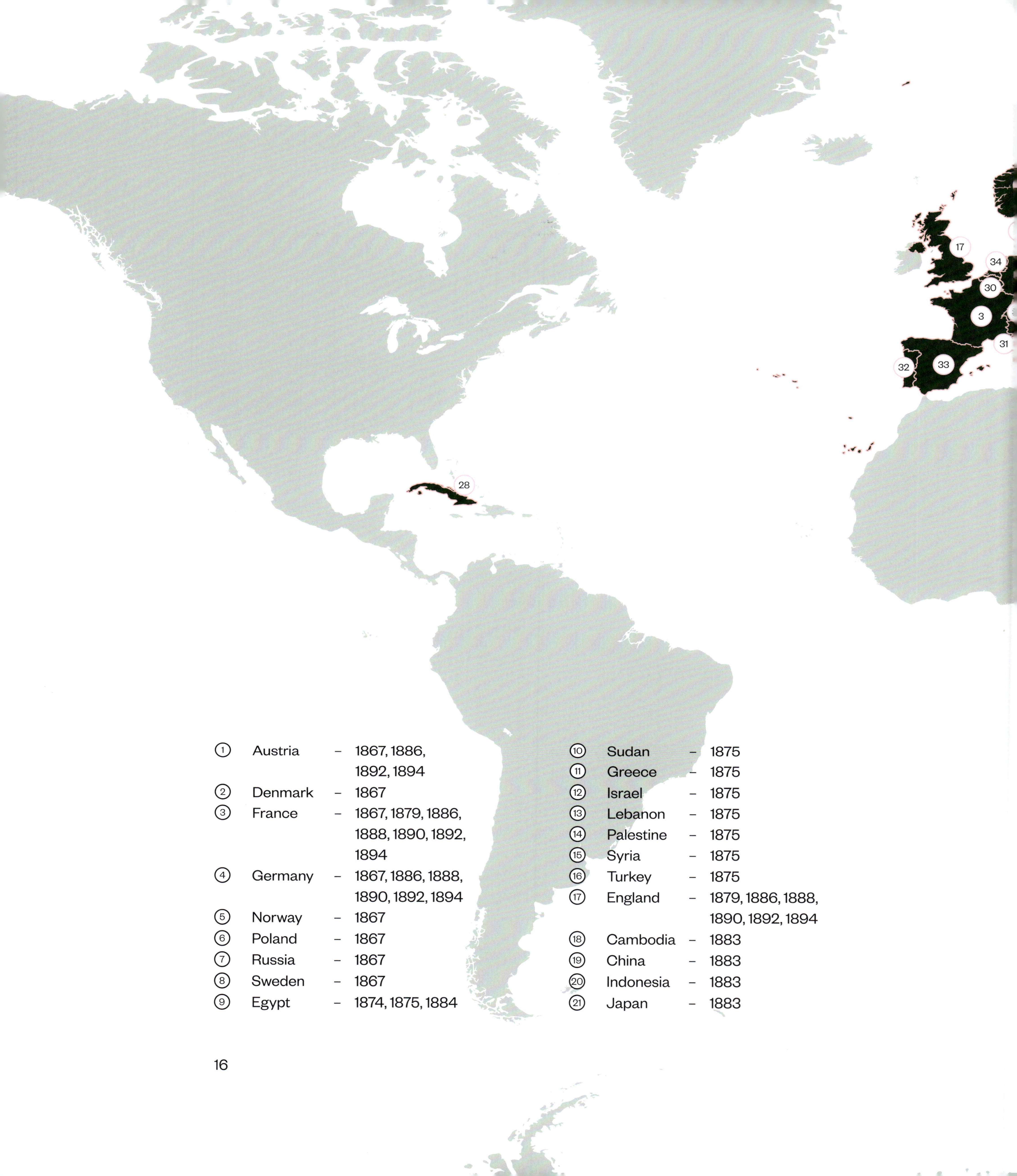

①	Austria	–	1867, 1886, 1892, 1894
②	Denmark	–	1867
③	France	–	1867, 1879, 1886, 1888, 1890, 1892, 1894
④	Germany	–	1867, 1886, 1888, 1890, 1892, 1894
⑤	Norway	–	1867
⑥	Poland	–	1867
⑦	Russia	–	1867
⑧	Sweden	–	1867
⑨	Egypt	–	1874, 1875, 1884
⑩	Sudan	–	1875
⑪	Greece	–	1875
⑫	Israel	–	1875
⑬	Lebanon	–	1875
⑭	Palestine	–	1875
⑮	Syria	–	1875
⑯	Turkey	–	1875
⑰	England	–	1879, 1886, 1888, 1890, 1892, 1894
⑱	Cambodia	–	1883
⑲	China	–	1883
⑳	Indonesia	–	1883
㉑	Japan	–	1883

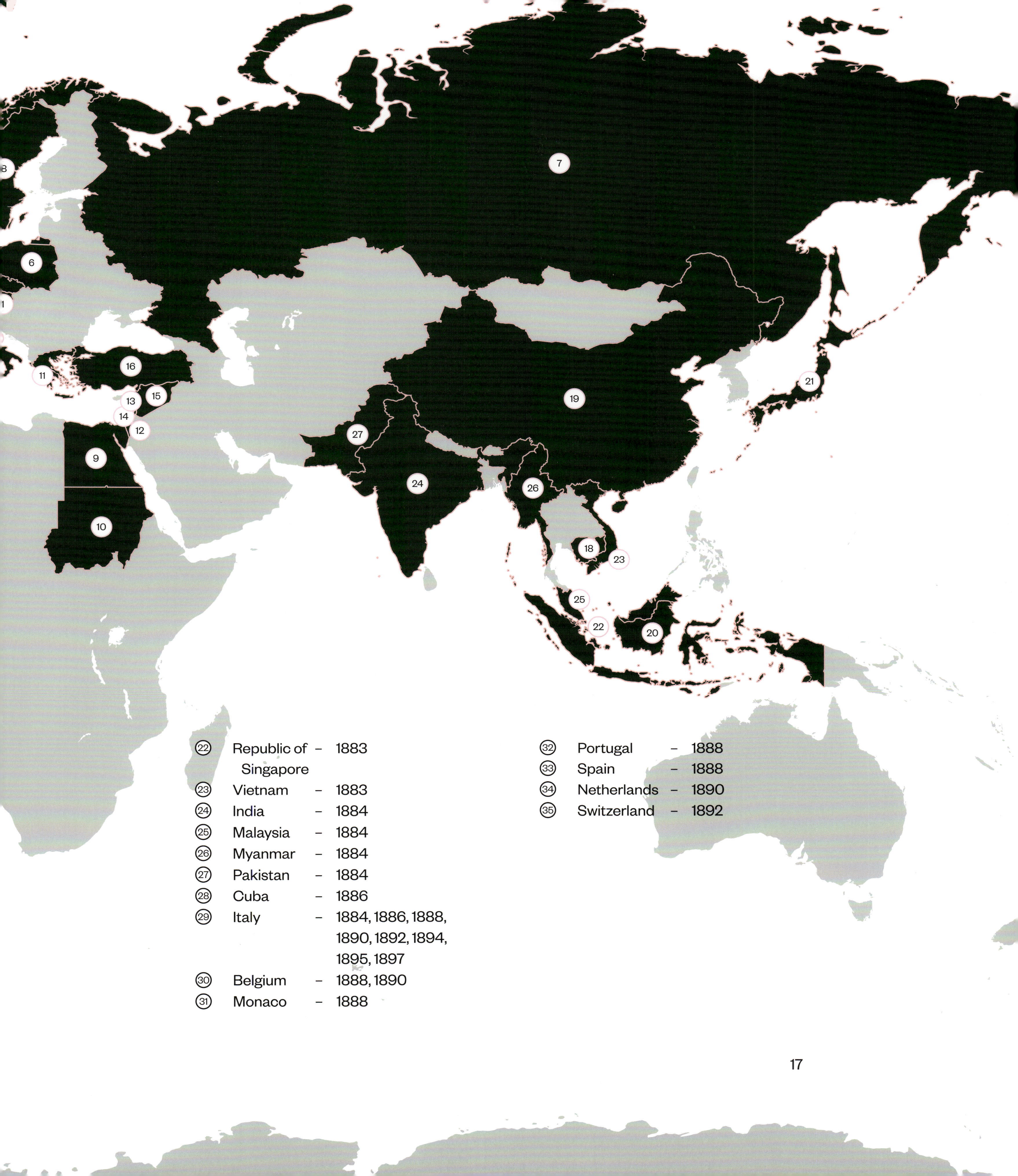

㉒ Republic of Singapore – 1883
㉓ Vietnam – 1883
㉔ India – 1884
㉕ Malaysia – 1884
㉖ Myanmar – 1884
㉗ Pakistan – 1884
㉘ Cuba – 1886
㉙ Italy – 1884, 1886, 1888, 1890, 1892, 1894, 1895, 1897
㉚ Belgium – 1888, 1890
㉛ Monaco – 1888
㉜ Portugal – 1888
㉝ Spain – 1888
㉞ Netherlands – 1890
㉟ Switzerland – 1892

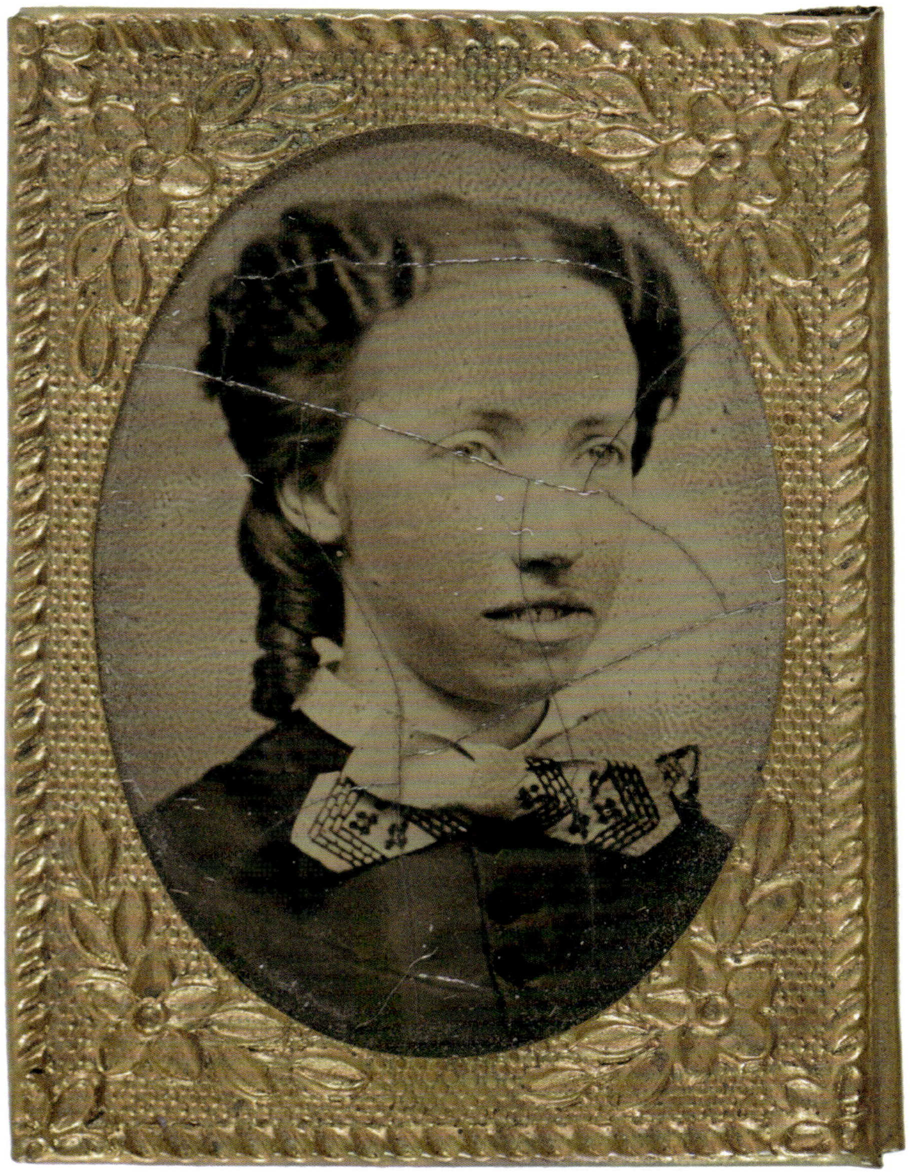

Fig. 1 *Isabella Stewart Gardner*, about 1860. Tintype, 2.54 × 1.91 cm. Isabella Stewart Gardner Museum, Boston

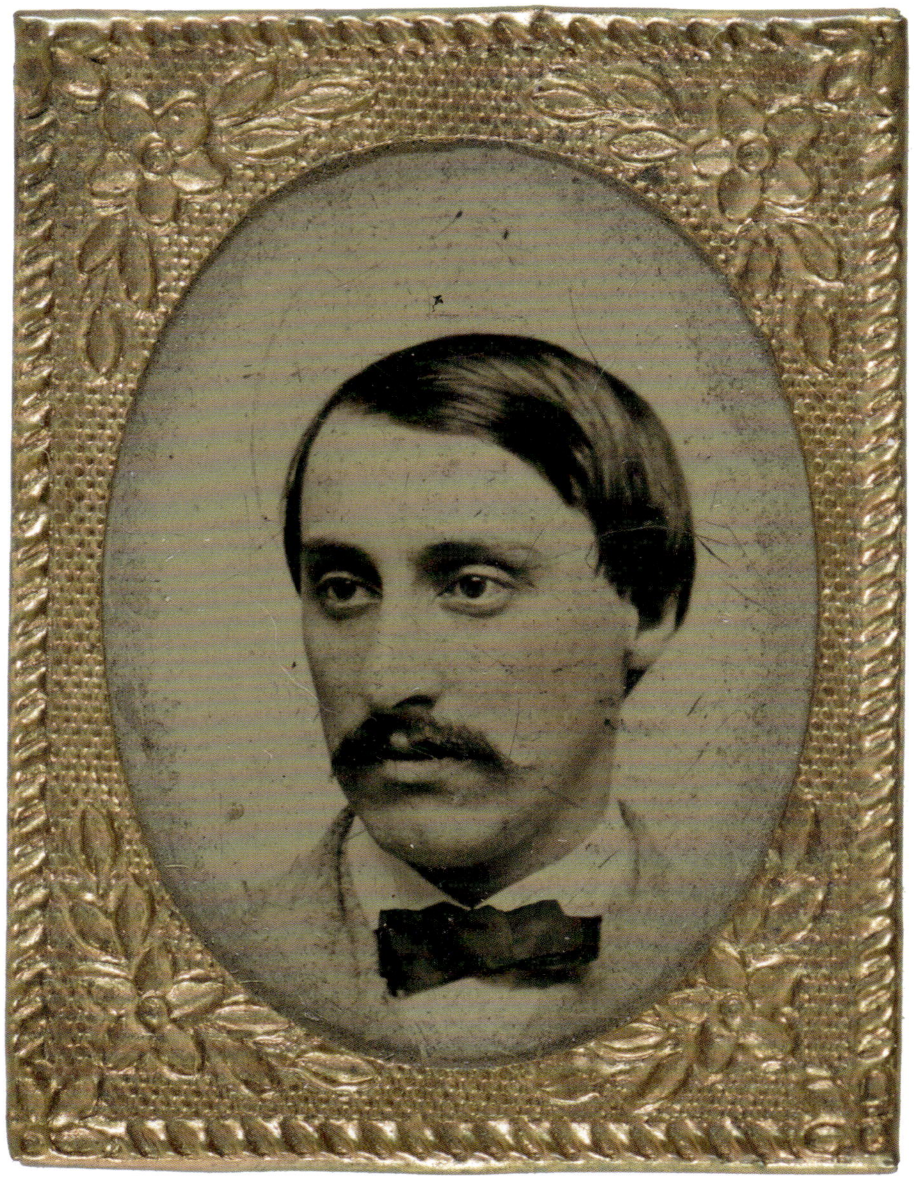

Fig. 2 *John L. Gardner Jr.*, about 1860. Tintype, 2.54 × 1.91 cm. Isabella Stewart Gardner Museum, Boston

Introduction

Diana Seave Greenwald & Casey Riley

Entering the historic building of the Isabella Stewart Gardner Museum, Boston's gray skies and chaotic traffic fade into memory, replaced by the sounds of bubbling fountains and the humid scent of the garden flowering at the heart of the museum. In April, the month of Isabella Stewart Gardner's birth, vines of brilliantly hued nasturtiums cascade down the pink-stuccoed Courtyard's walls, a show of horticultural splendor reminiscent of tales set in faraway palaces. The striking contrast between the museum's dun-colored exterior and its transportive interior heightens visitors' sense of having traveled a great distance by merely crossing a threshold. To describe the fairy-tale effect of the Gardner's spellbinding interior is to verge upon the cliché—and yet, the historical bridge between its enchanting atmosphere and the global travels of the museum's founder and namesake is a complicated one, and mandates restoration. This essay—and the volume it introduces—begins that process through a comprehensive engagement with some of the most vital documents compiled by Isabella Stewart Gardner (1840–1924): the extraordinary photographic travel albums that she herself assembled between 1867 and 1897.

Although many wealthy Americans, male and female, traveled internationally in the decades following the Civil War, Gardner traveled more—and to more far-flung places—than many women of her era and status. Her childhood as the privileged daughter of a successful merchant in New York City did not indicate this would be the case. In her teenage years, however, the horizon widened. Seeking an education for their daughter, Isabella's parents enrolled her in a Protestant "finishing" school for girls in Paris.[1] When she completed her studies, the Stewart family traveled to Italy, where they followed an itinerary broadly based on the highlights of the Grand Tour, including historical and religious sites in Milan and Rome, as well as a visit to Venice.[2] When she returned home, Isabella Stewart entered into a courtship with Boston's most eligible bachelor, John Lowell Gardner Jr. (known as Jack) (figs. 1 and 2). Their relationship blossomed rapidly, and the native New Yorker wed the son of one of New England's elite families within the year, in 1860. After a honeymoon trip to Washington, DC, the newlyweds moved to the Back Bay neighborhood of Boston—an enclave of wealth and prestige in the mid-nineteenth century, and a historical center of cultural power.

This sheltered life could have become Isabella's universe, but her family circumstances were increasingly complicated. Jack and Isabella struggled to have children; she suffered recurrent pregnancy loss.[3] Their only son, Jackie, was born in

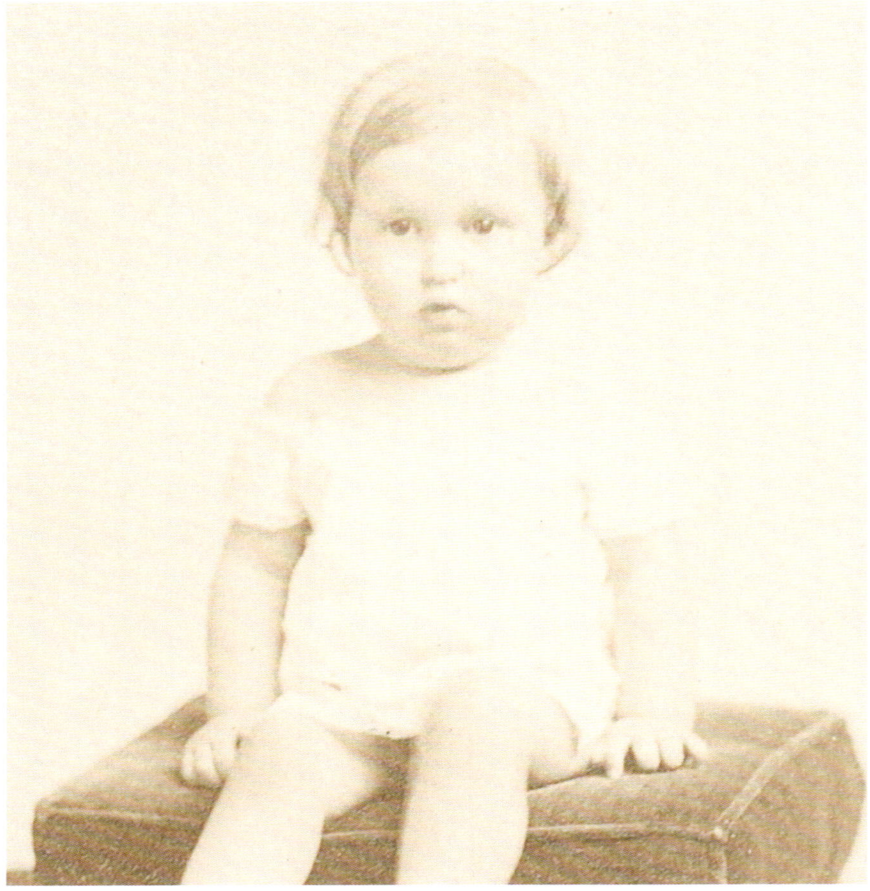

Fig. 3 *"Jackie," John Lowell Gardner III*, 1864–65. Albumen print on card, 10.1 × 6.2 cm. Isabella Stewart Gardner Museum, Boston

1863 and occupied the center of their lives, but died of pneumonia just before his second birthday (fig. 3). This tragic loss, combined with other deaths in the family—including a possible miscarriage—sent Gardner into a profound depressive episode.[4] After months of suffering, and at the recommendation of her physician, Jack and Isabella planned an extensive itinerary through Northern Europe to begin the process of healing from her emotional trauma. In 1867, the pair sailed across the Atlantic to embark upon a journey throughout Scandinavia and parts of the Russian Empire (fig. 4). After months of bedridden despair, the trip had a restorative effect upon Gardner's wellbeing, and she returned to the United States a changed woman. This journey through some of Europe's least temperate zones inspired many elaborately planned journeys taken by the couple over the next three decades. Jack and Isabella Gardner visited approximately thirty-nine countries (using twenty-first-century borders) across Asia, the Middle East, Europe, and North America, often on months-long itineraries. Gardner was an intrepid and stoic traveler who moved across continents and countries by train, steamer, sailboat, as well as her own feet—and occasionally on the back of a mule or elephant.

Far from idle in her staterooms and coaches, Gardner maintained an active correspondence with family and friends throughout her extensive social networks, kept written diaries, and—most significantly—created over two dozen original compendia to remember these voyages. From 1867 until 1919, when a stroke left her largely wheelchair bound, Gardner pursued the practice of photographic album-making,

Fig. 4 Copenhagen, June 1867 from Isabella Stewart Gardner's *Travel Album: Northern and Eastern Europe, Russia, the United States, Mexico, and Cuba*, page 3

Copenhagen –
June – 1867.

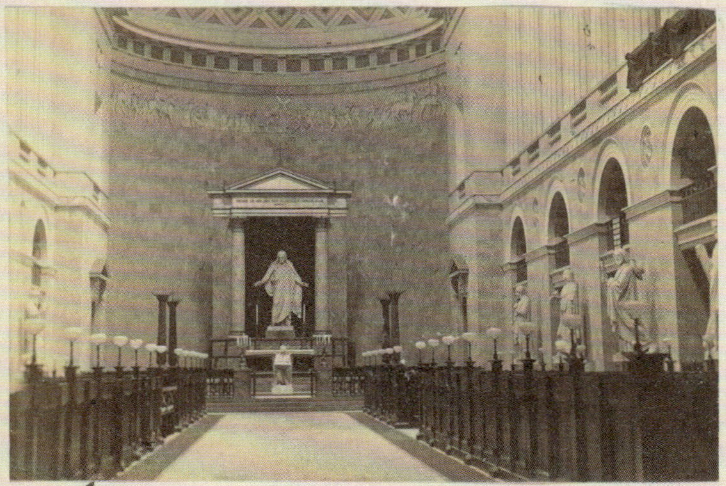

Interior of the Frue Kirke.

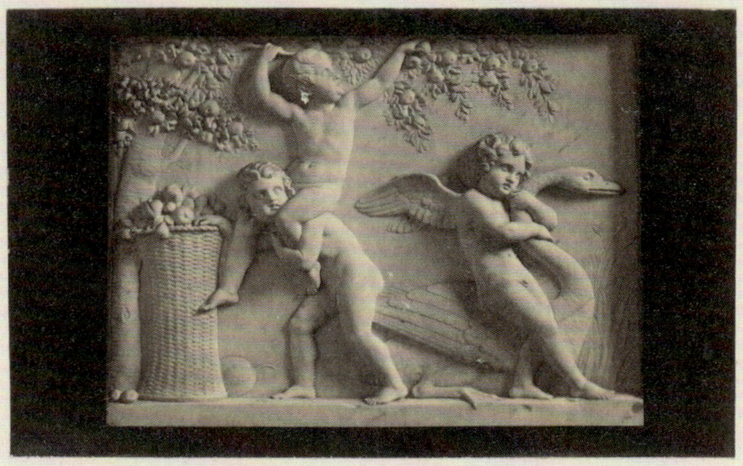

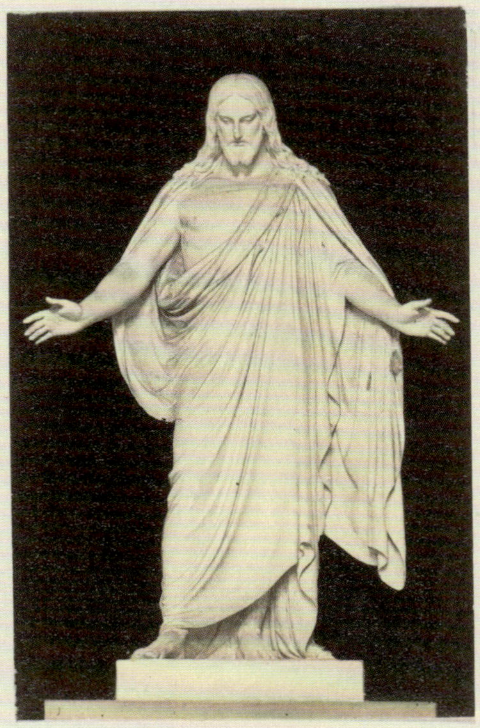

The Christ.

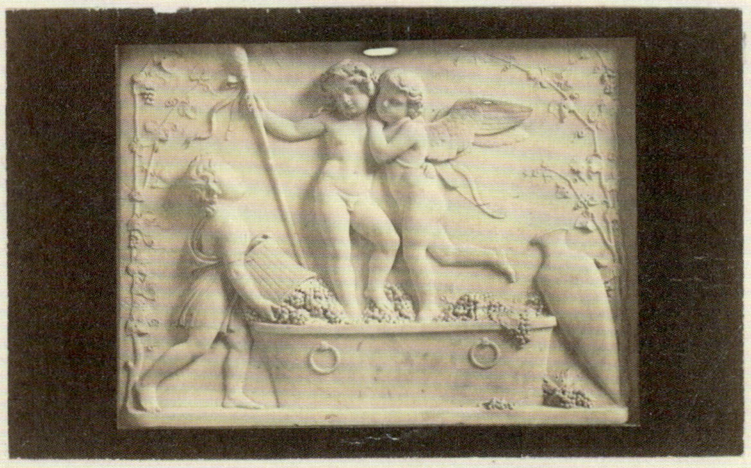

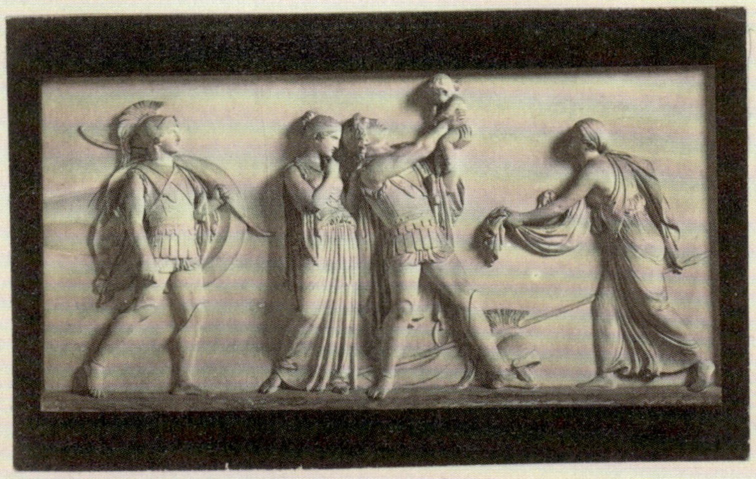

Bas Reliefs of Thorwaldsen.

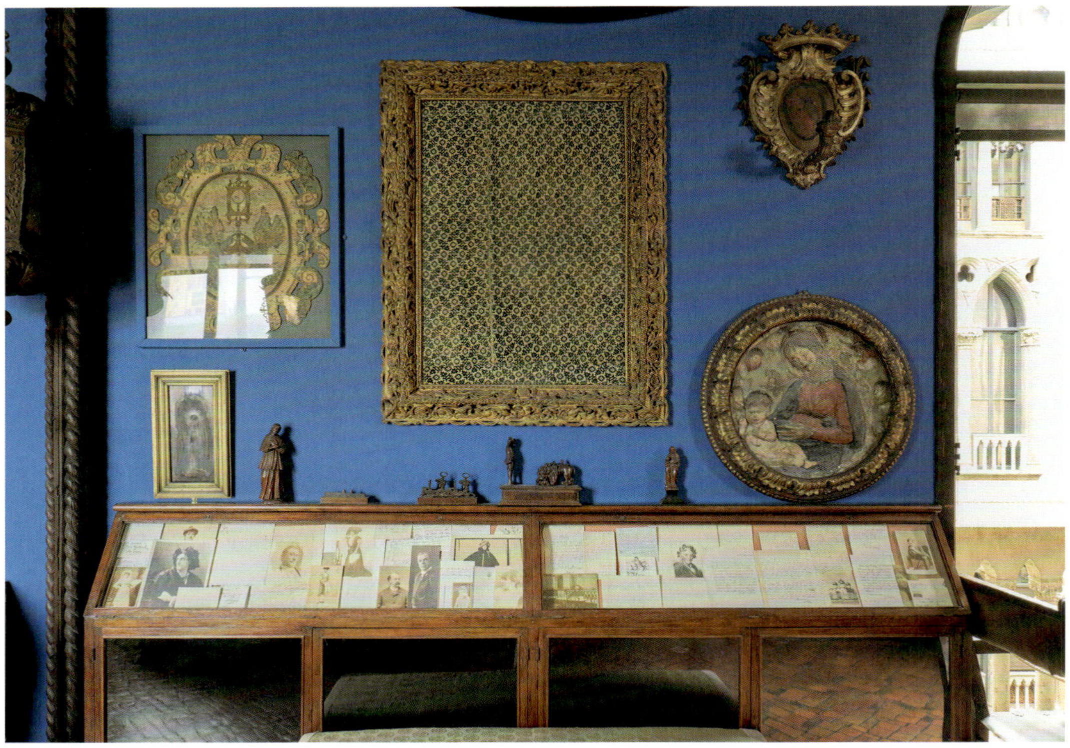

Fig. 5 Modern Actors Case and French Authors Case in the Long Gallery, Isabella Stewart Gardner Museum, Boston

intentionally collecting and arranging a wide variety of print photographs and other ephemera as she traveled (Plate 1). Her activities were not necessarily unusual: many wealthy European and American travelers in the nineteenth century compiled similar volumes in tribute to their own journeys across the globe, to serve as memory aids, didactic showcases, and family records. Still, Gardner devoted significant energy to the collection, creation, arrangement, and display of photographic documents throughout her life; in fact, she produced a broad range of photographic books over the course of her eighty-four years.[5] These compendia provide unparalleled insight into the mind of the woman who would become one of Gilded Age Boston's most celebrated—and misunderstood—civic innovators.

Today, the archives of the Isabella Stewart Gardner Museum maintain twenty-eight photographic travel albums compiled by Gardner herself. As visual and material narratives, her albums contain a range of landscape and ethnographic photographs from the nineteenth-century tourist trade, as well as photographic reproductions of art and architecture, revealing her ongoing acquisition of works produced by important commercial photographers, including Francis Frith (1822–1898), Antonio Beato (about 1835–1906), Pascal Sébah (1823–1886), and many others (Plate 4). The assemblages she devised within them also incorporate a striking variety of materials beyond photographic documents, including watercolor and ink sketches, captions and other handwritten texts, news clippings, botanical samples, and travel souvenirs. These imaginatively collaged albums, created by Gardner's own hand, represent the most important and understudied archives of her experiences, relationships, and ambitions.

When read closely, Gardner's albums reveal her increasingly conscious efforts to curate a public identity and civic legacy. Perhaps most crucially, they contain vital information regarding the curatorial practices that underpin her museum. Because they were assembled in concert with her traveling, Gardner's albums represent her immediate responses to a range of world cultures, training her for the collection and arrangement of the disparate works she would eventually display in her eponymous museum. At the same time, her travel albums serve as material evidence of the evolution of her design strategies for the same: a distinctively layered, syncretic architectural environment composed of elements—both live and inanimate—drawn from across the globe (fig. 5).

Isabella Stewart Gardner was a notable collector of rare books and manuscripts for much of her life—and acquired these materials during her trips. However, she collected very few works of fine art while traveling abroad between 1867 and 1897. Most of her journeys across continents occurred before she became an active participant in the international art market of the mid-1890s. Moreover, most of the acquisitions she made during this period of travel reflect her engagement with other, less refined marketplaces, including photographic studios, stationers' shops, and tourist kiosks. She bought fans in Spain, and opium bottles in China; in Japan, she made the wise decision to acquire two screens depicting *Spring* and *Autumn* (see Odo essay figs. 1 and 2), perhaps because of her interest in books and libraries.

Gardner's decision to create illustrated travel documentaries may be contrasted with the practice of using photograph albums to capture the rituals of home and family, as many women of her era did. As numerous scholars have documented, nineteenth-century women—from a range of class backgrounds—frequently created such albums as sites of private identity formation and the confirmation of familial ties.[6] Seen with the benefit of history, the photographic and material practices within Gardner's albums can be read as evidence of her curatorial labor. In short: her travel albums are important records of the ways in which one woman used photographic collage, and bookmaking, to realize her public and professional ambitions. No mere souvenirs to be relegated to the edges of her museum's history: when viewed in context—and especially as compendia that were circulated within her elite social networks—Gardner's travel albums are intrinsic to the formation and promotion of her identity as a cosmopolitan civic leader.

For reasons that have been lost to the historical record, Gardner was deliberately opaque about her curatorial and design strategies for Fenway Court, the name given to her museum during her lifetime. Near the end of her life, she actively destroyed much of her personal correspondence, erasing evidence pertinent to the creation of her museum. The archival record related to the formation of the institution is therefore largely pictorial, with few first-person written sources to explain her methodologies, collecting practices, or many contemporary political and social issues.[7] Her albums represent the most important compilation of source material for the museum; significantly, they were installed upon bookshelves in two separate galleries within the museum by Isabella herself. She meant for us to engage with her albums. And they

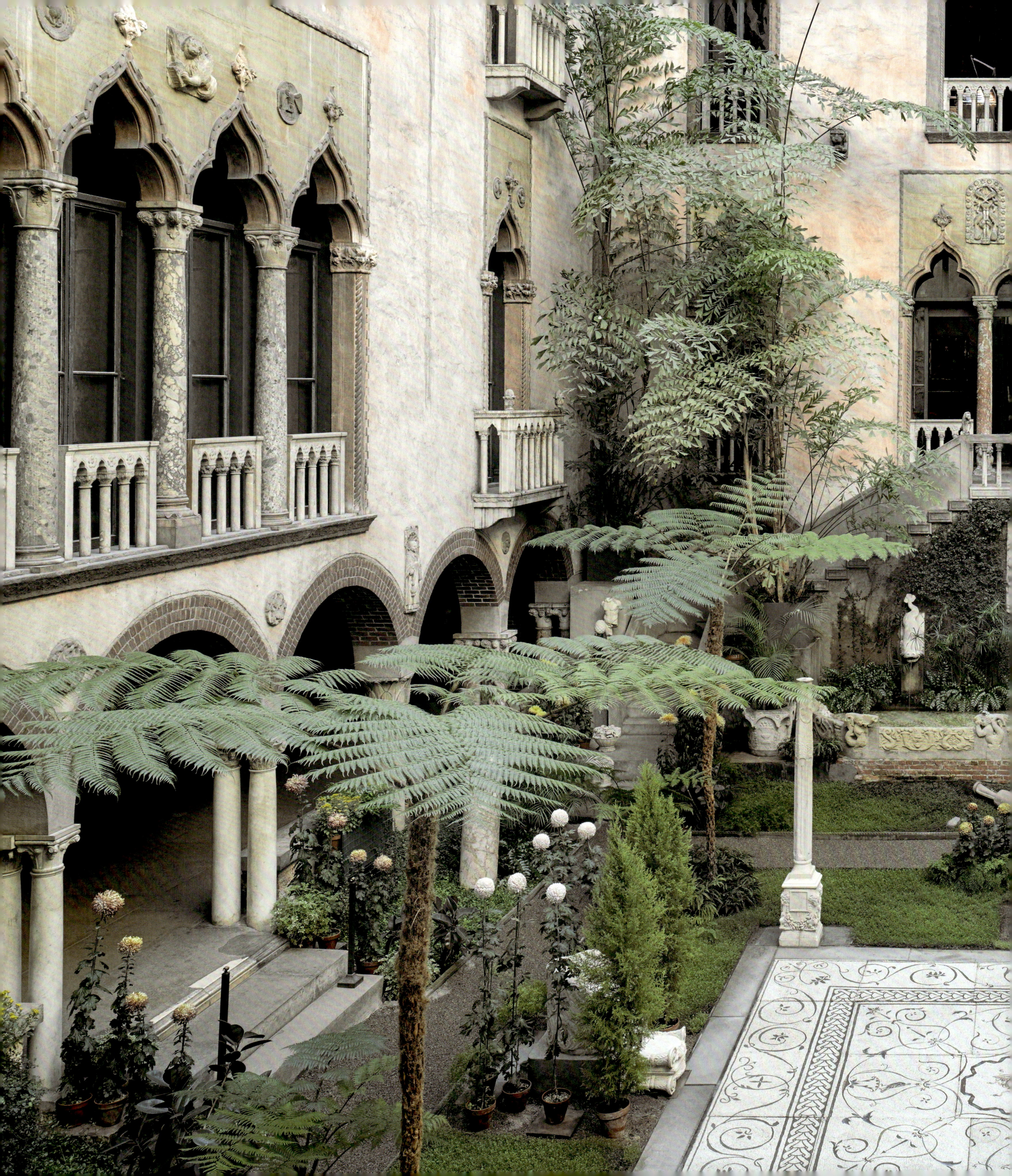

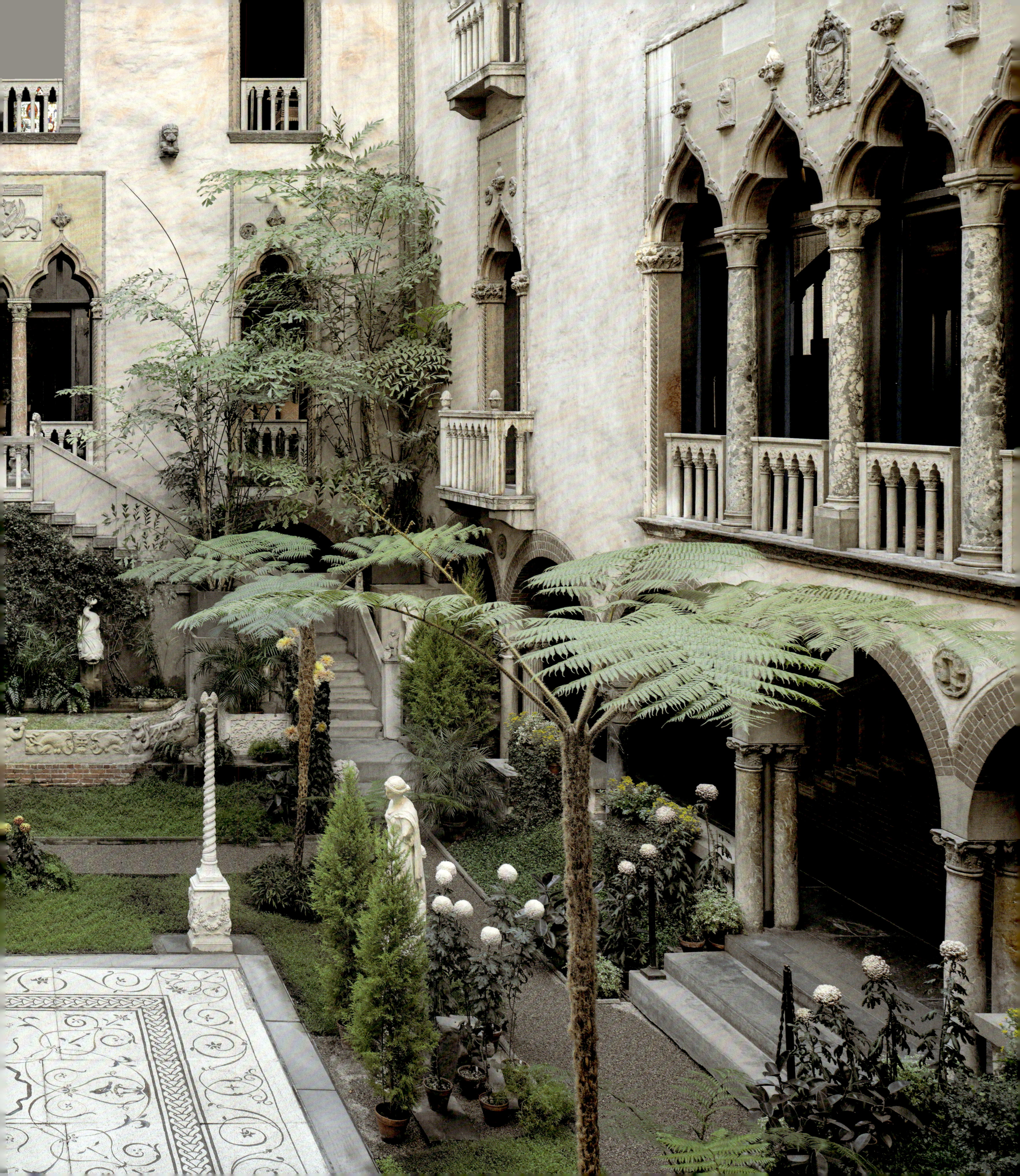

help today's visitors to access the nuances of what might otherwise appear to be a self-serving mélange of attributes deemed "foreign" by a privileged white woman during the early twentieth century.

For much of the Gardner Museum's history, Italy has been identified as the primary source for the foreign atmosphere often attributed to it. Gardner's passion for Italian cities and people, the art and literature of the Italian Renaissance, and her decision to live for several months at a time in the Palazzo Barbaro on Venice's Grand Canal have been well-documented in the literature about and produced by the institution.[8] Indeed, and in a kind of flattening of the idiosyncrasies that characterize her cultural bequest, Italy has become synonymous with the history of her museum. Even as Fenway Court was under construction, newspaper articles speculated that Isabella had purchased Florence's grandest palace, the Palazzo Pitti, to reconstruct it in America. Her first biographer, Morris Carter, reinforces the rumor by writing that it was in Venice that she "seemed to have discovered for the first time the glory of painting."[9] Not coincidentally, her travel album of Venice (Plate 6) has received the most attention in museum publications—from her first biographer's discussion of the photographic reproductions of Venetian paintings to an exhibition catalogue essay by photographic historian Elizabeth Anne McCauley.[10]

Though Italy was inarguably a potent force within Isabella's travel itineraries and albums, her photo albums illuminate a more worldly cultural perspective and set of interests. When taken together, these painstakingly created volumes reveal the breadth of her aesthetic and intellectual engagement with the world: its people, architecture, plant life, traditions, faiths, and more. Notably, many of these volumes showcase her experiences outside of Italy. Of the twenty-eight travel albums she produced, six are dedicated to her trip around Asia from 1883 to 1884; two are focused on the Middle East; Spain receives two volumes. Her 1879 British albums (Plate 5) were formative to the development of her collage-making practices, as were separate journeys throughout Northern and Central Europe.[11]

Perhaps less well-known and underemphasized is Isabella Stewart Gardner's 1874–1875 "Egypt Diary," an important early project representing the origins of her artistic practices in album-making (Plate 3). This album and diary also provides foundational insight into the germination of her curatorial methodologies. Over a five-month tour of Egypt centered upon a cruise up and down the Nile River, Gardner developed her acumen for assemblage, gradually allowing visual elements to convey key elements of her travel narrative. This trip was only her second voyage abroad with Jack, and the diary's first-person narrative, written in Isabella's fluent longhand, records her responses to the sights and people of Egypt. At the same time, through accomplished watercolor sketches of the landscape, her diary parallels the descriptive content of contemporary travel literature, including her own well-worn guidebook of Egypt.[12] Most importantly, the selection and arrangement of photographs of canonical sites within the Nile Valley reveal the degree to which Gardner relied upon photography in the formation of her vision. The assembling of photographic documents within

Fig. 6 Courtyard, Isabella Stewart Gardner Museum, Boston

the "Egypt Diary" effectively certified her comprehension of the diverse cultural landscapes that comprised Egypt in the 1870s, and reveals a newfound interest in the built environments of preindustrial societies. At the same time, her watercolors of the many archaeological sites along the Nile demonstrate a growing fascination with the design and ornamentation of ancient structures beyond Europe.

The "Egypt Diary" also reveals Isabella Stewart Gardner's absorption of the work of the many scientists, artists, and literary figures whose expeditions preceded her own voyage as preparation for her journey. The pictorial and written elements of her narrative illuminate her face-to-face interactions with Western scholars, missionaries, and contemporary Egyptians, and demonstrate the ways in which these engagements enhanced her self-presentation (Plate 2). Moreover, the album demonstrates her astute transformation of this independent research and fieldwork into an attractively designed book—whose pages approximate miniature exhibits—for a future audience. When taken together, Gardner's aesthetic, intellectual, and social activities in Egypt represent a consolidation of creative energies that would determine the course of her future travels and, eventually, fuel her civic ambitions.

Nowhere is her leadership more evident than in the creation of her museum at the turn of the twentieth century. As a well-traveled American woman of elite status and means, Gardner had access to cultural capital afforded to few other Americans, especially those who were Black and Indigenous, as well as many recent immigrants to the United States, including from Asia and the Middle East. While the photographic albums she created in the late nineteenth century are evidence of this privilege, they also illuminate the degree to which her global journeys shaped the character of her vision, and the goals underpinning her synesthetic curatorial regime. One may see this most clearly in the design of her museum's courtyard, the light-filled space around which her institution revolves (fig. 6). With its distinctively blush-colored stucco walls and fenestrations taken directly from the façade of a Venetian palazzo—the Ca d'Oro—the space at first appears to be a relatively straightforward transposition of Venice into Boston.[13] Her travel albums, however, provide grounds for an alternative interpretation of this space. Years after her adolescent sojourn to the city, Gardner visited Venice as a married woman, at the tail end of a year-long trip through East, Southeast, and South Asia. Rather than entering the city from Europe, as many American travelers did while following the Grand Tour, Isabella came to Venice by way of India—where she had spent three months touring a vast range of cities and sites seen by few of her acquaintances.

The two albums Isabella produced of India devote a great deal of attention to architecture. She selected numerous photographs of a courtyard in Northern India that, upon further reflection, bears a striking resemblance to the museum's central Courtyard (fig. 7). Moreover, the choice of pink for the Courtyard's wall color—something typically linked exclusively to the Palazzo Barbaro (fig. 8)—can also be understood in an Indian context. Isabella traveled to Jaipur, where the buildings have been painted pink routinely since 1877 (fig. 9). Other aesthetic allusions and potential

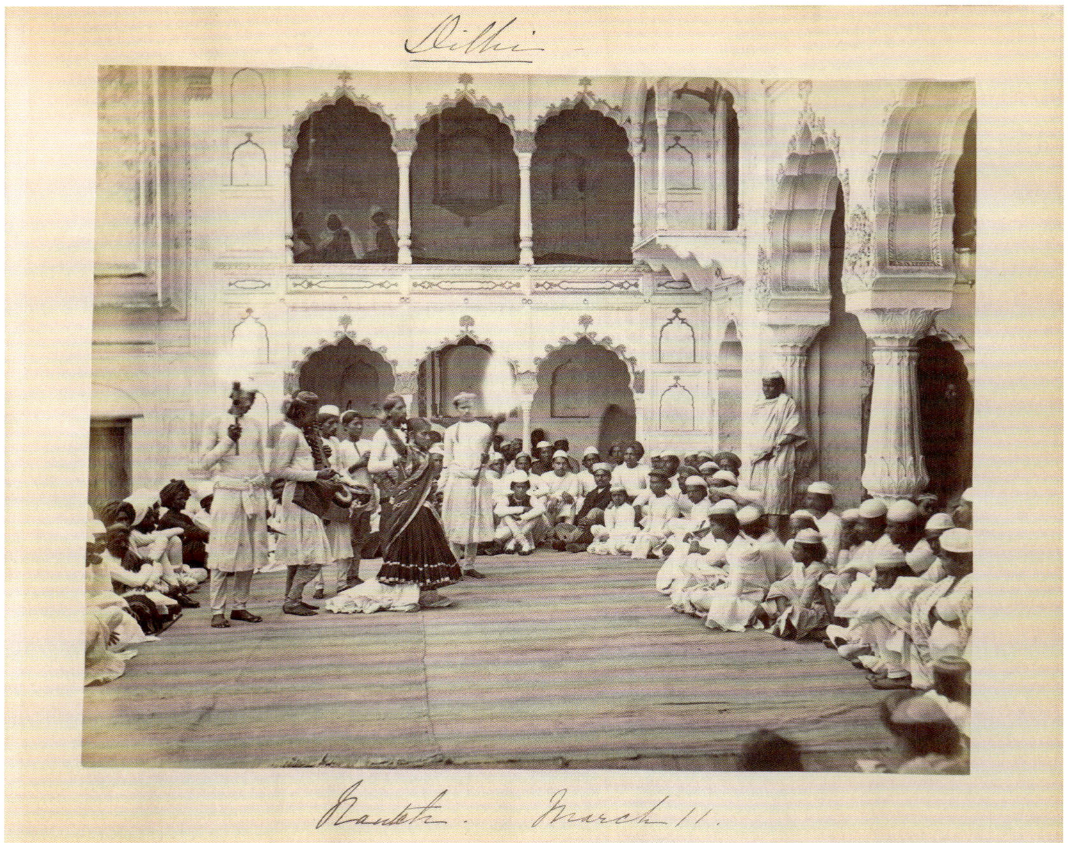

Fig. 7 Delhi, March 1884 from Isabella Stewart Gardner's *Travel Album: India, Pakistan, Yemen, and Egypt, Volume VI*, page 38

cultural transpositions may be traced to her assemblages within her albums of India. Her decision to install a stained-glass window from Soissons Cathedral low enough to cast colors on the tiled floor of her museum's Long Gallery (fig. 10) echoes the low placement of cut-glass fenestrations in a palace near Jaipur—a palace she selected for photographic documentation within the pages of her album (fig. 11).

None of the above is meant to deny the broader influence of European sources, especially those in Italy, upon the collection and design of the Isabella Stewart Gardner Museum. Rather, it is to restore the institution—and the far-ranging intellectual and civic intentions of its founder—to a network of global influences. Doing so not only illuminates her individual accomplishments and acumen, but reflects most accurately her engagement with an increasingly interconnected and technologically advanced world. Moreover, it suggests an awareness of the demographic and cultural realities of early-twentieth-century Boston—an American city increasingly defined by waves of immigration and national migration—and perhaps an effort to greet them.

It is also evident from the albums that Isabella cultivated sincere respect for, and a passionate interest within, a broader range of cultures than the literature surrounding her museum would suggest. While her indebtedness to colonial enterprise is inarguable, her engagement with a world beyond Boston's innermost circles was sustained, deep, and intellectually driven. Seen in this way, her museum might have reflected a desire on her part to create an homage to the cultures she knew she was privileged to visit.

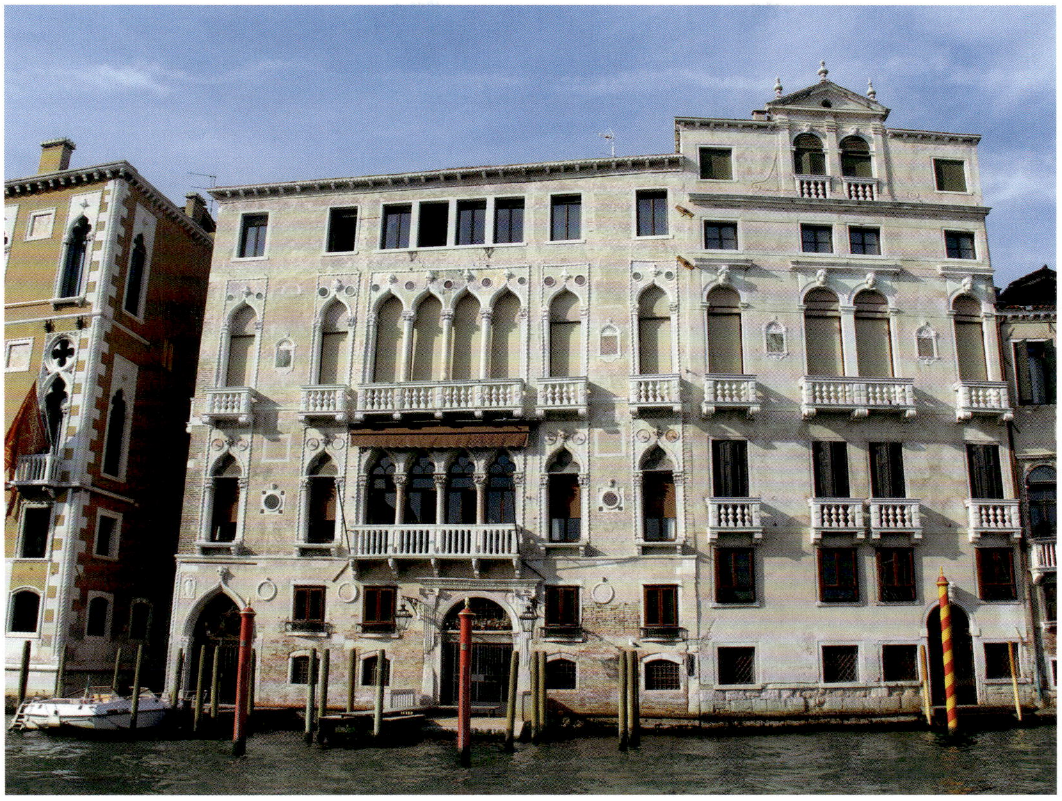

Fig. 8 Palazzo Barbaro, Venice

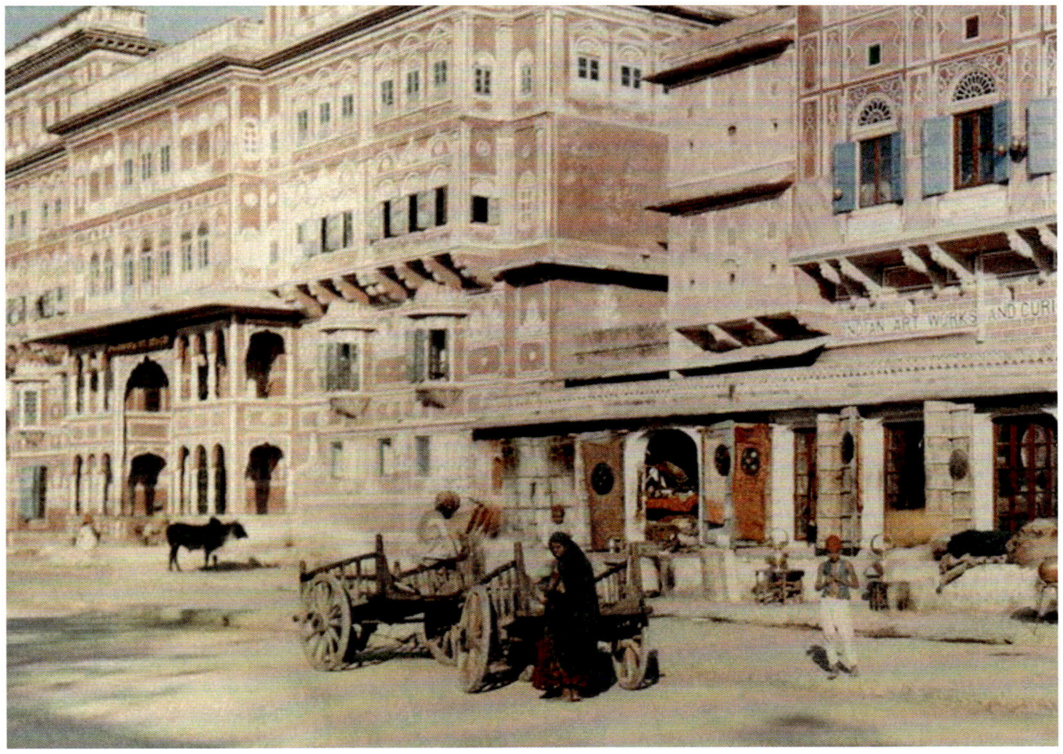

Fig. 9 Jules Gervais-Courtellemont (French, 1863–1931),
A view of the streets of Jaipur, 1926

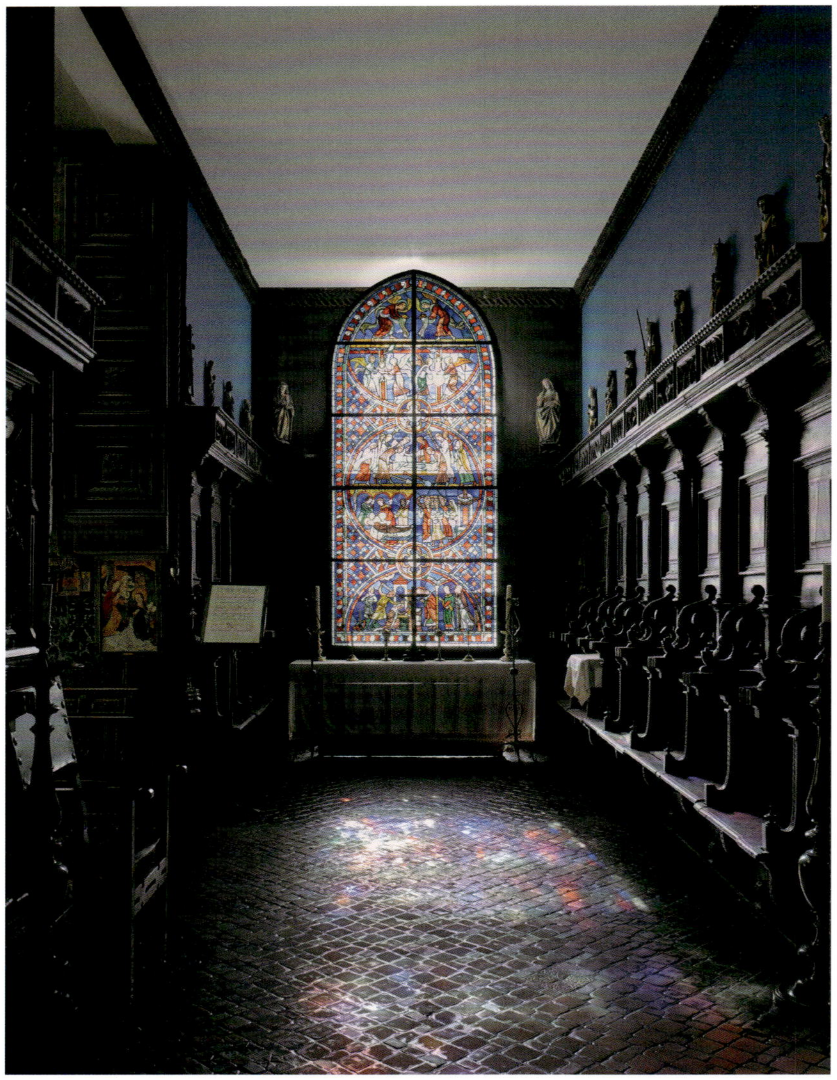

Fig. 10 French, Soissons, Window with Scenes from the Lives of Saint Nicasius and Saint Eutropia, about 1205 (with later restorations). Pot metal glass, 366 × 157.5 cm

At the same time, Isabella Stewart Gardner's travel albums cannot be separated from the complexities of her position and the power in which she was imbricated. Despite historical reports of her respectful interactions with the foreign cultures and people, Isabella's voyages were facilitated by extractive and repressive colonial enterprises, particularly in Asia. Consider the album pages in which she introduces Angkor Wat, the exceptional temple complex in Cambodia (Plate 8). It seems that few American visitors had traveled to the complex in the late nineteenth century, making her visit highly unusual. How had she come to travel there? Through a chance meeting on the steamship in China where she describes meeting "De Baré, Durant de Fontmagne, a naval Frenchman our only companions."[14] These two French officers appear throughout the next few weeks of her diary from this period; according to the text, de Baré convinced Jack and Isabella Gardner to visit Cambodia, which was then a colonial protectorate of France.[15] The fact that French colonial forces enabled their visit is prominently visible in her assemblages related to the site. In one example, she collages a photograph of the road to Angkor alongside an elaborate menu handwritten in French (Plate 7). The title above the menu suggests that the Gardners dined upon

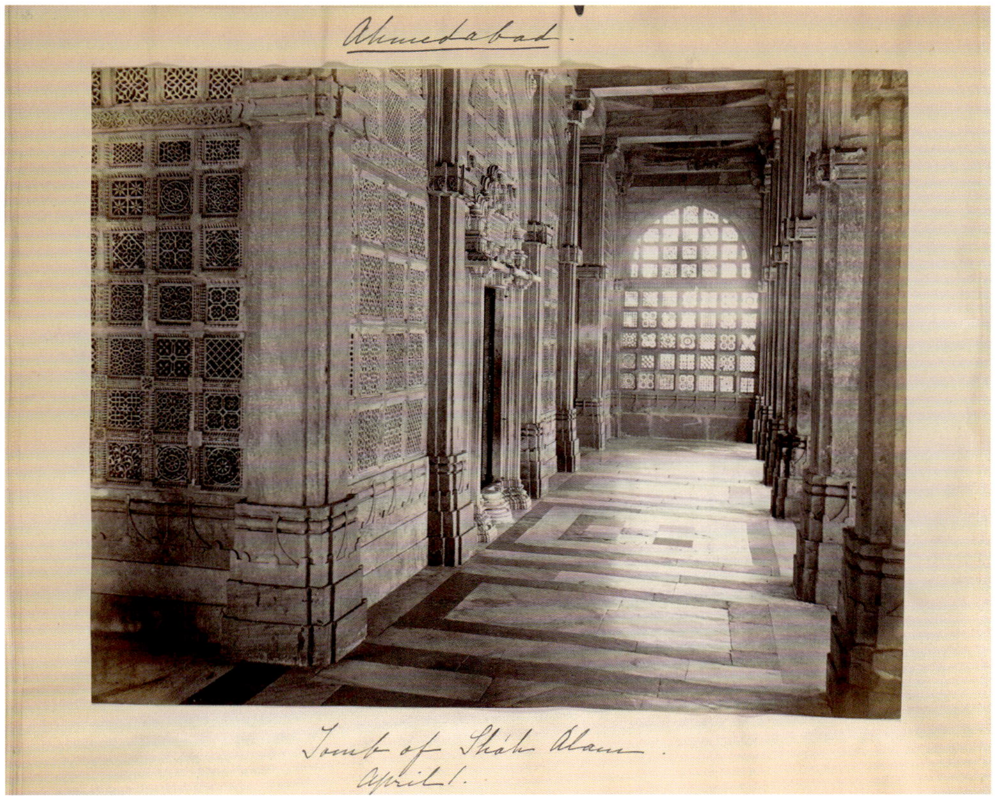

Fig. 11 Ahmedabad, April 1884 from Isabella Stewart Gardner's *Travel Album: India, Pakistan, Yemen, and Egypt, Volume VI*, page 94

French cuisine, enjoying a multicourse meal amid the ruins of the temple complex—a fitting metaphor for the all-consuming appetites and ravages of colonial power. These scenes cannot and should not be extricated from any reading of her albums, as they also reflect the mindset and perquisites of their creator.

The photography within her albums further illuminates the imperial underpinnings of her project. Like many white American and European travelers of the nineteenth century, Gardner collected and arranged photographic "types"—stereotyped constructions, largely produced in commercial photography studios, of people wearing the supposed costumes and props of their cultural heritage. So-called "street" and "peasant" types were popular depictions of racial stereotypes, which reflected and reinforced the harmful, racialized fantasies of their intended audiences, and were marketed by the same purveyors who sold architectural photography, travel books, tourist souvenirs, and more. As cultural ephemera, they attest to the flowering of racial pseudoscience in the nineteenth and twentieth centuries, the theories of which held sway in American and European academic, legal, and professional contexts for decades, to their detriment. Viewing such portraits within her albums requires a clear-eyed reckoning with the violence and exploitation thrumming beneath the surface of her comfort, and with her subjective view of the world.

The collection and remixing of material objects from diverse cultures is, at its heart, a colonial enterprise. Displaying the works she purchased around the world in a Boston museum—thousands of miles away from the contexts in which each was created—represents a form of violence in its displacement of objects from the people

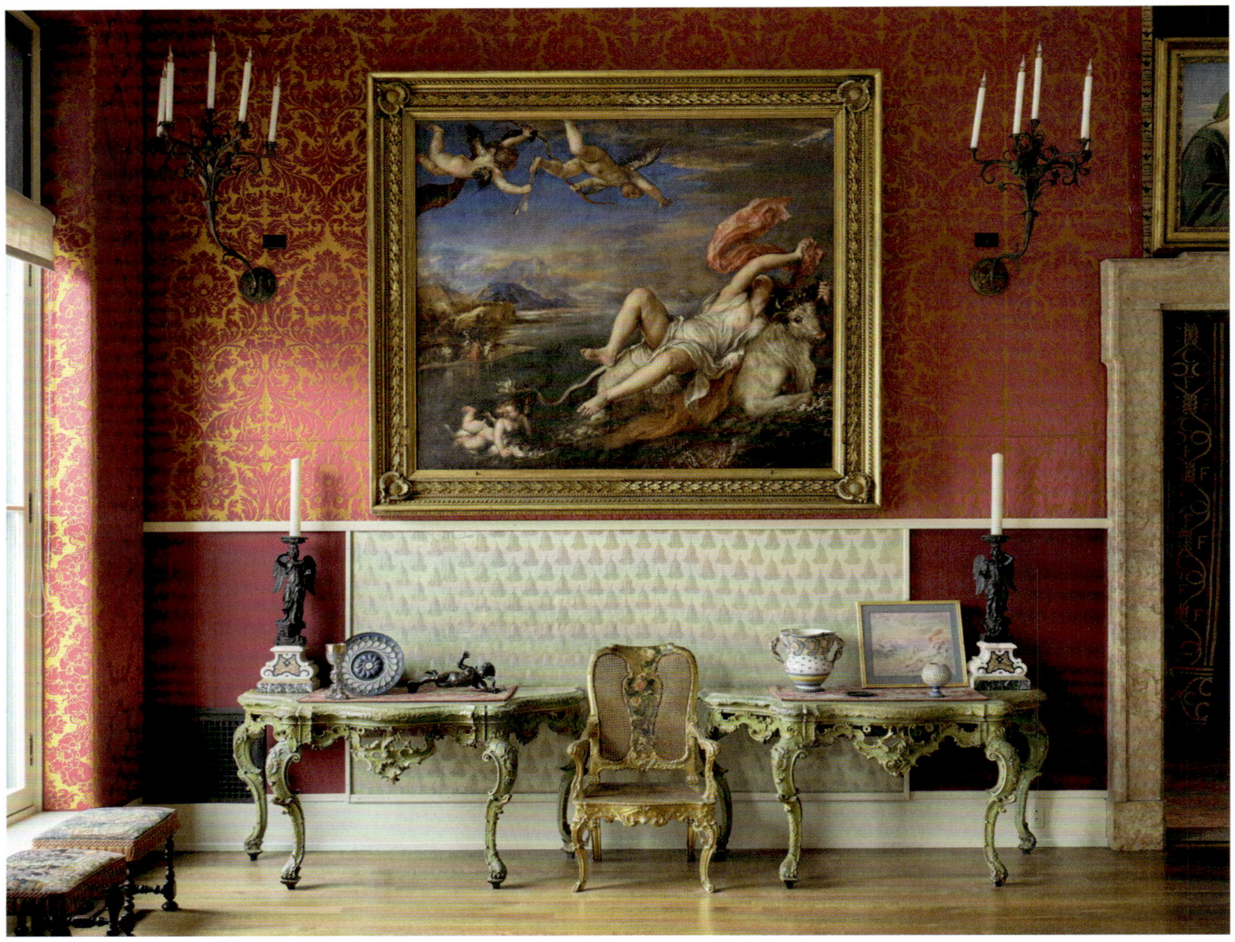

Fig. 12 Titian's *Rape of Europa*, 1559–62 in the Titian Room, Isabella Stewart Gardner Museum, Boston

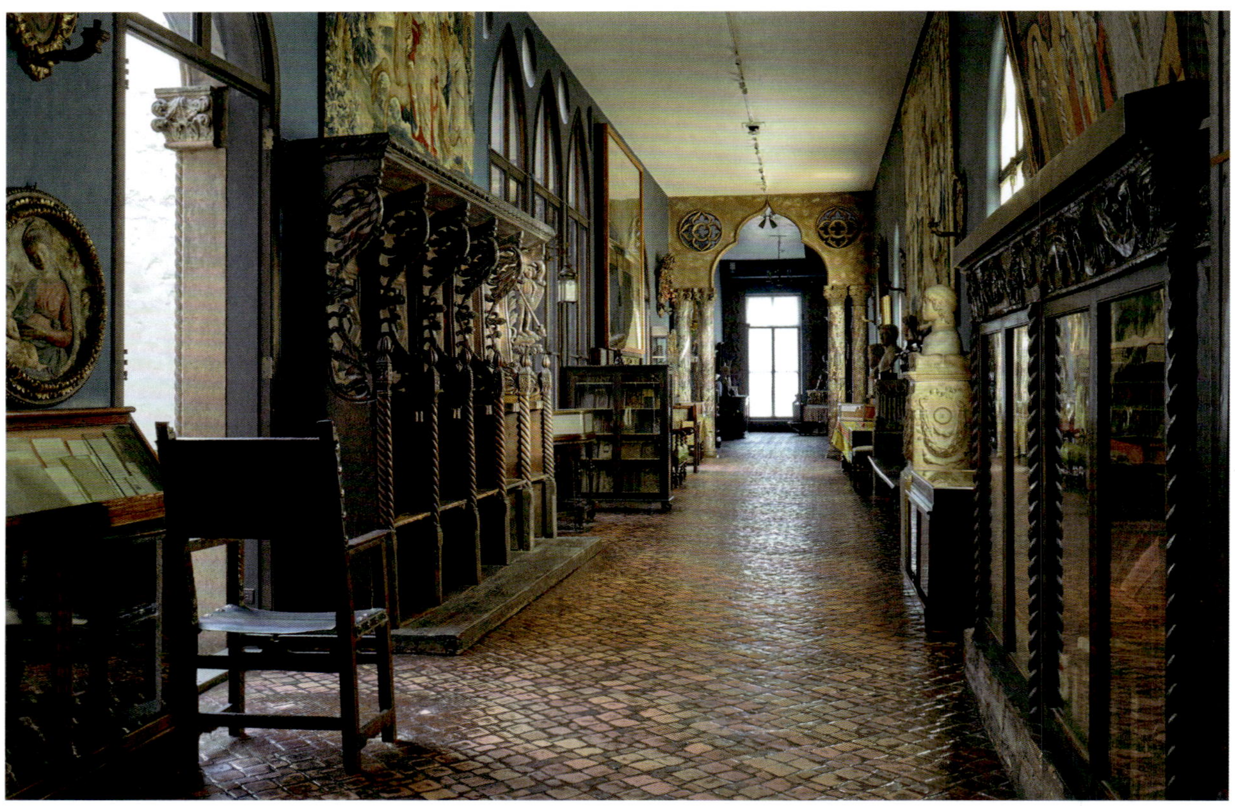

Fig. 13 Long Gallery, Isabella Stewart Gardner Museum, Boston

who cherished them. The idiosyncrasies of her signature arrangements of objects from disparate cultures and eras can further distance these artworks from their cultural origins. In our collective reading of her photographic travel albums, this volume balances appreciation and exploitation as uneasy partners in a complex creative dynamic. As such, we build upon the work of many colleagues within the field of art history, and of museums, who seek a decolonized perspective upon institutions like the Gardner Museum in the twentieth-first century.

Within this context, Gardner's albums can be read as templates for her actions as an art collector and museum founder. Photographic album-making anticipated the strategies of research, collection, installation, and preservation that she would put into use when she amassed the objects that would establish her as a major patron and institutional leader at the turn of the twentieth century. Her travel albums illuminate her actions as the dedicated chronicler of a peculiarly American aesthetic and intellectual evolution, one made possible by the imperial ambitions of the same. Moreover, to evaluate Isabella's photographic travel albums as evidence of her *labor*—as opposed to her leisure—reveals the ways in which she used album-making to realize her public ambitions. As such, they illuminate the ways in which a woman of her race, class, and national origin could become the founder of a museum, a rarity not only in Boston, but around the world.

In her analysis of Titian's *Rape of Europa* within Fenway Court (fig. 12), art historian Anne Higonnet contends that "the Gardner Museum's installation reminds us not only that museums create visual meanings, but that those meanings have had a history."[16] The material construction of installations precedes the visual meanings they convey, and these processes of creation are just as endowed with historical significance and cultural signifiers. To achieve an "integrated analysis" of the Gardner Museum as advocated by Higonnet requires the consideration of sources drawn from the social and material life of its founder, a woman who traveled widely, networked avidly, and collected scrupulously. These sources—correspondence, print ephemera, souvenirs and memorabilia, but above all the photographic albums that she crafted and archived inside the Long Gallery (fig. 13) and the Vatichino (fig. 14)—were not incidental to the life of her museum: they illuminate the evolution of her grand civic project. Crucially, her curatorial strategies arose from her longstanding engagement with photography. While the acquisition of the paintings, statuary, and furniture within Fenway Court is a testament to her extraordinary connoisseurship, Isabella's installation practices reveal a humbler set of practices centered upon her scissors, her paste pot, and the camera's eye. It is our hope that this volume provides a more comprehensive examination of her work as an assembler of albums, and the pivotal role of photography in the formation of her institution and its links to the foreign cultures from which she acquired works of art. We hope not only to contribute to the burgeoning discourse surrounding scrapbooks and photograph albums, but also to present a novel analysis of the material and visual practices that shaped Isabella Stewart Gardner's gift to the education and enjoyment of the public—forever.

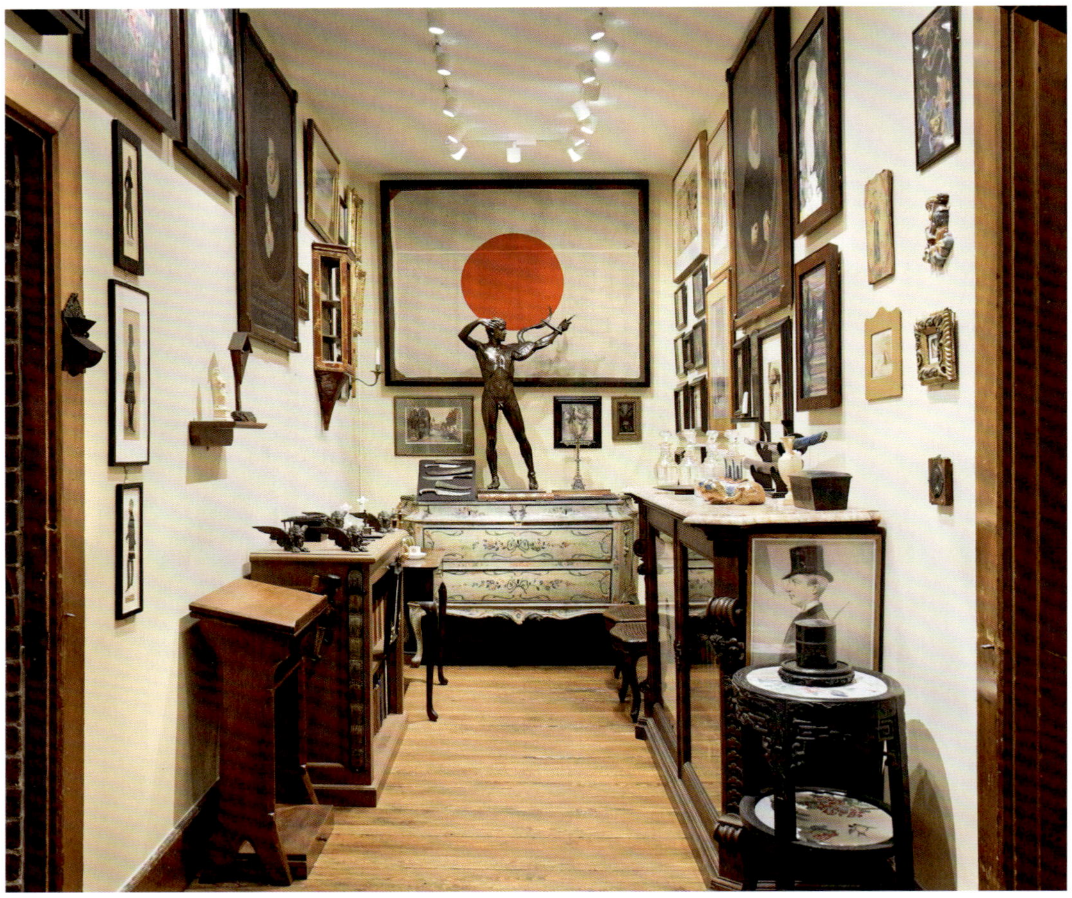

Fig. 14 Vatichino, Isabella Stewart Gardner Museum, Boston

1. Silver and Greenwald 2022, 28–29; "finishing school" is a private school where wealthy girls are "prepared" for an elite social life and marriage by learning languages, music, dance, watercolors, and other skills.
2. Ibid., 29–30.
3. Ibid., 32–33.
4. Ibid., 36–39.
5. For further discussion of Gardner's photographic practices, see Riley 2015b, 2020a, 2020b.
6. Numerous studies of women's photographic albums in the nineteenth century assert a vital relationship between album creation and the domestic sphere. Anne Higonnet states that women's albums drew upon pictorial traditions rooted within the home, as "home was a world to be invented and managed by women." See Higonnet 1992, 177. Patrizia Di Bello suggests that the album is a "feminine space" in which women used photography "to give power to their fantasies and validate their experiences" within the confines of home and family; see Di Bello 2007, 23. Similarly, Elizabeth Siegel contends that women's "[album] art depicted activities within an accepted domestic sphere, representing a range—though a limited one—of women's experiences and important social and familial functions"; see Siegel 2009, 16. While Marilyn Motz's case studies of women's photograph albums from 1880 to 1920 suggest that the album served as a place for women to establish "a personal identity apart from one's family," she demonstrates that "the ideal of the domestic sphere" remained a constant framework for these albums' construction. See Motz 1989, 89.
7. The notable exception to this is the Bernard Berenson letters, which were discovered at Berenson's Italian home—Villa I Tatti—in 1965. They have since been published in a much-cited compendium produced by the museum. The museum archive contains twenty-eight travel albums, fifteen guestbooks that are full of snapshot photography of Gardner and her friends, as well as clippings and notes, and five albums containing photography by T. Marr & Son. There are also two albums of travel snapshots that Gardner took and kept.
8. McCauley et al. 2004.
9. "Mrs. 'Jack' Gardner's Palace Mystifies Back Bay Bostonians," *Philadelphia Press*, 17 March 1901 (ARC.100017); Carter 1925, 86.
10. McCauley 2004.
11. Riley 2015a, ch. 2, 111–80.
12. For further discussion regarding this album, see ibid., ch. 1, 32–110. Gardner's guidebook to the region is held in the archives of the Isabella Stewart Gardner Museum; see Murray 1873.
13. Appolonia 2004, 183–84.
14. Isabella Stewart Gardner, entry from Tuesday, 30 October 1883, *Travel Diary: China, Vietnam, Cambodia, Indonesia, India, Singapore, Myanmar, Egypt, and Italy, 1883–1884* (ARC.009111), Isabella Stewart Gardner Museum Archives, Boston.
15. Muller 2006.
16. Higonnet 2003, 145. Higonnet's longer analysis of the Gardner Museum asserts the fixity of the installations as a characteristic feature of "collection museums" such as the Frick Collection, the Musée Condé, the Huntington Library, and Dumbarton Oaks; see Higonnet 2009. David Carrier offers a reading of Gardner's installation of Titian's *Europa* as a point of contact between the visitor, Gardner, and Bernard Berenson (1865–1959), who helped Gardner to acquire the painting; see Carrier 2006.

Pages 6–7, August 1867

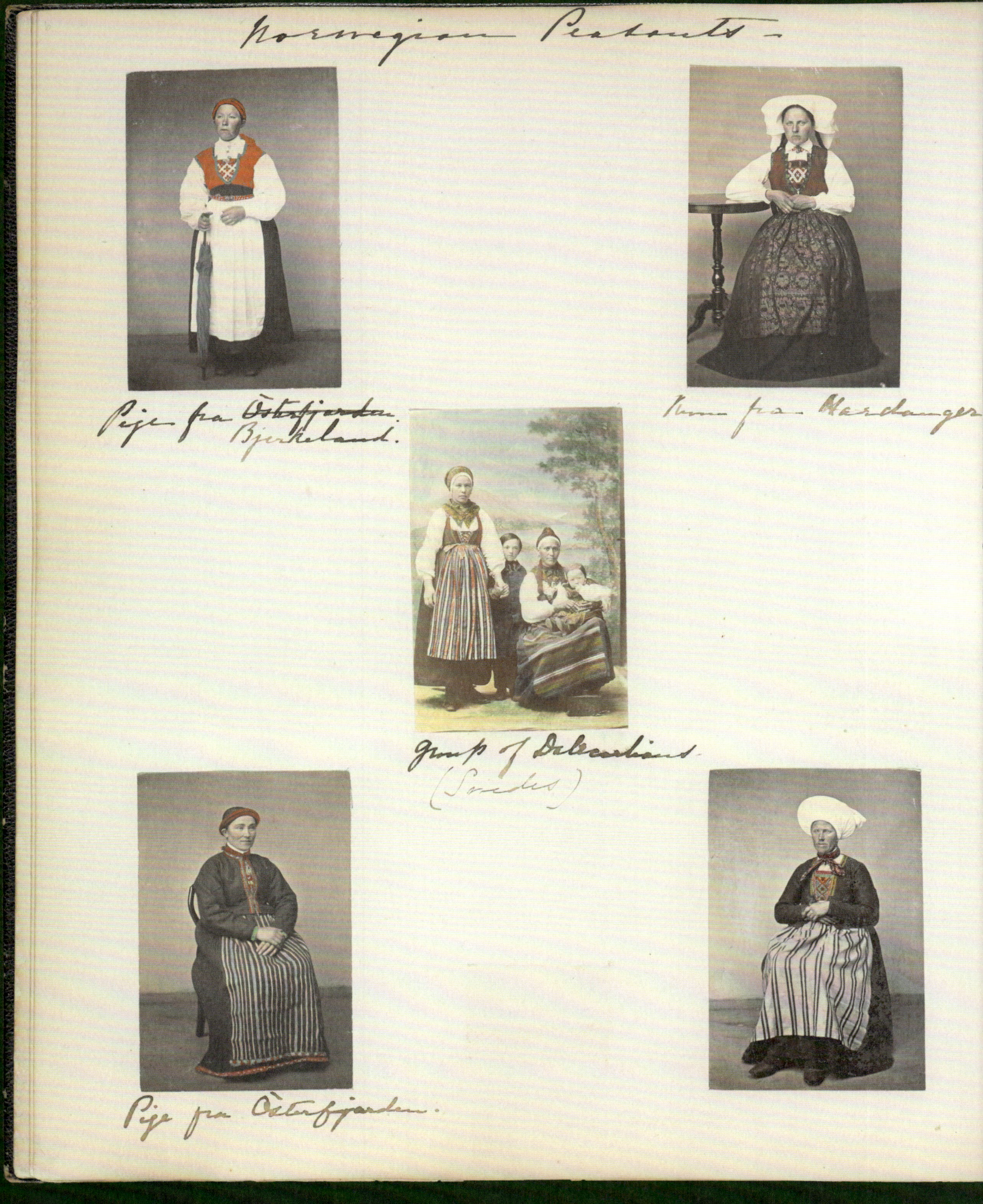

Norwegian Peasants.

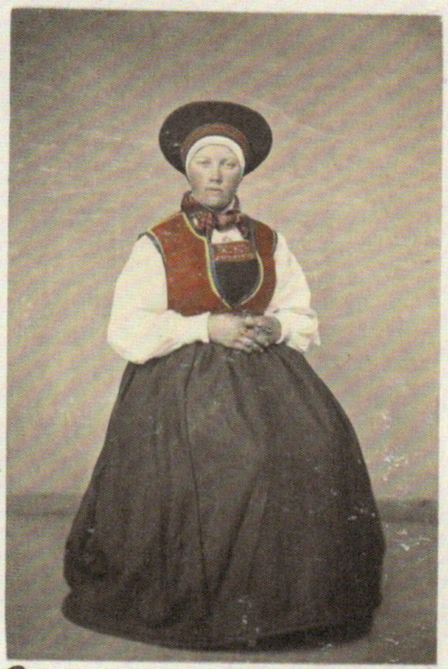

Pige fra Arendhavred.

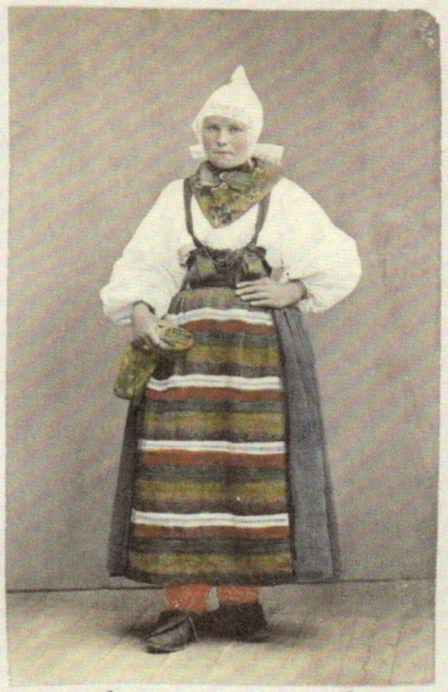

Dalecarlian.
(Swede.)

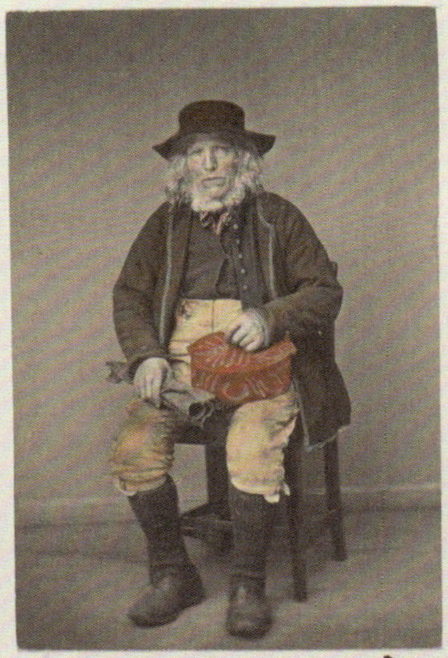

Mand fra Österfjorden.

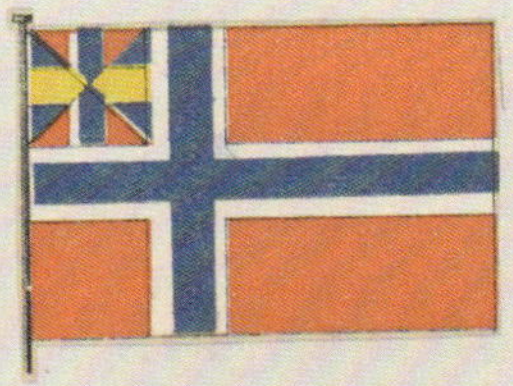

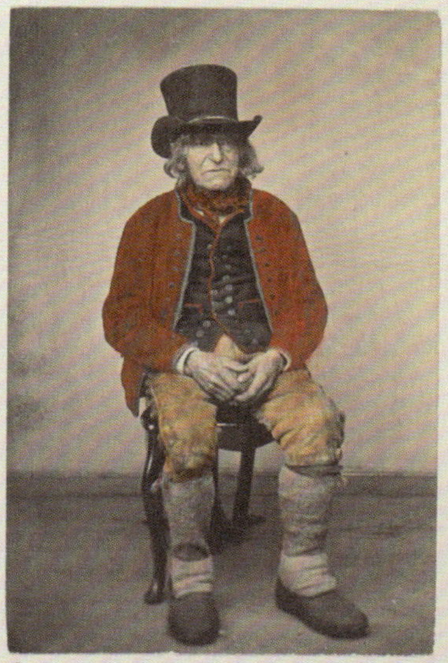

Mand fra Hardanger.

Left Norway Saturday August 17.

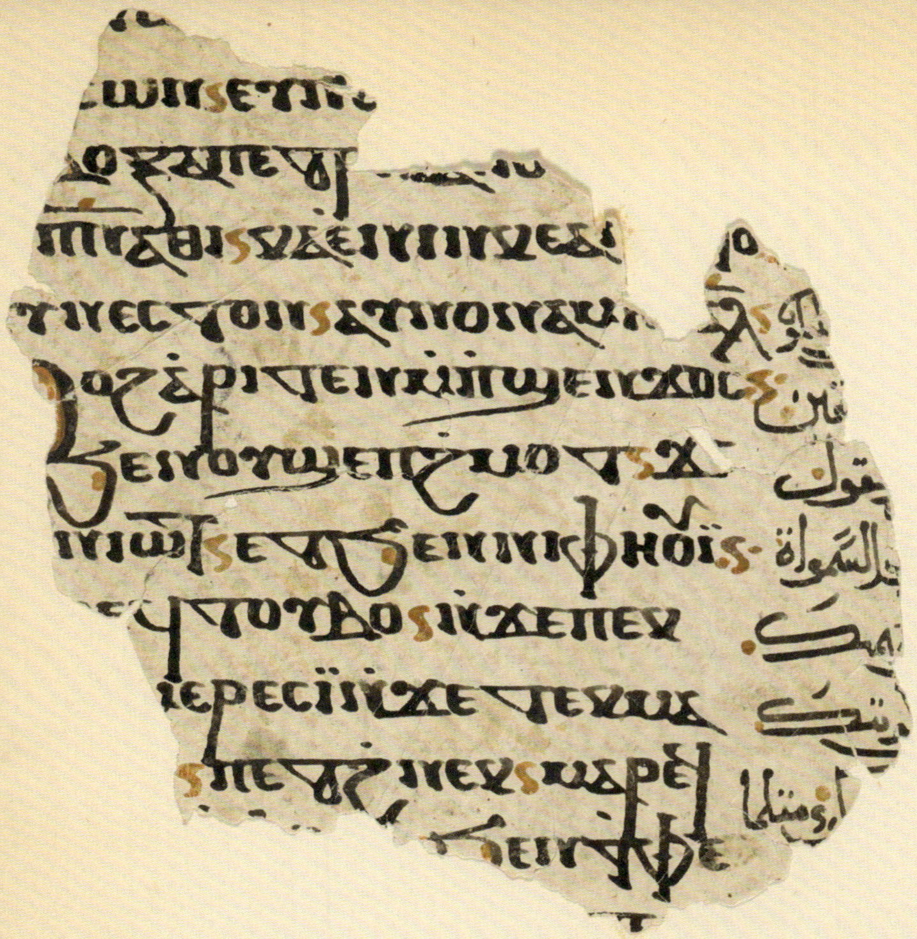

A specimen of Coptic, from the old Empress Helena Convent — Arabic on the other side.

poor fellaheen at the Shadoofs did sing their sad song that sounds like a longing for some-thing of which they have no hope — We moored for the night near 3 little steamers & a mysteriously shrouded Dahabeah. We went for a walk on the bank - the blaze of the charcoal tents in the distance & the moon so soft & full - We picked cotton in the pods & Baucci met the cook of one of the steamers & found out that

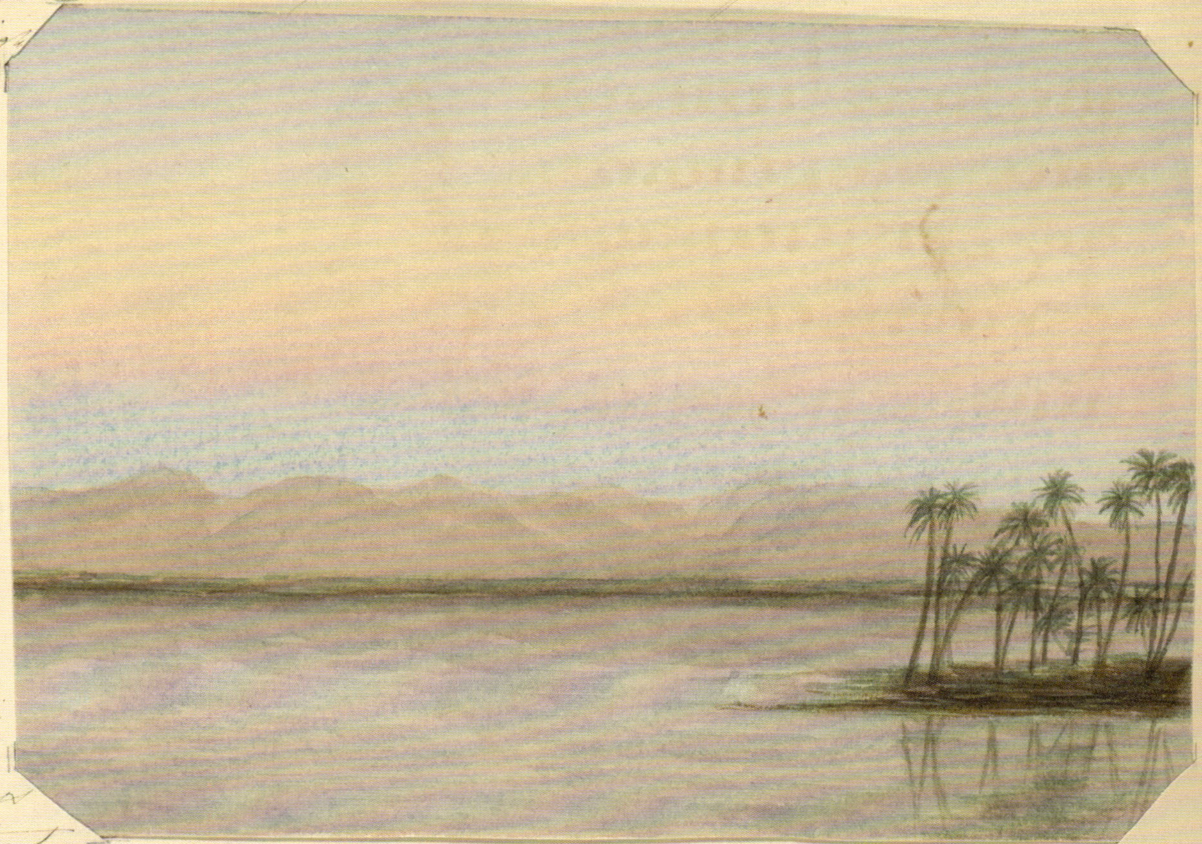

the Dahabeah & steamers were on a secret railroad expedition to Kartoum. Saturday Jan. 23. Getting nearer Nubia & having perfect weather. The character of the scenery changing. Sunday Jan. 24 We were called early by Baucci, with the news that "the Rachel" was coming, so we hurried on deck & had a yell with our A. Skelton. I was really disappointed not to have met him at some stopping place, & hear his experiences. As the sun set the clouds were purple & silver, different from any other sunset. The old Reis suddenly jumped up & called out "Temsah" & we saw our 1st Crocodile sprawled out on a sand bank - a loath-some creature. He slumped himself slowly into the river, leaving a wet slimy track behind him.

Monday Jan. 25 I was awakened by a stirring on the boat about 4 1/2 a.m. & looked out. Venus as large as a moon, & a fleet of Dahabeahs in full sail — a race for

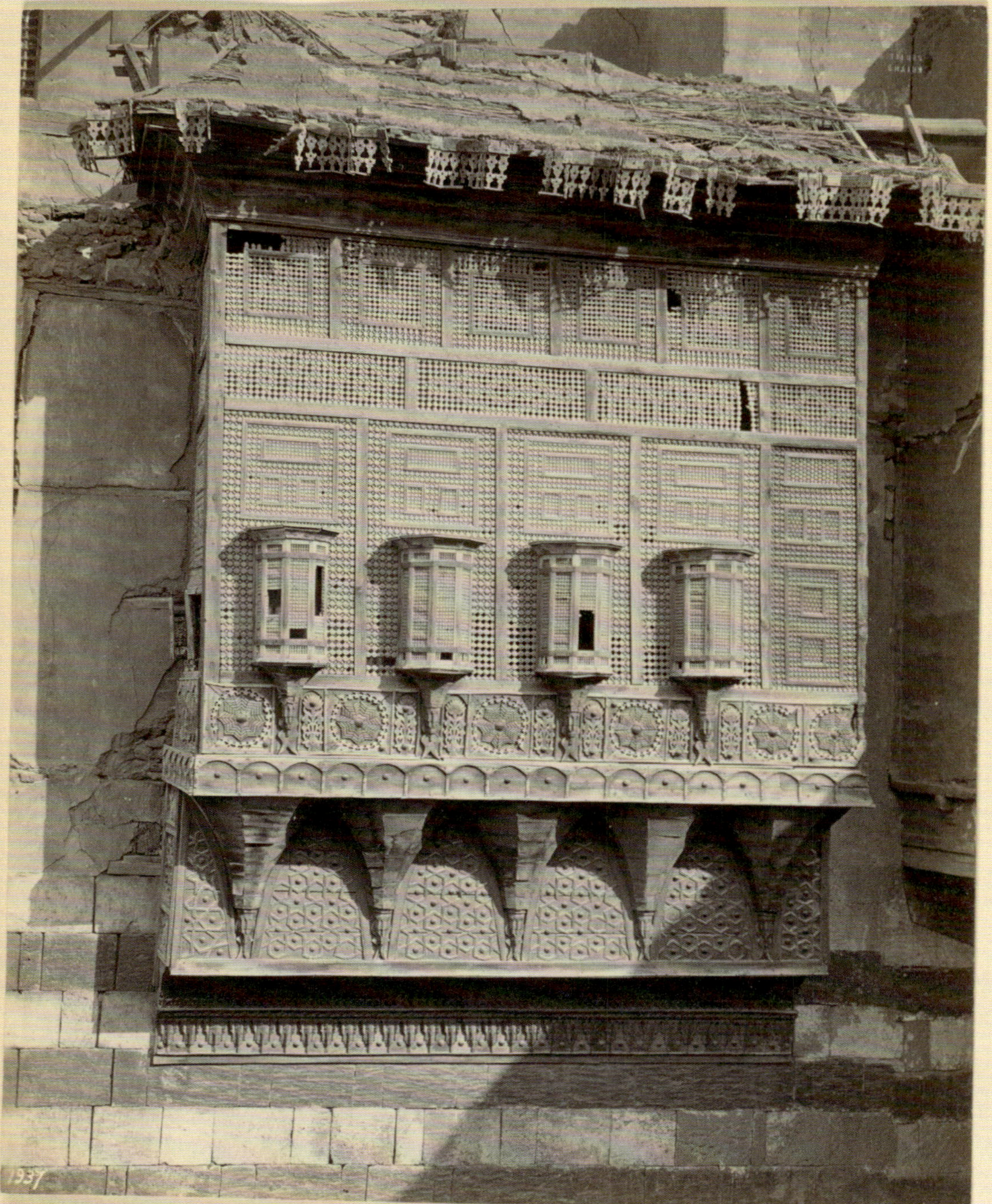

the sky parlour with smoking coffee & plumcake. In the evening just before dinner, the Sheikh arrived with an escort, he & his head man came into the little parlour & had cigarettes & insisted upon going outside & waiting from declining dinner, insisted upon going outside & waiting until we had dined – which we did & went out, to find them squatted in a row. Our new Sung, our guests had coffee & then departed with salaams &c – Wednesday, Thursday & Friday we sailed along, quite near two English Dahabeahs. The Nile shores dotted with Shadoofs, the pole & bucket manner of irrigation. Arrived on Saturday Jan. 9 at Girgeh & the 24 hours bake of the crew's bread began. They were in great spirits, asked for backsheesh & "a mutton" was granted. After breakfast we went ashore, mailed our letters & went to the Latin Convent. It was good to see the I.H.S. & the ✕ over the very picturesque door of the Chapel. The one monk of the brotherhood lives at the top of the house & we climbed up quite a flight & out onto the roof & then looked onto others. And there the monk lived. It was a charming place with a delicious air

ruined mosque Girgeh Jan. 9.

view. He had a deewan & awning there & opening on to it were his two rooms. One a nice little parlour with a mushrebeeyeh (sort of bow window). He gave us coffee & told us of the schools that were connected with the church to which many Copts belonged. The Copts & Latins, he said, were now together "con amore". Sunday Jan. 10 The crew had been ashore all night baking their bread at the public bakery & they were ready to start about 10 & we sailed off with a fine wind. The Gebel Girgeh was still on our left & it was rather dangerous

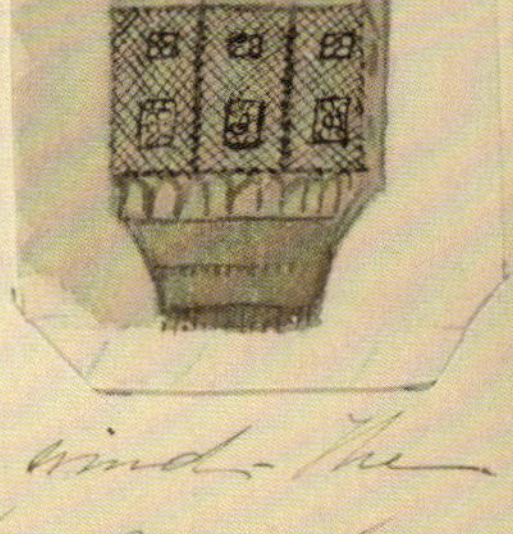

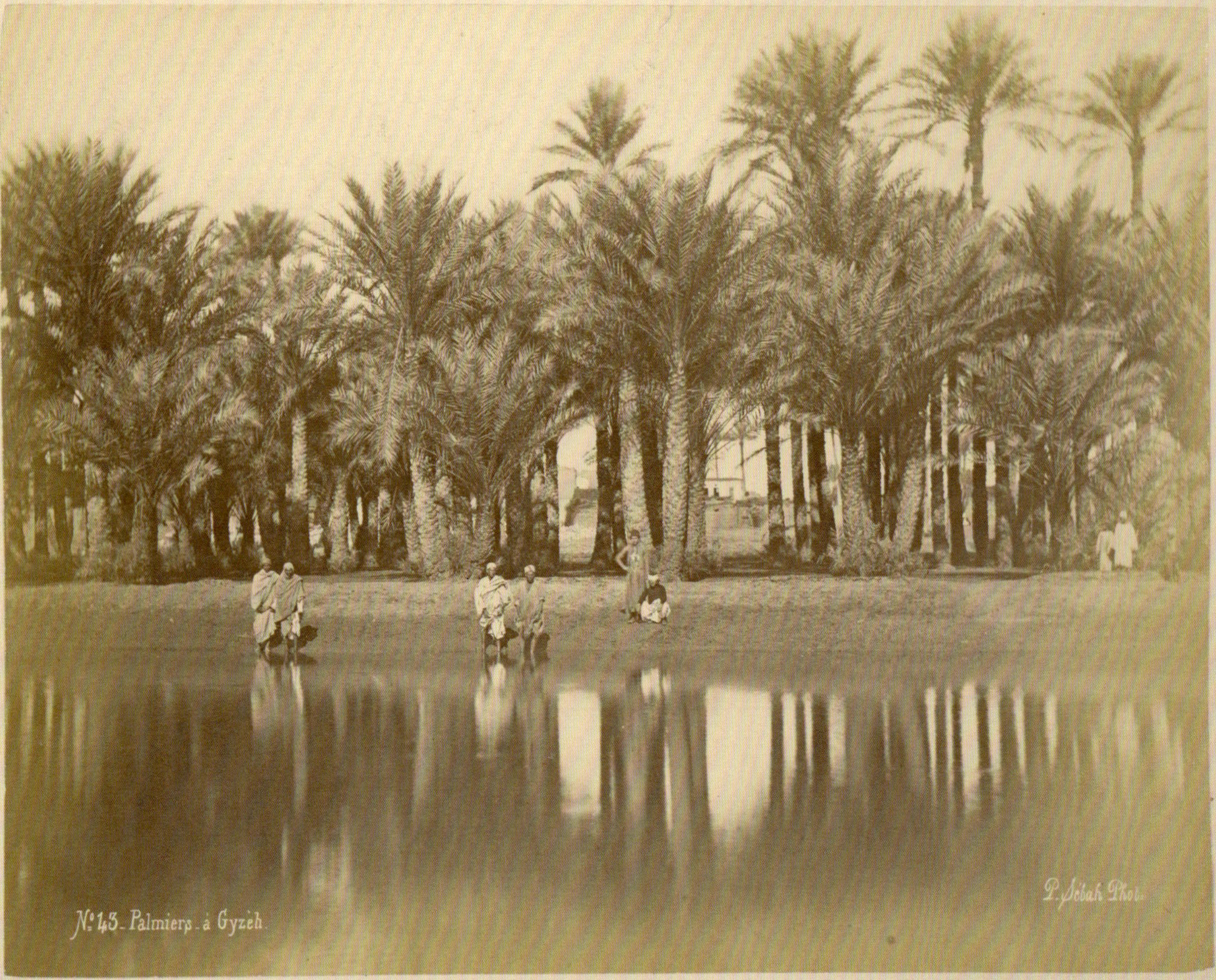
N°43_Palmiers à Gyzeh
P. Sébah Phot.

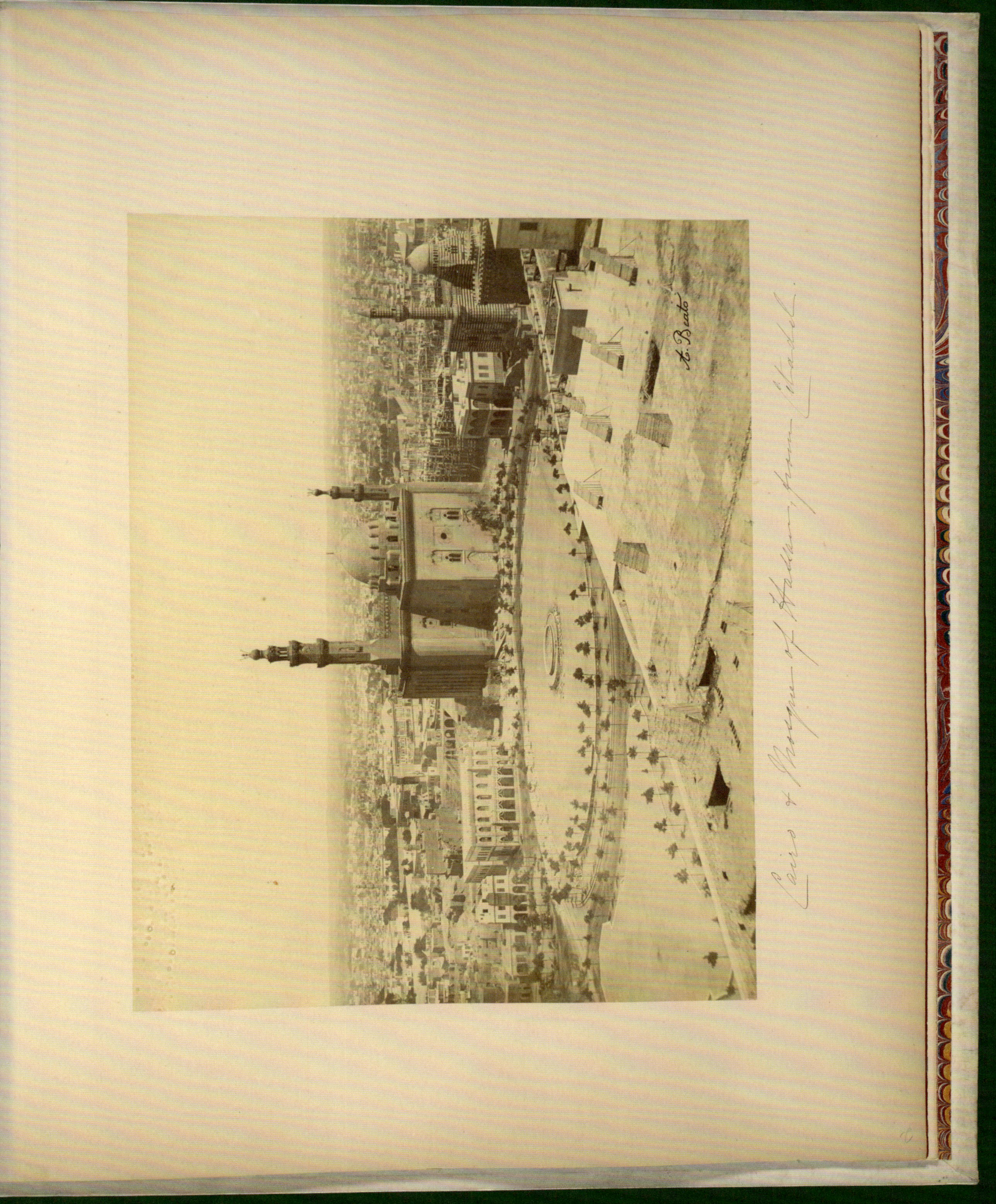
A. Beato
Cairo & Mosque of Hassan from Citadel

Volume I, pages 26–27, June 1879

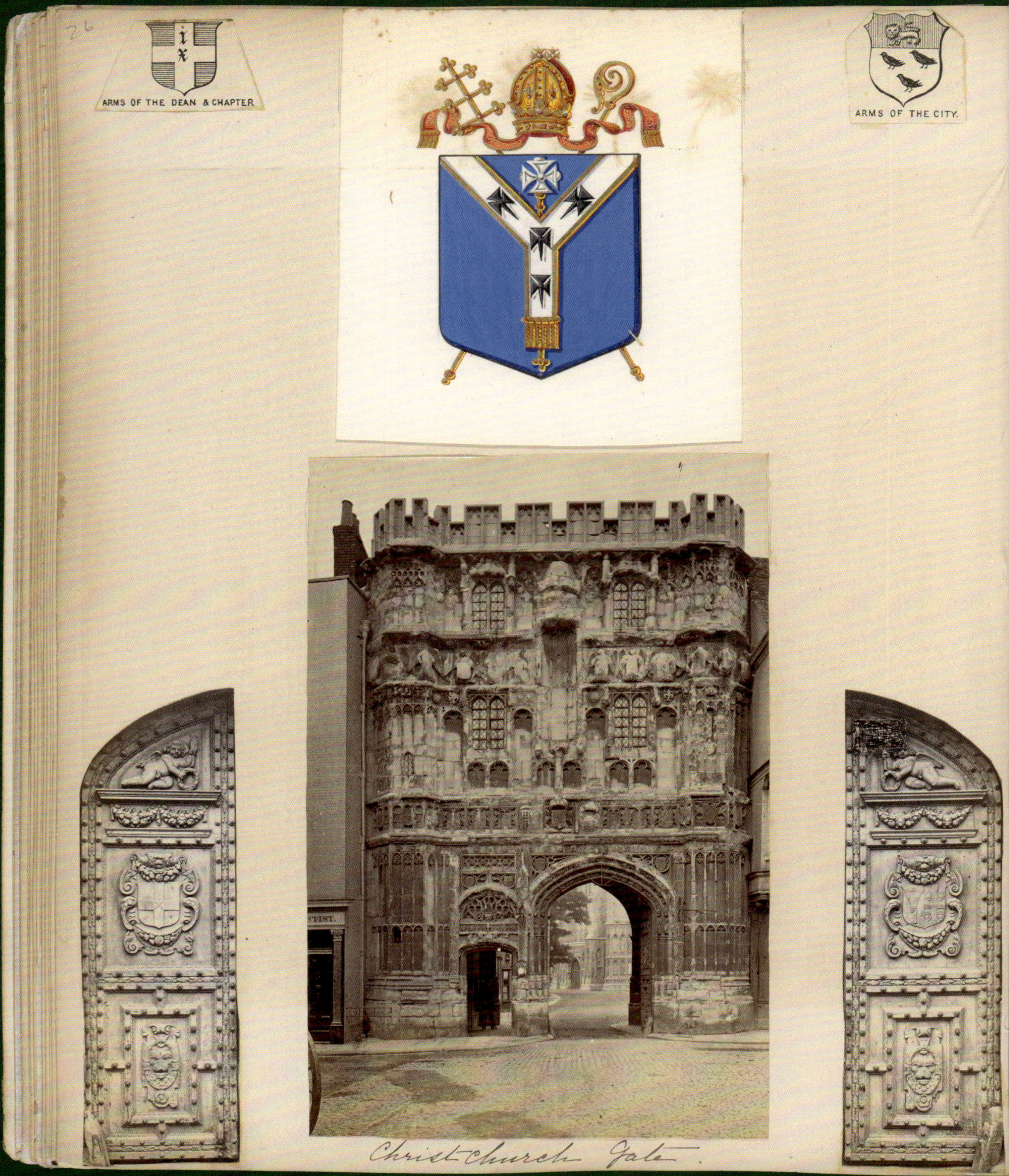

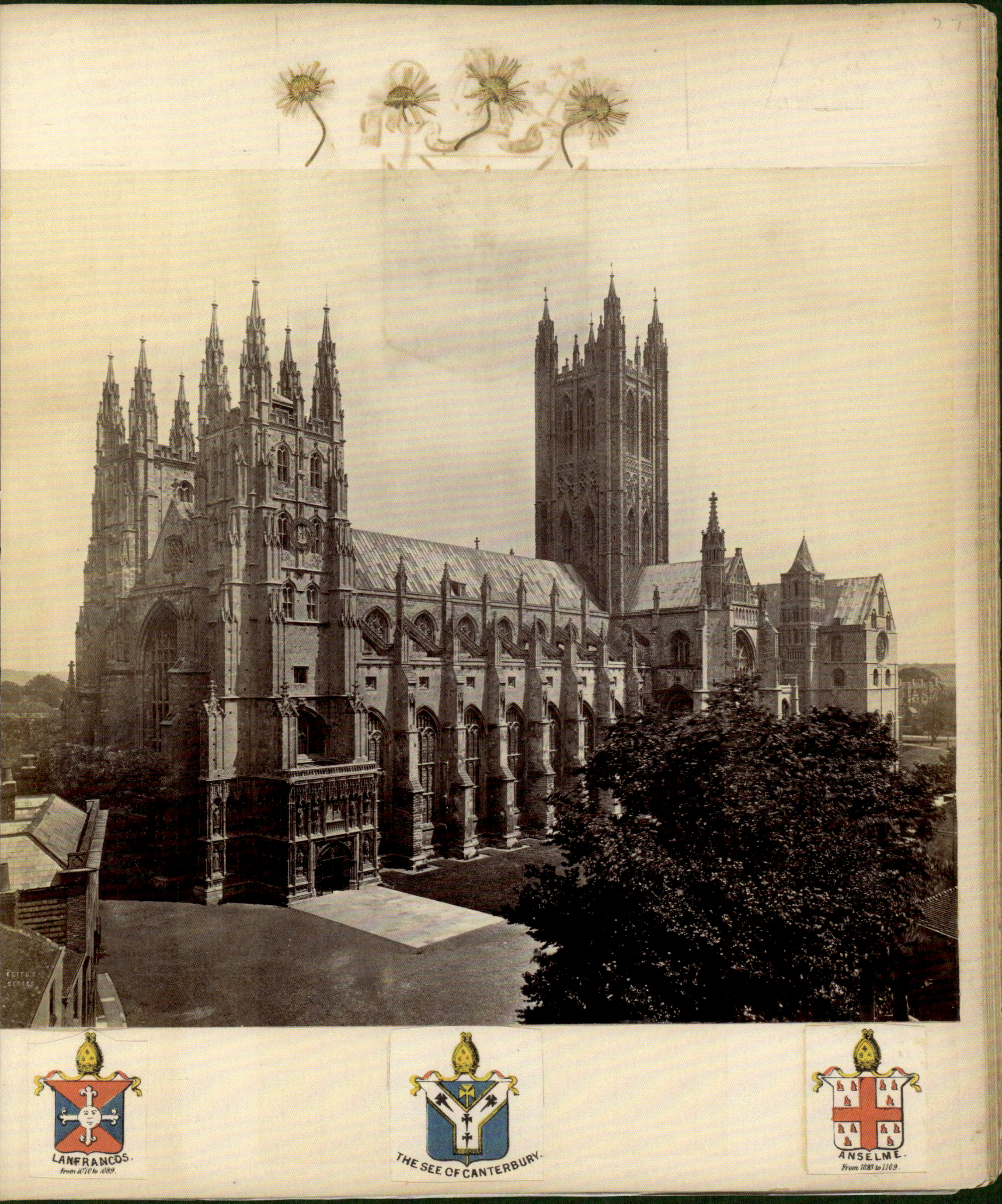
LANFRANCOS.
From 1070 to 1089.
THE SEE OF CANTERBURY.
ANSELME.
From 1093 to 1109.

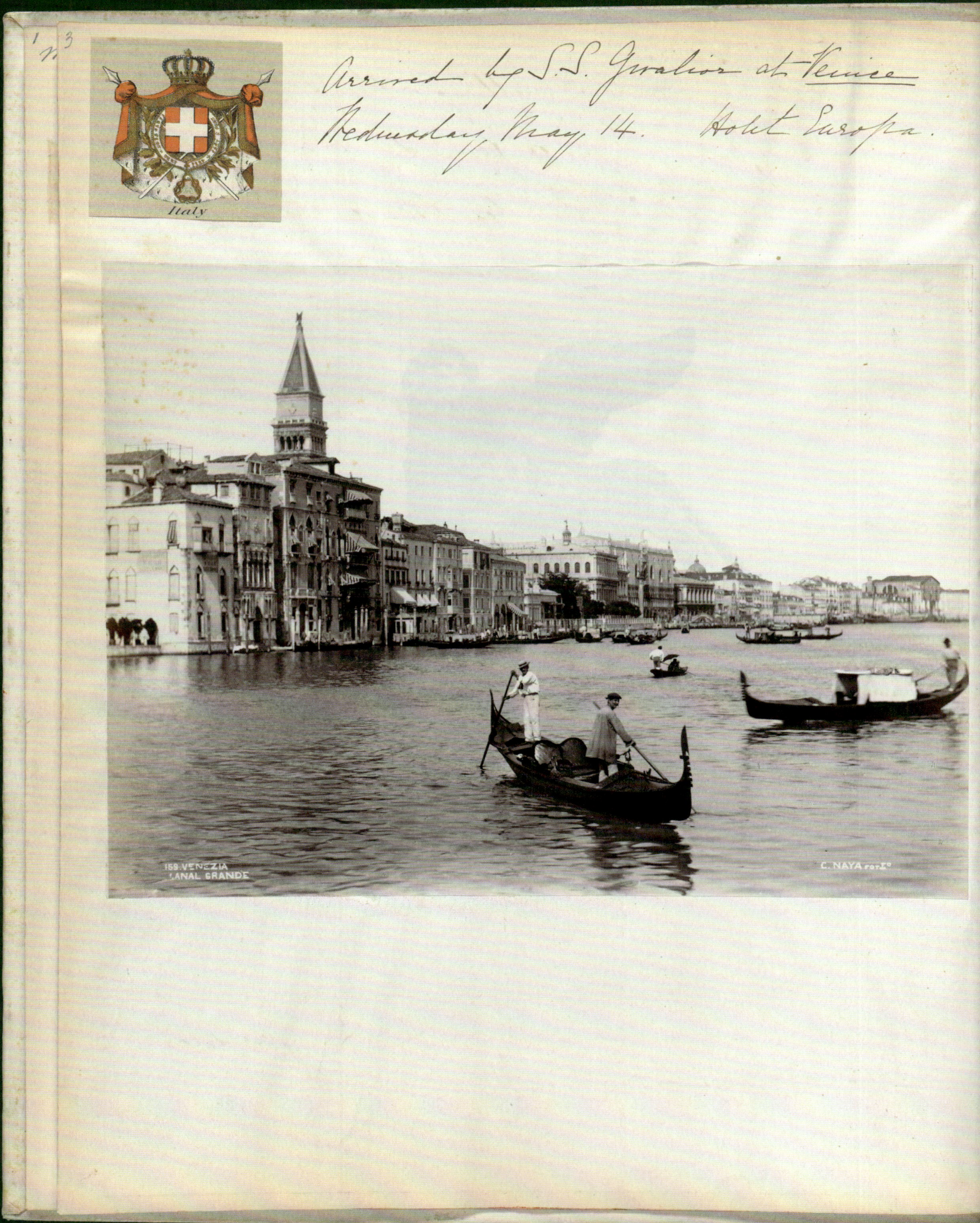
Arrived by S.S. Gwalior at Venice
Wednesday May 14. Hotel Europa.
Italy
159. VENEZIA
CANAL GRANDE
C. NAYA

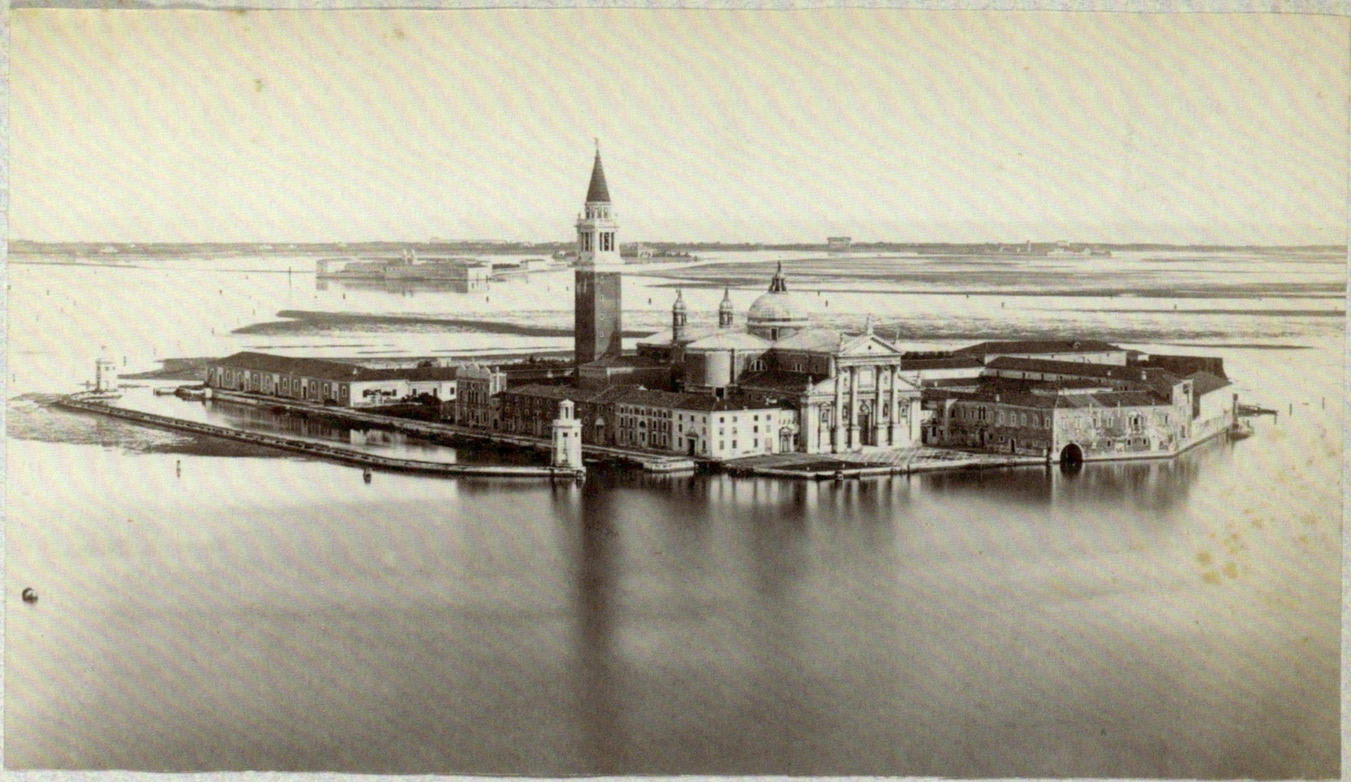

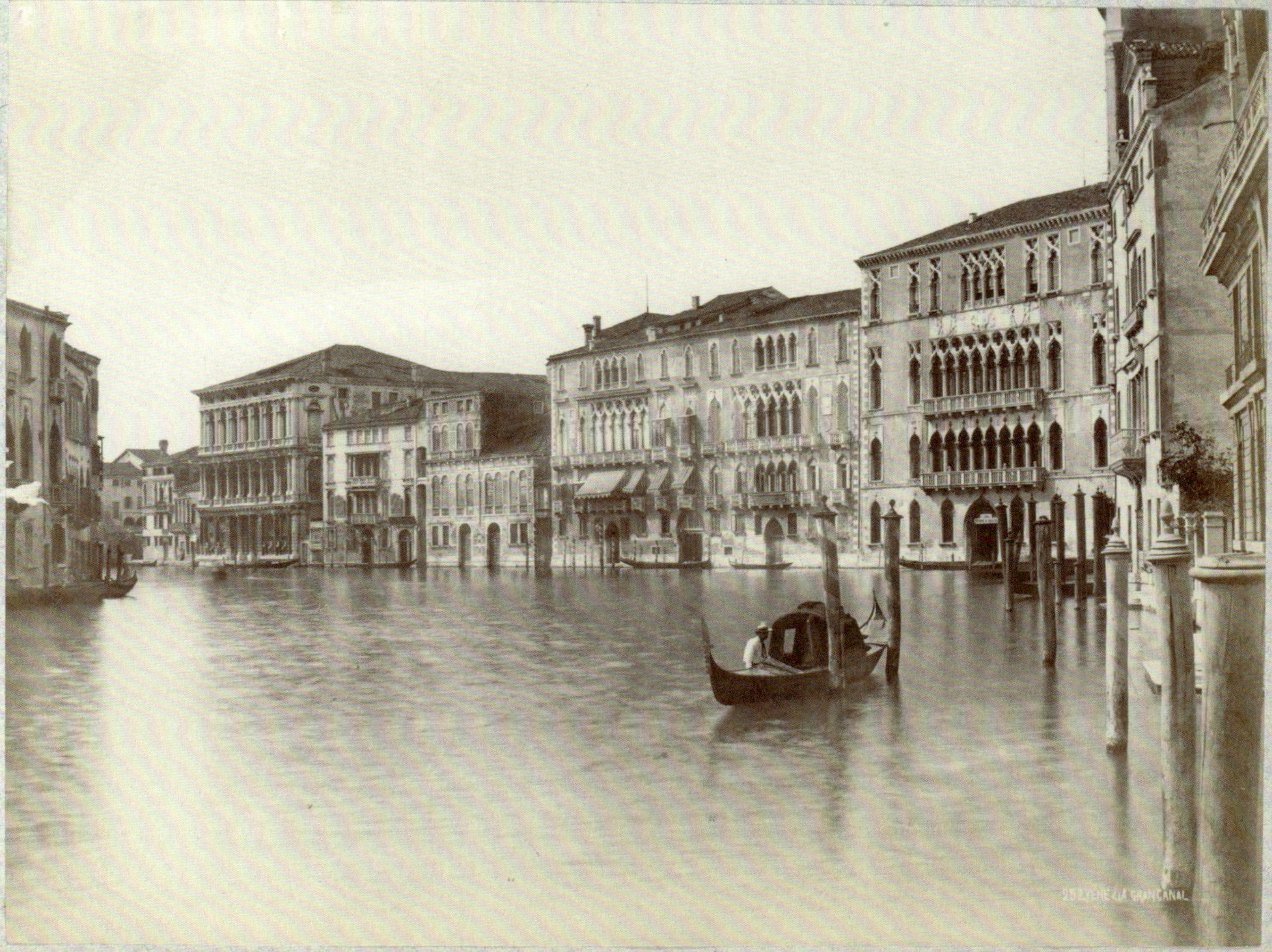

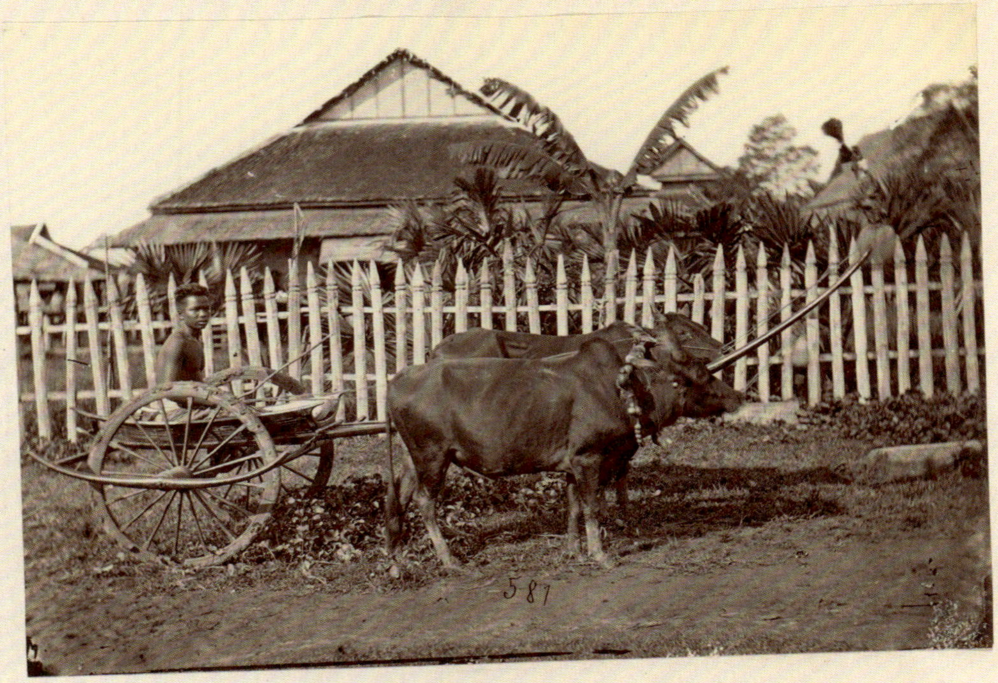

Ruines d'Angkor-Wat
Siam
Diner du 17 novembre 83

Sardines - Pot au feu Beurre

Mortadelle -

Paté de perdreaux.

Homard

Asperges

Canard
Haricots verts

Confitures - Fraises

Vermouth Vin de Médoc Sancerre Champagne

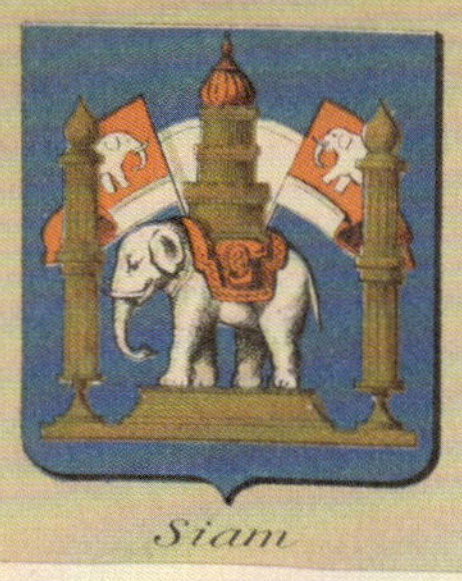

Angkor Wat. Nov. 17 – 19.

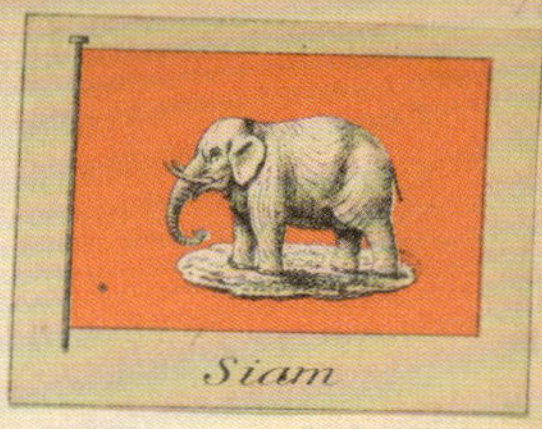

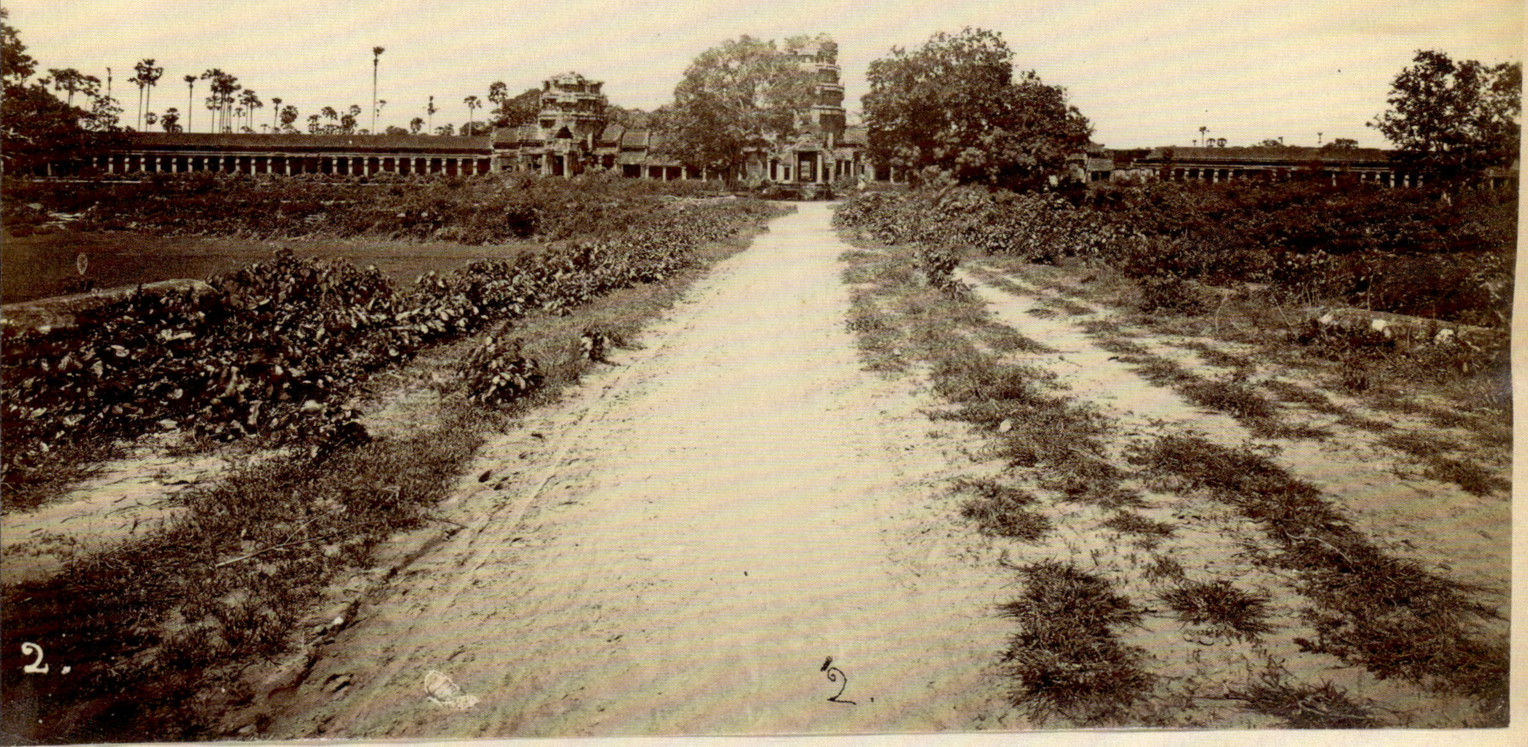

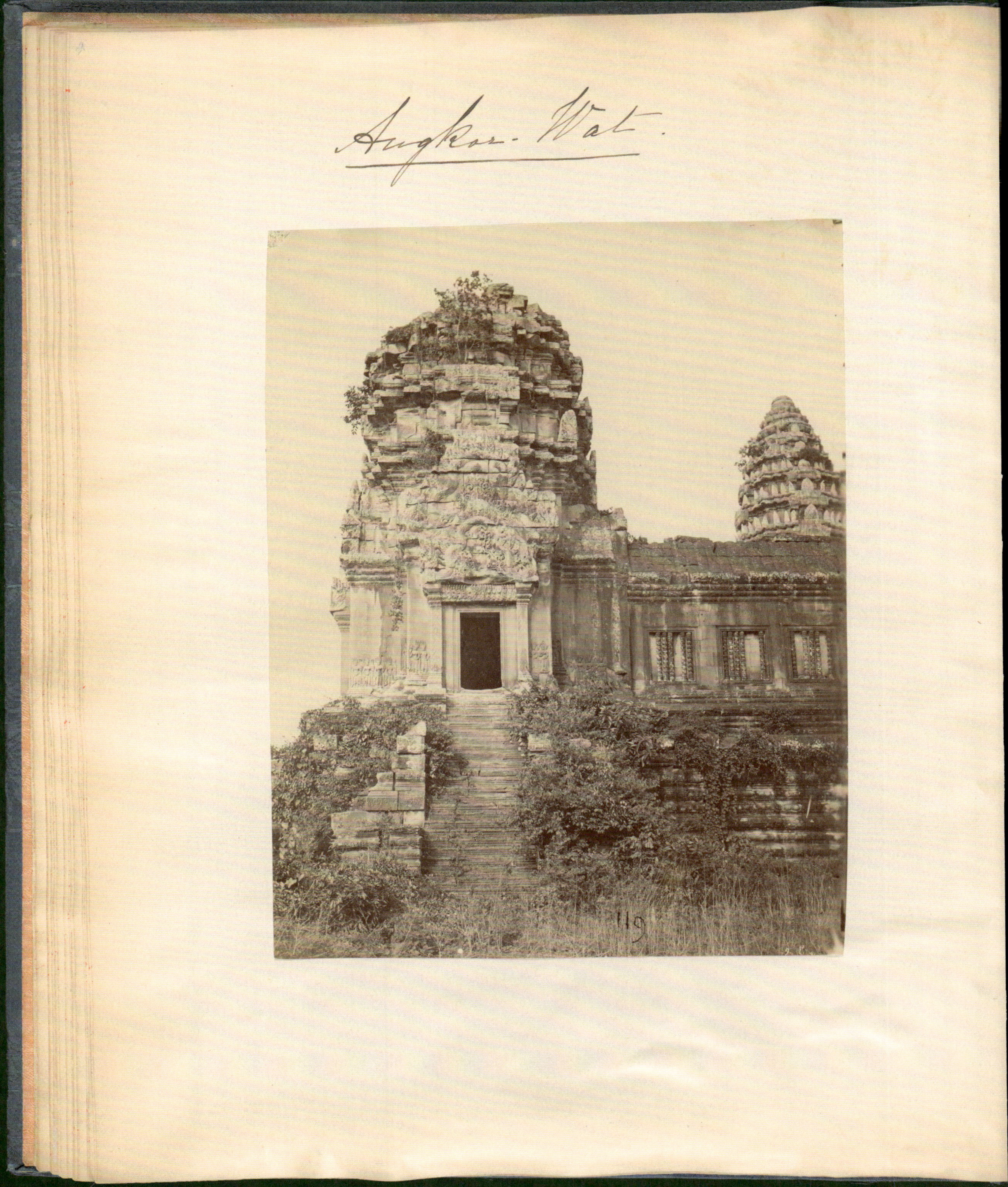
Angkor-Wat.
119

Angkor-Wat.

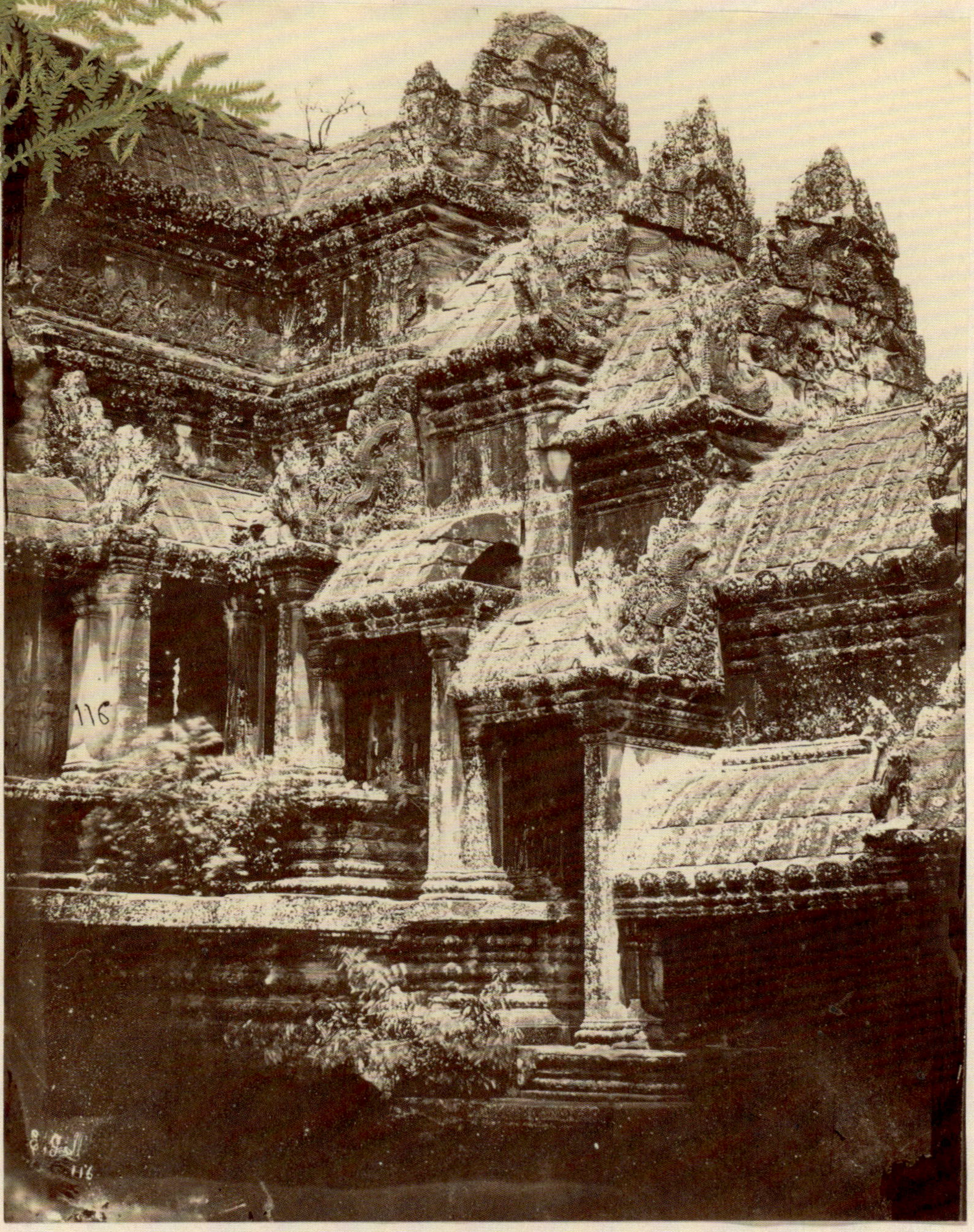

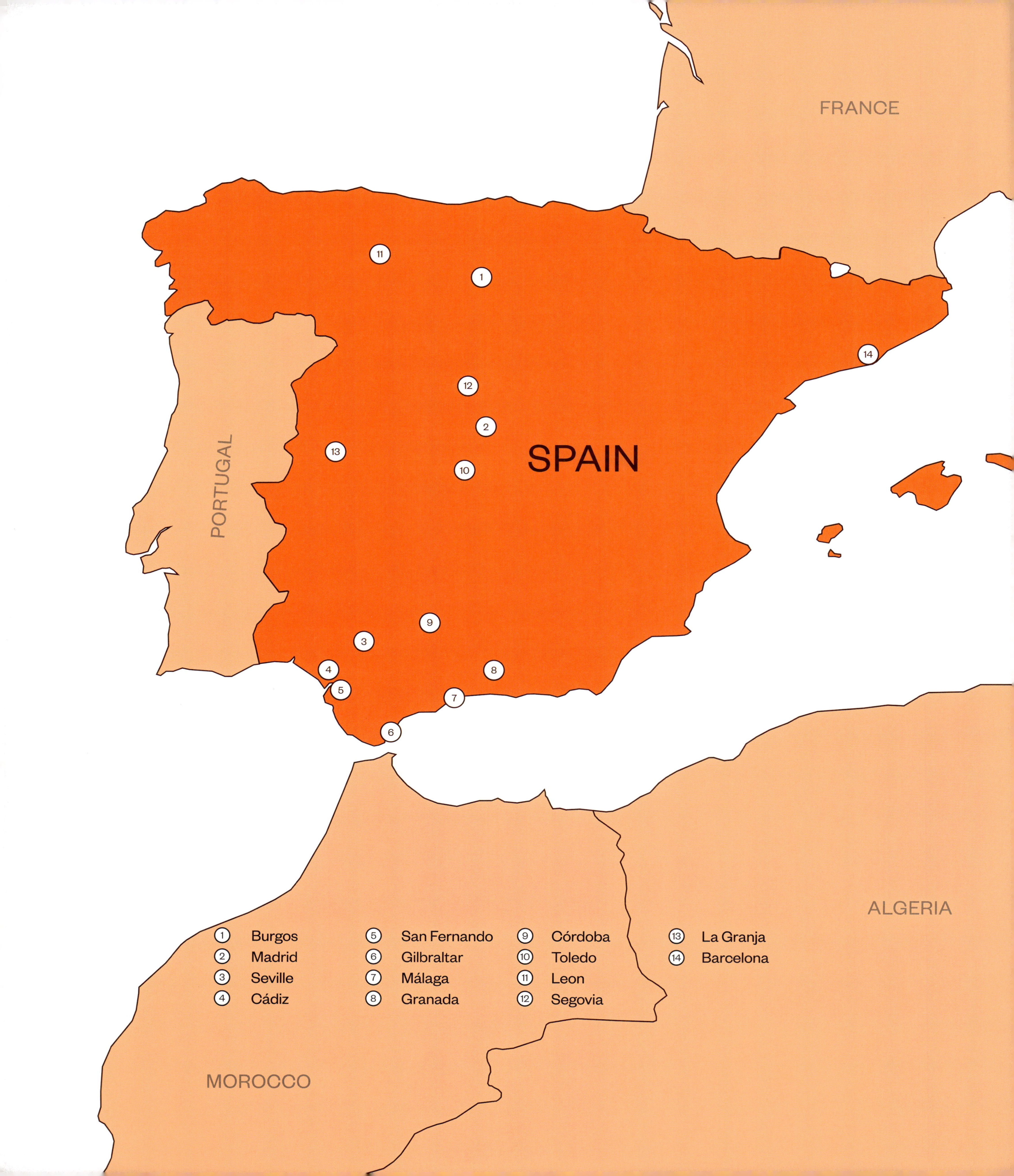
FRANCE
SPAIN
PORTUGAL
ALGERIA
MOROCCO
1 Burgos
2 Madrid
3 Seville
4 Cádiz
5 San Fernando
6 Gilbraltar
7 Málaga
8 Granada
9 Córdoba
10 Toledo
11 Leon
12 Segovia
13 La Granja
14 Barcelona

Spain

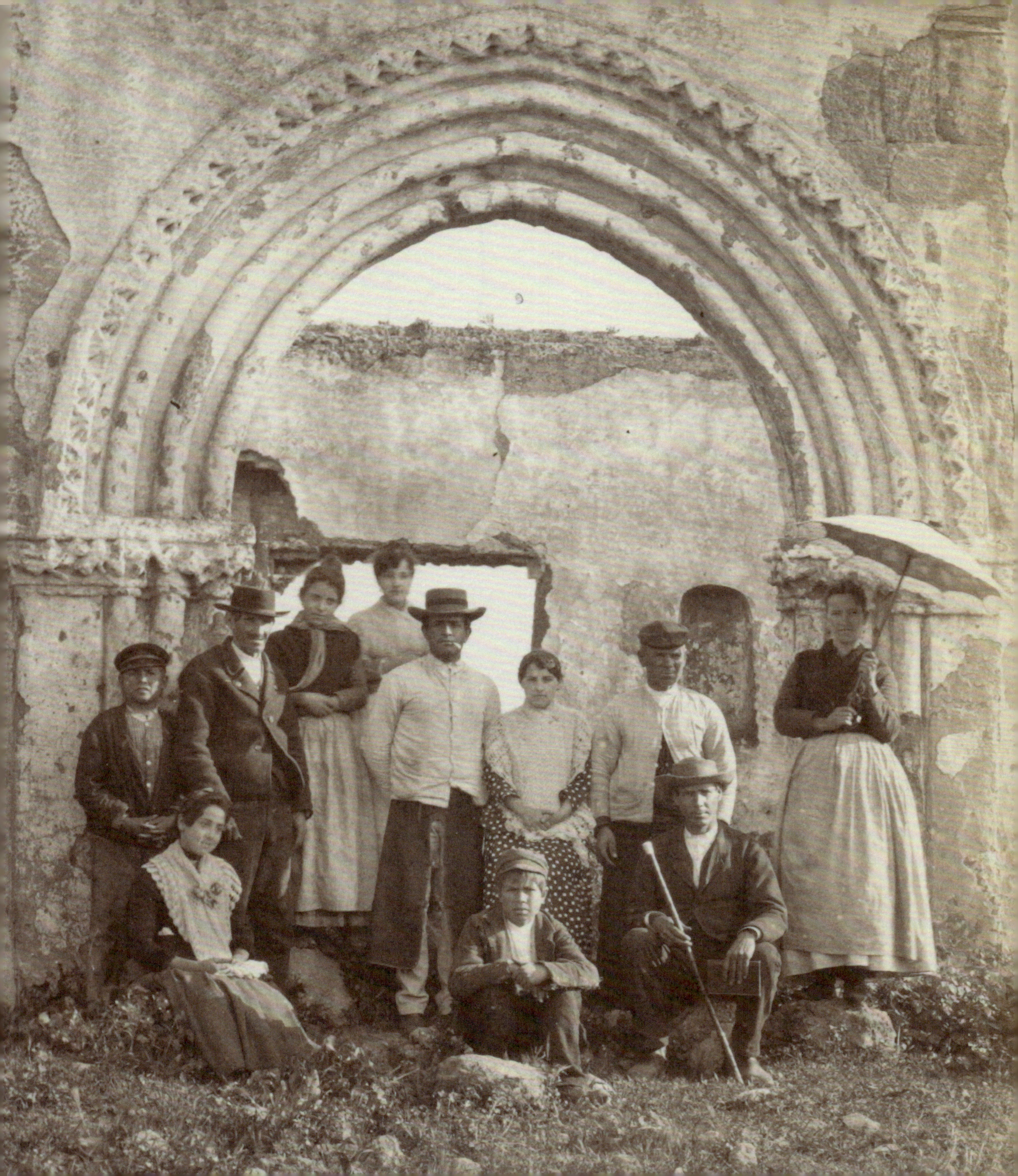

"Spain Says It All" Isabella Stewart Gardner's 1888 Travel Albums of Spain

Madeleine Haddon

In 1888, Isabella Stewart Gardner and her husband, John Lowell Gardner Jr., traveled to Europe to escape a fanfare of rumors and gossip circulating around them in Boston. The couple made plans to visit Spain for the first time, an encounter that forever impacted Isabella Gardner in its cultivation of her love for Spanish art and culture. In commemoration of this trip, Gardner compiled two travel albums that allowed both her and its future readers to retrace the steps of the journey. Her impressions of the country captured within these albums shaped the formation of her growing art collection. They also impacted the very walls of her museum with the construction in 1914 of the Spanish Cloister to house John Singer Sargent's (1856–1925) *El Jaleo* (1882) (fig. 1)—the painterly embodiment of Sargent's own trip to Spain—which Gardner had admired since first laying eyes on the picture shortly after it was completed.[1] Gardner's fascination with ideas of "Spanishness"—complex fictions developed within the nineteenth century around Spain's history and culture—is preserved within the museum through *El Jaleo* and the Spanish Cloister but through her own hand in the two albums that she compiled from her 1888 trip. For Gardner, and many others of her generation, Spain represented a country with a multicultural past and present that synthesized and blended cultures much as Gardner achieved within her own museum. As Gardner wrote as the epigraph to the second volume of her Spanish travel albums, "Spain says it all."[2]

Gardner's decision to travel to Spain stemmed from several weeks she spent in Sargent's company as he painted her portrait in January 1888 (fig. 2). Sargent and Gardner most likely discussed his travels to Spain and love for Spanish art during the many hours of their sittings, when the two lifelong friends first started to get to know one another. Gardner enjoyed the sessions so much that she did not want the portrait to be completed for it meant their time together would end. Sargent redid the painting at her insistence nine times.[3] Sargent had traveled to Spain in 1879 after completing his artistic training in Paris and left enthralled with a nation that to him teetered on

Seville, April 1888 (detail) from Isabella Stewart Gardner's *Travel Album: Spain and Portugal, Volume II*, page 14

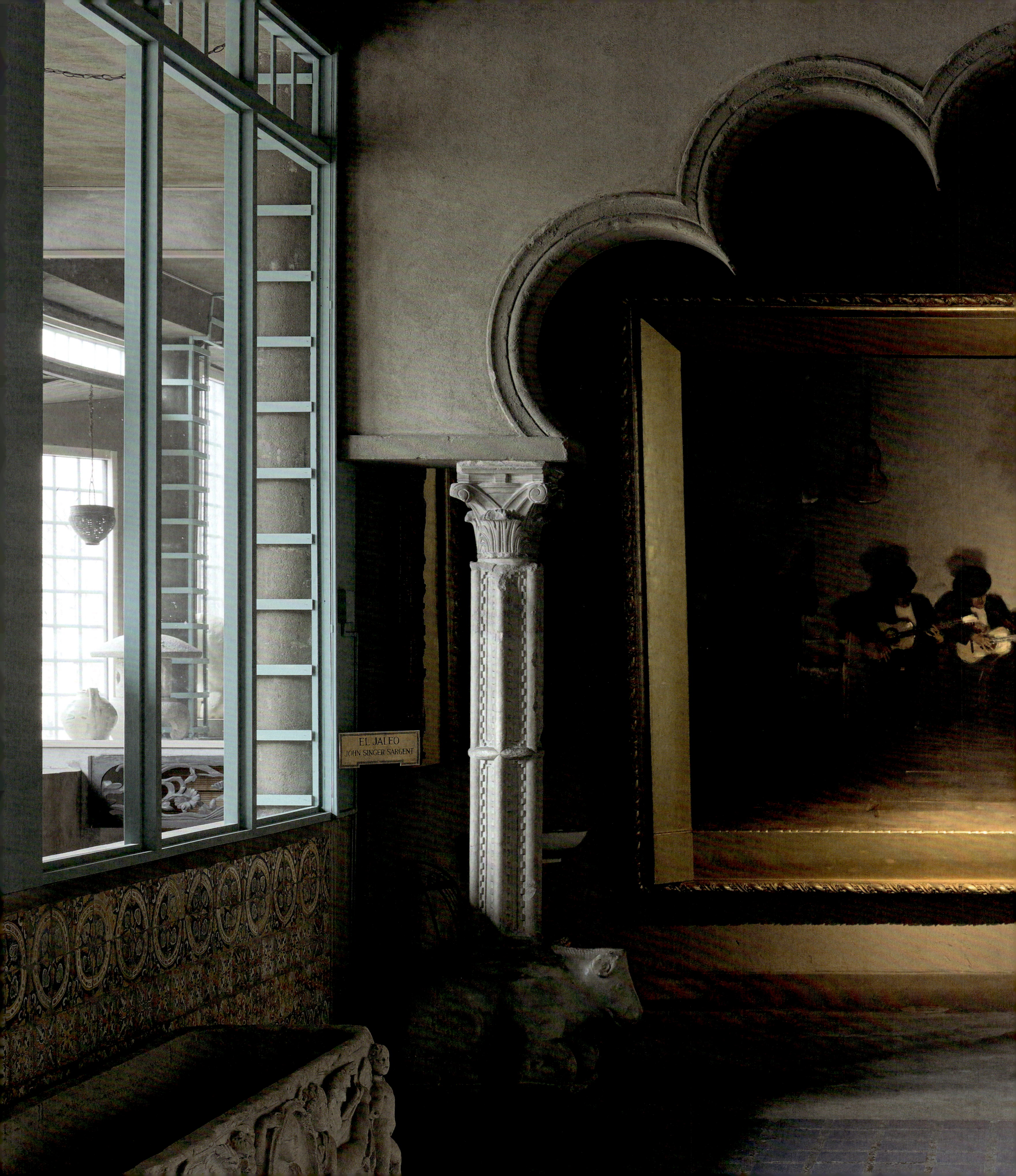
EL JALEO
JOHN SINGER SARGENT

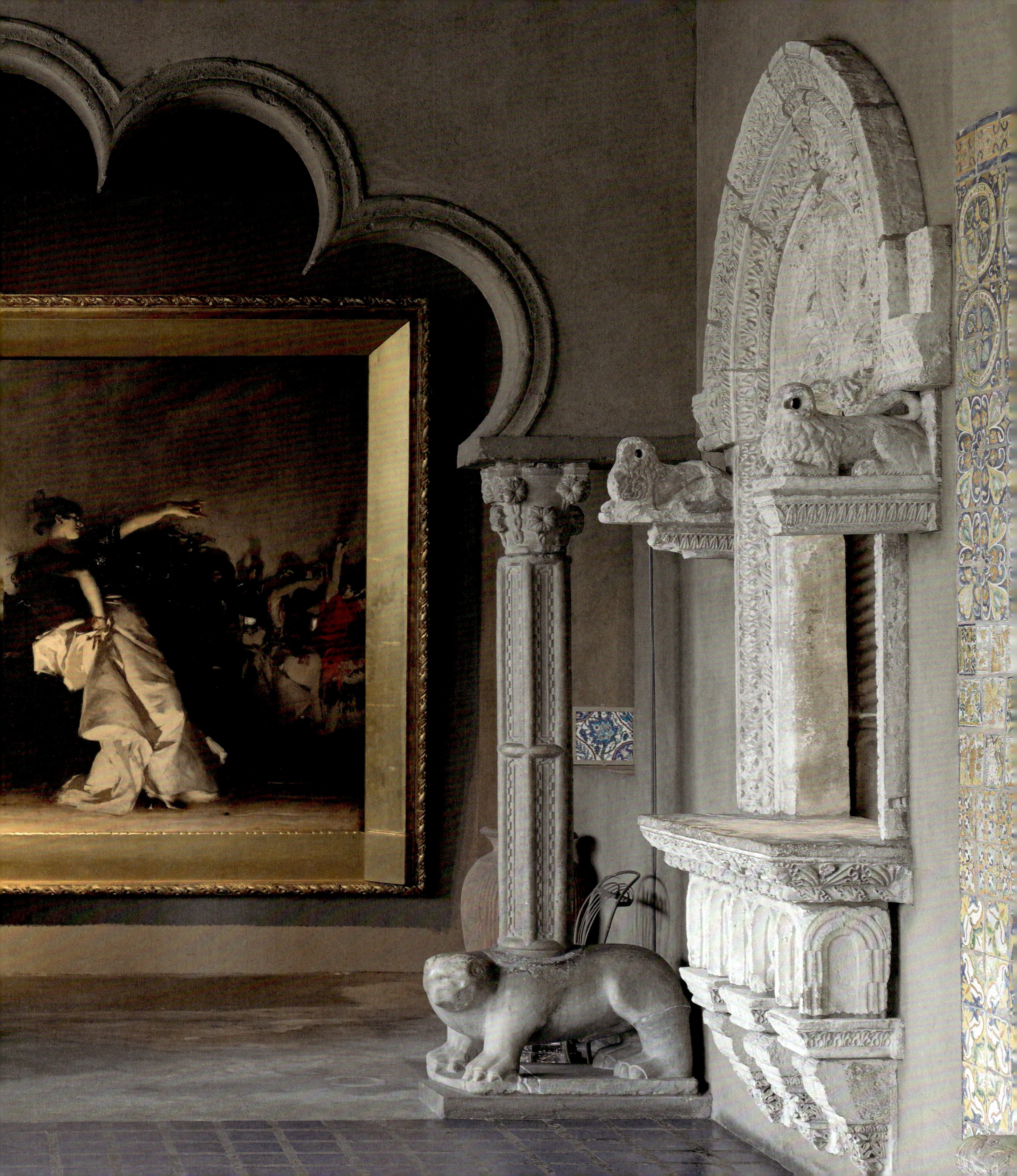

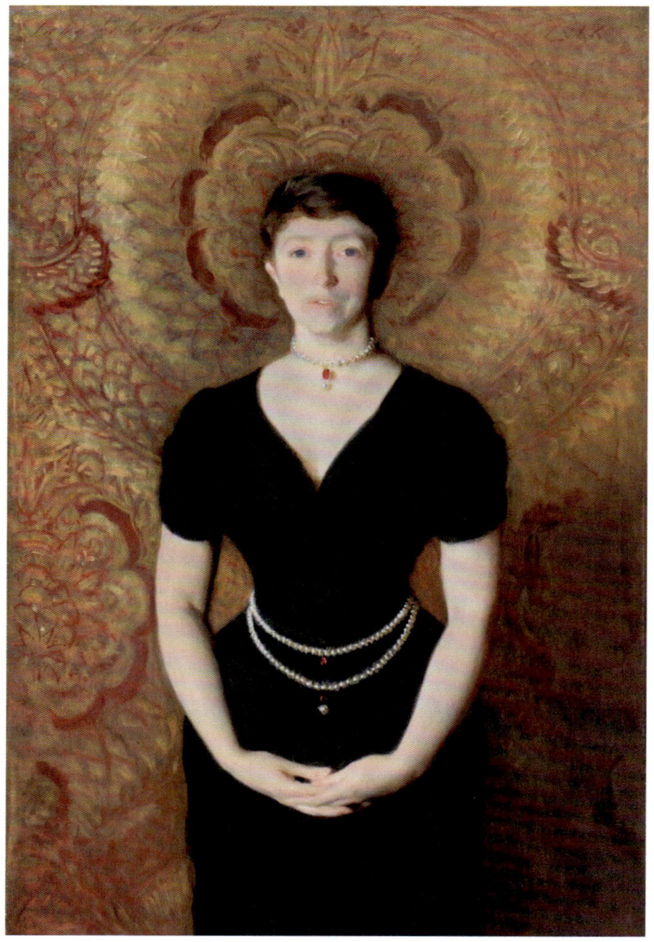

Fig. 2 John Singer Sargent (American, 1856–1925), *Isabella Stewart Gardner*, 1888. Oil on canvas, 190 x 80 cm. Isabella Stewart Gardner Museum, Boston

and blended the borders between Europe and North Africa. Following his return to Paris, Sargent spent two years attempting to create a painting that captured what he had experienced in Spain.[4] This sustained effort eventually resulted in his creation of *El Jaleo* (fig. 1), which caused such a sensation when it debuted at the 1882 Salon that one critic called it a depiction of "strange excrescences [who] purport to be human beings playing the guitar," an expression of an "uncouth blackness."[5] Gardner was captivated by the painting and eventually achieved her goal of acquiring it in 1914.[6]

Sargent's portrait of Gardner—a critical moment within the development of their relationship—was unveiled on 30 January at his first one-man exhibition at the St. Botolph Club. The painting provoked such a scandal surrounding the revealing cut of the dress in which she posed and Gardner's relationships with young artistic men that Jack Gardner demanded it be removed from the exhibition. It was not publicly exhibited again until after his death.[7] The Gardners decided they needed to escape Boston, and the extended period that Gardner and Sargent had spent together is presumably why the Gardners chose Spain as their destination. Sargent most likely advised the Gardners on their itinerary. Gardner was also aware that her close friend the artist Ralph Wormeley Curtis (1854–1922) intended to be in Seville through April, based on a letter that she received from him in February.[8] Curtis joined the Gardners in Seville when they arrived.

Fig. 1 Spanish Cloister with John Singer Sargent's *El Jaleo*, 1882. Isabella Stewart Gardner Museum, Boston

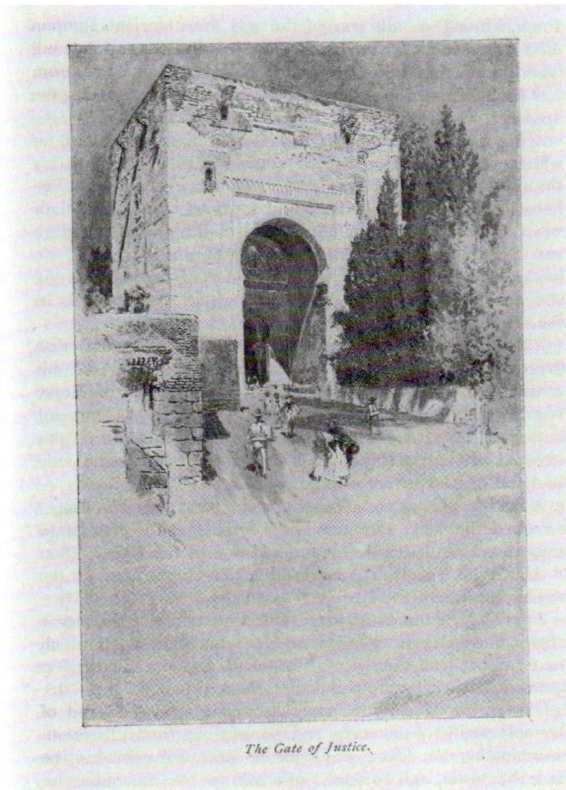

Fig. 3 Washington Irving (American, 1783–1859), *The Alhambra: A Series of Tales and Sketches of the Moors and Spaniards*, pages 182–83, 1896. Library of Congress

Fig. 4 Sir William Stirling Maxwell (British, 1818–1878, author) and Nicolaas Henneman (Dutch, 1813–1898, artist), Talbotype Illustrations to *Annals of the Artists of Spain*, plate 31, 1840s. Salted paper print. Metropolitan Museum of Art, New York

Both Gardner and Sargent had been swept into a much broader cultural fascination with Spain, and attendant ideas of "Spanishness" that developed during the mid- to late nineteenth century. The word "Spanishness" first became part of the English language during the 1830s, around the same time that the equivalent term began to appear in French and Spanish.[9] It is still employed frequently without much clarification for what it means beyond an attempt to describe the shifting stereotypes concerning the qualities of being Spanish. Relatively few Americans traveled to Spain, compared to other European destinations, but these trips increased significantly during the nineteenth century.[10] Many American art collectors and artists, like Gardner and Sargent, visited in order to seek out the masterworks of Spanish art by Diego Velázquez (1599–1660), El Greco (1541–1614), and Francisco de Goya (1746–1828) housed in the Prado Museum in Madrid.[11] However, for many American travelers, the Iberian Peninsula also represented intoxicating exoticism. Such depictions of Spain were popularly consumed through publications such as Washington Irving's (1783–1859) best-selling book *The Alhambra: A Series of Tales and Sketches of the Moors and Spaniards* (1832) (fig. 3), inspired by his travels through Andalusia in 1828–1829, in which he describes Spain as "a country apart; severed in history, habits, manners, and modes of thinking, from all the rest of Europe."[12] From the beginning of the nineteenth century onward, Spain was described as "the doorway to the Orient," an open hunting ground for those searching for easy access to the exotic, while Spaniards were perceived as a "half-European and half-African race of men."[13] Americans were enthralled by Spaniards not only as exotic fantasies of the "other," but also as people who threatened their established notions of what it meant to be European. The racial ambiguity of the Spanish, rooted in the Muslim conquest of Spain from 756 to 1492 and the country's large Romani population and cultural influences, embodied American fears of miscegenation.[14] American art critics responded to *El Jaleo*, in particular, with hostile invocations of the period's discourse around Spain's racial ambiguity.[15]

The Gardners, like many before them, were also in search of the romanticized and exoticized experiences of Spain that dominated American popular culture. Arriving in Spain on 18 March 1888, they started their trip in Burgos and remained in the Iberian Peninsula for three months, until 12 June. They toured the full breadth of the country, traveling to Madrid, Seville, Cadiz, San Fernando, Gibraltar, Malaga, Granada, Cordoba, Toledo, Leon, Segovia, and Barcelona, an itinerary that is unique from that of other foreign travelers, who were primarily interested in seeing only Madrid and Andalusia. They also made a brief stop in Portugal to visit Lisbon and Sintra. The Gardners consumed all aspects of the cultures of the Iberian Peninsula, from east to west, north to south, with particular focus on splendors of historic art and architecture, as if it were a place frozen within the past. Gardner's albums are powerful recreations for her memories and future readers of her perceptions of what defined Spanishness.

Gardner's albums document each of the stops along their trip in chronological order (Plate 9). Each section introduces a new city, beginning with the dates she and Jack were there and where they arrived from, and a quotation that relates to that location. It also includes a colorful lithograph of the city's coat of arms and small illustration of a man and a woman in the region's traditional dress that stand on either side of a plaque containing factual information in Spanish about the details of that region—including its physical size, number of inhabitants, and number of cities and towns. The coats of arms resemble ones that Gardner pasted in her 1879 album from England and France, as if produced by a related publisher. However, the fact that the coats of arms are similar typographically to the figure illustrations which contain Spanish text suggests that these both come from a collection that Gardner purchased while in Spain.

The two small male and female pairs in these stamp-like illustrations are among the only people who populate the pages of the first album. Gardner fills most of the space with photographs of architecture, monuments, and works of art. The first volume focuses primarily on her time in Madrid and over fifty pages are taken up with photographs—often trimmed to fit within her collages—of her favorite works from Madrid's Museo Nacional del Prado and Real Academia de Bellas Artes de San Fernando by artists such as Velázquez, Murillo (1617–1682), Titian (c. 1488–1576), and Goya (Plate 10). Gardner creates a textbook-like reference of historic monuments and celebrated artworks that presumably informed her own future purchases of Spanish art.[16] These portions of both albums resemble William Stirling Maxwell's (1818–1878)

Annals of the Artists of Spain (1847) (fig. 4), the first chronology of the history of Spanish art to be published in English and the first book of Spanish art to include photographs created through the early photographic process known as "talbotypes" invented by William Henry Fox Talbot (1800–1877).[17] Gardner owned both 1848 and 1891 editions of the *Annals of the Artists of Spain*. These were purchased at auction on 20 November 1891, but she was most likely familiar with this landmark book even before then.

Most of the photographs that Gardner included in her albums were taken by the French photographer Juan Laurent Minier (1816–1886) (Plate 11). Laurent moved to Madrid in 1843 and spent his life working in Spain, taking photographs of tourist and historic sites throughout the country that he sold in Spain and abroad. In 1879 Laurent was appointed the official photographer for the Prado and his works could be purchased at the museum as well as from his studio, J. Laurent y Cia, located at Calle de San Jeronimo, No. 39. Laurent also photographed the collections of Madrid's Real Academia de San Fernando and the Museo de Bellas Artes in Seville.[18] Laurent's photographs are not listed among Jack Gardner's detailed inventory of what the couple acquired in Spain but presumably these photographs were purchased during their trip. Sargent's scrapbook of photographs from his travels between 1874 and 1880 also contains a group by Laurent, of the Alhambra.[19] Sargent may have recommended Laurent's work to Gardner for her own albums and recordkeeping of her travels. Photographs within the album were also obtained from the studios of two other Spanish photographers—Emilio Beauchy (1842–1928) (Plate 15), who specialized in images of Seville and is considered one of Spain's first photojournalists, and Camino (d. 1888) (Plate 18) who was based in Granada and about whom very little is known. The fact that another significant portion of these images were acquired from photographers based in Seville and Granada further suggests that Gardner also purchased Laurent's photographs while in Madrid.

Gardner's handwritten captions that accompany these pasted photographs communicate her personal reflections on how she felt when beholding these monuments to the glory of Spanish art and architecture (Plate 12). She marks her arrival in Seville, where the Gardners spent six weeks of their trip, with the statement "Here one *feels* existence."[20] Being in Seville made Gardner feel more alive, as if the "Spanishness" that she encountered here helped her connect with her own essence and being. Gardner describes experiencing a similar sense of feeling alive within her Egypt diary from 1874–1875. Places that Gardner traveled in which she considered herself to be immersed in the so-called "Orient" produced within her sensations of feeling more connected to herself, with reality sharpening into vivacious color. Her language recalls that of many of her contemporaries who similarly felt that they experienced greater authenticity and sense of wonder when stepping into and consuming places that had been deemed part of their problematic fantasies of the "Orient."[21]

A number of the captions that Gardner writes in the album are quotations that she copied down from Richard Ford's (1796–1858) *Hand-Book for Travellers in Spain* (1848), which she had purchased in London on 6 February 1888. Ford's book, first published in

London in 1845, was based on his own travels through the country on horseback that same year and was the authoritative guidebook for nineteenth-century English and American travelers to Spain.[22] Just above Gardner's remark "Here one *feels* existence," she writes in Spanish, "One who has not seen Seville, has not seen wonder."[23] This is a direct quotation of Ford's opening lines on Seville in *A Hand-Book for Travellers in Spain*.[24] On the first page of her section on Madrid, she transcribes another quote from Ford, a Spanish proverb that translates as "the subtle air of Madrid, while will not extinguish a candle, will put out a man's life."[25] Gardner also brought along with her the 1882 edition of Ford's *Hand-Book for Travellers in Spain* and Henry O'Shea's *Guide to Spain and Portugal* (1885).[26] Ford's guidebook likely led Gardner to Laurent's studio. He lists it in his section on Madrid, along with its address, as the place where "photographic copies of the pictures in the Museo, and views from all Spain, may be obtained."[27]

Gardner introduces the second volume of her album with "Spain says it all" (Plate 13).[28] This epigraph she has adopted from the frontispiece of the 1882 edition of Ford's book and she centers the phrase under her name and the date. The page functions as the frontispiece to her own illustrated guide to Spain. Ford explains that the saying is the epitome of the much-used phrase "Españolismo," the belief held by Spaniards that all anyone could ever want is contained within their country.[29] Gardner wanted this thought to be at the very forefront of her readers' mind as they looked through the images in her album.

The subsequent pages mark a transition within Gardner's trip as she begins to live in Spain's present and not solely within its past. It is not until this volume that Gardner at last introduces photographs of Spaniards among those of Spanish art and architecture (Plates 16 and 17). Andalusia, especially Seville and Granada, was (and still is) seen as the epicenter of Spain's multicultural and interracial past and present, its deep history as a meeting point for European, North African, and Roma cultures. As Jack Gardner wrote to his nephew George of their travels: "The places are interesting but the Spanish ––! Oh!! In the South of Spain they are much more cleanly than in the north, supposed to be a relic of Moorish Civilization."[30] The photographs of Spaniards that Gardner selected for this volume perform the stereotypes of Spanishness prevalent within her time, exposed through their dark skin and posed in typologies of traditional dress.

Even when she is pasting in images of Spain's present, Gardner excludes any signs of its industrialization or modernization, much like other American and European travelers before her. Nor was this subject matter commonly captured by photographers such as Laurent, Beauchy, and Camino. Their mostly foreign clientele had little interest in consuming any aspects of a modernized Spain. Through her selection of photographs and creation of these two albums, Gardner herself edited out the elements of the modern age that she surely experienced. She is an active participant in authoring and maintaining the constructs of Spanishness that dominated her society.

In Seville Gardner attended the Holy Week processions on Palm Sunday and

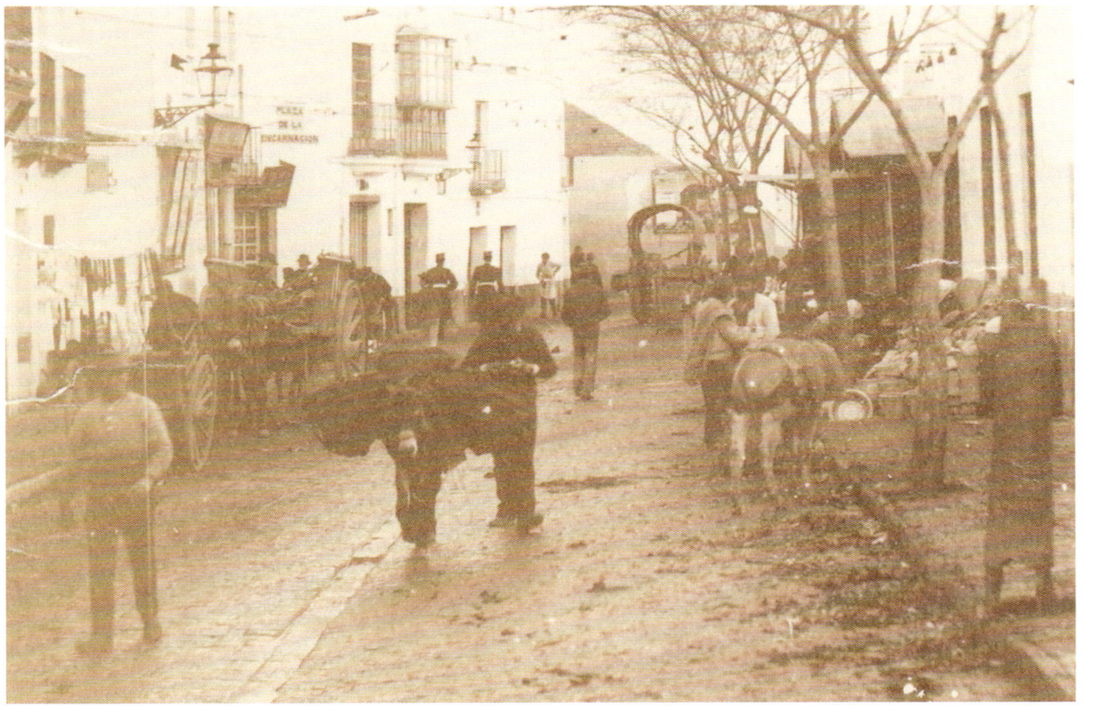

Fig. 5 John Lowell Gardner Jr. (American, 1837–1898), *Street in Seville*, Spain, 1888. Albumen print, 21 × 16.5 cm. Isabella Stewart Gardner Museum, Boston

Good Friday (Plate 13). The photographs she includes depicting these events show how many thousands of people filled the streets in celebration of these centuries-old sacred events. Using a mix of aerial vantage points and street-level views, they invoke the crowds in which Gardner herself would have been embedded and her readers are made to feel as if they themselves are able to watch the procession pass (Plate 14). Remarkably, while observing the processions in Seville Gardner suffered an attack of angina that subsided the moment one of the large crucifixes being carried came by, an episode that became a fabled story of divine intervention.[31]

Next, Gardner transports her reader to the bullfights in Seville's Plaza de Toros (Plate 15). The Gardners were seated in a front-row box on Easter Sunday; a bull was killed in Gardner's honor by the famous matador Luis Mazzantini (1856–1926). Although she was the guest of honor, Gardner, given her love of animals, was horrified by the violence and crouched on the ground, covering her eyes to avoid seeing the cruelty.[32] However, she selected for her album four photographs and one print all of graphic bullfighting scenes, even one in which blood visibly drips down the body of the dying bull. She also dedicated a page to images of bullfighters Mazzantini and his *cuadrilla* and Salvador Sánchez Povedano (1842–1898), known as Frascuelo (Plate 16). Gardner acquired another photograph of Mazzantini that she kept in her collection outside of the albums and inscribed at its top "I saw him kill 3 bulls." Despite her horror, Gardner takes pride in having witnessed what Americans perceived as one of the epitomes of experiencing Spanishness, while simultaneously emphasizing the "savage" and violent natures of the bullfight—Spain's national pastime.[33]

Among the few examples of candid photographs in Gardner's travel albums, she selected two snapshots of herself for her second Spanish album. The first was taken

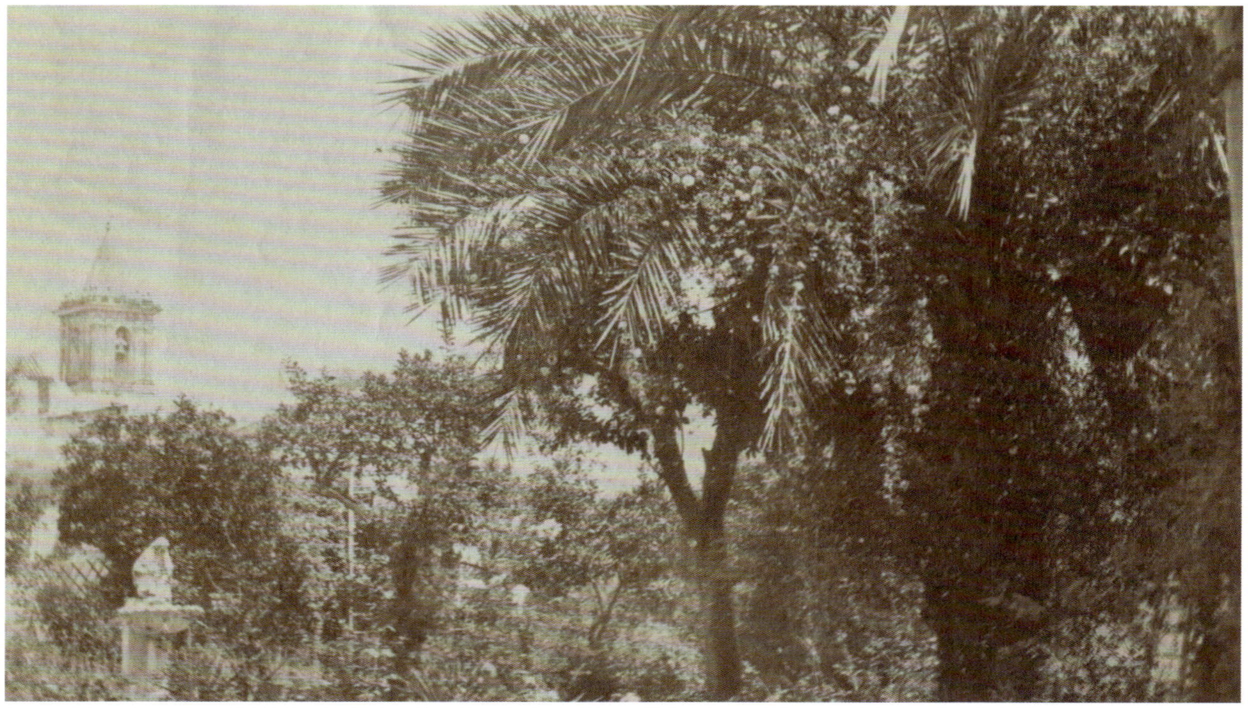

Fig. 6 John Lowell Gardner Jr. (American, 1837–1898), *Alva Garden, Seville, Spain,* 1888. Albumen print, 20.4 × 16.3 cm. Isabella Stewart Gardner Museum, Boston. Enclosed in Isabella Stewart Gardner's *Travel Album: Spain and Portugal, Volume II*

of her, Jack, Ralph Wormeley Curtis, and other friends at lunch on 19 April in a tent they rented in celebration of the Feria de Abril (Plate 17). Their glasses are raised and a vase of flowers has been placed on a chair in the foreground. Gardner angles her chair outward in order to face the camera, and she appears with a slight smile or as if in midspeech. Below this image she pasted another of herself taken the following day, 20 April, in Alva Garden. She is seated by the edge of a pool of water almost completely hidden by the luscious and overgrown vegetation, holding a parasol above her head as she gazes up at the sun. From the same day, Curtis produced a rapid sketch of Gardner appearing as if posed in a black lace mantilla with a sword through her black heart. Given how impacted Gardner was by Seville—a city in which she could feel her existence and to her was the embodiment of Spain saying it all—she wanted not only to remember and transport back with her the people and places she saw but also how she appeared when within the city, in order to preserve how being in Spain made her feel. She possibly had each of these aide-mémoire created with her albums specifically in mind.

Photographs taken by Jack Gardner in Seville have also been preserved within the museum's collection (figs. 5 and 6). His images depict a street in Seville; a bullfight most likely the one that the Gardners attended on Easter Sunday; and Alva Garden, where Gardner posed for the photograph featured in her album. This site is recognizable by the top of the statue she sits beside, which can be seen along the left-hand side of Jack's shot. The two photographs in the Alva Garden appear to be the same view taken at slightly different angles. It is likely that Jack took this photograph of Isabella as he did on other occasions throughout their travels together.

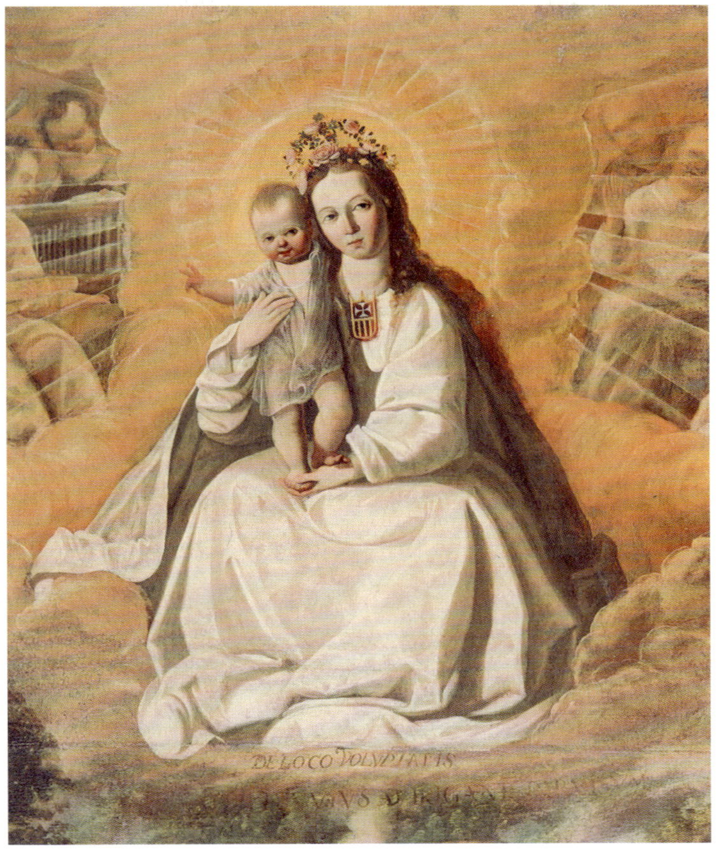

Fig. 7 Workshop of Francisco de Zurbarán (Spanish, 1598–1664), *The Virgin of Mercy*, about 1640. Oil on canvas, 139 x 42 cm. Isabella Stewart Gardner Museum, Boston

On 23 April, Gardner purchased her first European Old Master painting in Seville, the Workshop of Francisco de Zurbarán's (1598–1664) *The Virgin of Mercy* (1640), from the dealer Jacobo López Cepera (fig. 7). This acquisition also marked the beginning of Gardner's collecting of Spanish painting. Many of her contemporaries followed suit, for the first time beginning to see Spanish art on par with its European peers.[34] Gardner continued to collect Spanish art for the rest of her life and ultimately acquired works by Velázquez, Bartolomé Bermejo (1436–1498), Pedro García de Benabarre (1455–1480), as well as a second Zurbarán. Her albums served as the ideal photographic resources needed to make informed decisions about her purchases. Gardner first exhibited *The Virgin of Mercy* in her bedroom at Beacon Street. When Fenway Court was built, the work remained in her private quarters until 1916. However, when she constructed the Spanish Chapel, Gardner had the painting moved there near an inscription "In Memoriam" carved on the wall. She felt that the Christ Child in the painting resembled her deceased son.[35]

Gardner was enthralled by Spain for the rest of her life, most famously iconized through her construction of the Spanish Cloister and purchase of *El Jaleo*, a painting that would have been another way for Gardner to relive her experiences there. Sargent and Gardner remained in dialogue about Spain throughout their friendship, especially about the music and dance that fascinated them both. They exchanged flamenco records through the early 1920s.[36] In 1890, when the infamous Spanish dancer La Carmencita (1868–1910) (of whom both Sargent and William Merritt Chase [1849–1916] would paint portraits) came to New York, Sargent organized a private performance at Chase's studio in Gardner's honor. Gardner traveled to Spain again in 1906. She stayed only a few days but insisted on making the trip—"I must see the Prado now or never, so all of my time and attention in these two days has been to arranging a hurried visit to Spain."[37] Archer Milton Huntington (1870–1955), founder of the Hispanic Society of America, appointed Gardner as the first female board member of his museum. She proudly displayed Huntington's letters and the Hispanic Society Membership Medal that he presented her in the Long Gallery.[38] Through Gardner's albums, she could return to her time in Spain and the diverse layers of Spanishness that she had come to emulate within her own museum.

This exhibition and catalogue dedicated to Gardner's travel albums occurs at the museum in conjunction with an examination of legendary American artist Betye Saar's (b. 1926) production of sketchbooks and assemblages in response to her own travels from the 1970s through the present day. In a lecture that Saar delivered in 1994 entitled "The Art of Assemblage: Two Views" about the Gardner Museum and its parallels with her own work, she concludes with the Spanish Cloister:

> *The tiles are Mexican. I wasn't able to find out where this window was imported from.... But, for me, this piece sort of symbolizes how I felt about the Museum. About how Mrs. Gardner loved the pattern and the exotic. And history and religion and nature. How she liked the old and the beautiful, the exotic. And how all of these things sort of interwoven to her life. And her passion for collecting and arranging. This window sort of symbolizes that, for me, like this sort of dark unknown quality about her, that there are all these things that are left. And you can just sort of invent your story about them. But you really don't know who the real woman is, that it's a mystery that lies behind this window. And that's what makes us curious about her. That's what attracted me about her.*[39]

The notions of Spanishness that Gardner was able to create based on her 1888 trip to Spain within her museum were to Saar the embodiment of what made Gardner's museum a space that she felt drawn to. As she says, it is what the museum itself symbolized for her. Saar traveled to Spain in the late 1960s but did not produce any sketchbooks that have survived from her trip there. The Spain that Saar experienced would have been radically different from the one Gardner knew at the turn of the century from the perspectives of time periods but also from those of race and class. However, Saar as an artist was also interested in the multicultural aesthetics of Hispanic identity—an expanded conception of Spanishness that has come to include Latin America (another place that Saar traveled)—evidenced in assemblages such as *La Bonita* (1989), a work that Saar also included in her 1994 lecture at the museum. Saar's narrative of found objects recalls the imagery that Gardner assembled in her travel album from Spain—two photographs of women covered with black veils, an image of the Virgin and Child, a black miniature fan, and a small black heart. For both Gardner and Saar, the elusive idea of the Hispanic embodied "the old and the beautiful, the exotic" and the stories that could be told from cutting, pasting, and layering these together all in one place.

1. For more on Gardner's acquisition of *El Jaleo* from her cousin Thomas Jefferson Coolidge (1831–1920), see Volk et al. 1992. For more on Gardner's construction of the Spanish Cloister, see Prokop 2012.
2. Isabella Stewart Gardner, *Travel Album: Spain and Portugal, Volume II*, 1888, 1, Isabella Stewart Gardner Museum, Boston.
3. Carter 1925, 104; Prokop 2012, 103.
4. These were all compiled together for the first time in the National Gallery of Art's exhibition and catalogue on *El Jaleo*, Volk et al. 1992.
5. Cartwright 1883, 6.
6. Although Gardner and Sargent did not meet until 1886, between 1882 and 1883 Gardner saved a clipping of an article entitled "Sargent's New Picture" published about *El Jaleo* when the painting was first brought to the United States.
7. Carter 1925, 105; Prokop 2012, 103.
8. Ralph Wormeley Curtis to Isabella Stewart Gardner, 9 February 1888, Isabella Stewart Gardner Museum Archives, Boston.
9. Haddon 2021, 3n5.
10. For more on America's fascination with Spain, see Boone 1998, 2007, 2019; DeGuzman 2005; and Kagan 2019.
11. The best-known collectors of Spanish art from the period besides Gardner include Archer Milton Huntington (1870–1955), Charles Deering (1852–1927), Henry Clay Frick (1849–1919), Louisine Havemeyer (1855–1929), and William Randolph Hearst (1863–1951).
12. Irving 1837, 13–14. Gardner owned a copy of Irving's *Manuscript of the Chronicle of the Conquest of Granada* (1828–29) and a copy of a letter Irving sent from Madrid in 1842.
13. Numerous books, periodicals, travel journals, and letters written in the United States and France from the early nineteenth century onward characterize Spain as half-European and half-African. See Haddon 2021, 10.
14. See Dearing 2010, 166–67, for a discussion of the important distinctions between the terms "Romani," "Gypsy," and "gypsy."
15. Haddon 2021, 56–66.
16. Collecting photographs of works of art was a critical part of Gardner's approach to amassing her own collection. See Riley 2020a, 96; Prokop 2012, 104.
17. For more on Stirling Maxwell and his production of the *Annals of the Artists of Spain*, see Macartney and Matilla 2016.
18. For more on Laurent and his production in Madrid, see Jiménez Díaz, Muñoz, and Teixidor Cadenas 2019.
19. This scrapbook is now at the Metropolitan Museum of Art. Boone 2007, 240n19.
20. Isabella Stewart Gardner, *Travel Album: Spain, Volume I*, 1888, 58, Isabella Stewart Gardner Museum, Boston.
21. For more on these perceptions within the context of Gardner's travels in Egypt, see Riley 2015a, 69–71.
22. Both William Stirling Maxwell and his assistant Nicolaas Henneman (1813–1898), who produced the photographs for the *Annals of the Artists of Spain*, were close friends with Ford. He assisted as well with the development of their book.
23. "Quien no ha visto Sevilla no ha visto maravilla." Gardner 1888a, 58.
24. Ford 1845, 241.
25. "El aire de Madrid es tan sotil / Que mata á un hombre, y no apaga á un candil." Ibid., 34.
26. The English publisher John Murray published Ford's *Hand-Book for Travellers in Spain*. Gardner also used a number of Murray's other travel guides for her trips abroad such as to Egypt as well as England and Wales.
27. Ford 1882, 37.

28. "Quien dice España dice todo." Gardner 1888b, 1.
29. Ford 1847, 5.
30. John L. Gardner Jr. to George Peabody Gardner, 24 May 1888, Isabella Stewart Gardner Museum Archives, Boston.
31. Tharp 1965, 138; Prokop 2012, 104.
32. Carter 1925, 106; Tharp 1965, 137.
33. Gardner also purchased during her trip and brought back with her a folding fan decorated with bullfighters.
34. Kagan 2019, 261–62.
35. Tharp 1965, 138; Carter 1925, 107–8; Prokop 2012, 106–7.
36. Silver 2020; Llorens Serra and Sagredo Ezquerra 2006, 57.
37. Isabella Stewart Gardner to Henry Adams, 29 June 1906, Isabella Stewart Gardner Museum Archives, Boston.
38. Silver 2020.
39. Saar 1994, 8–9.

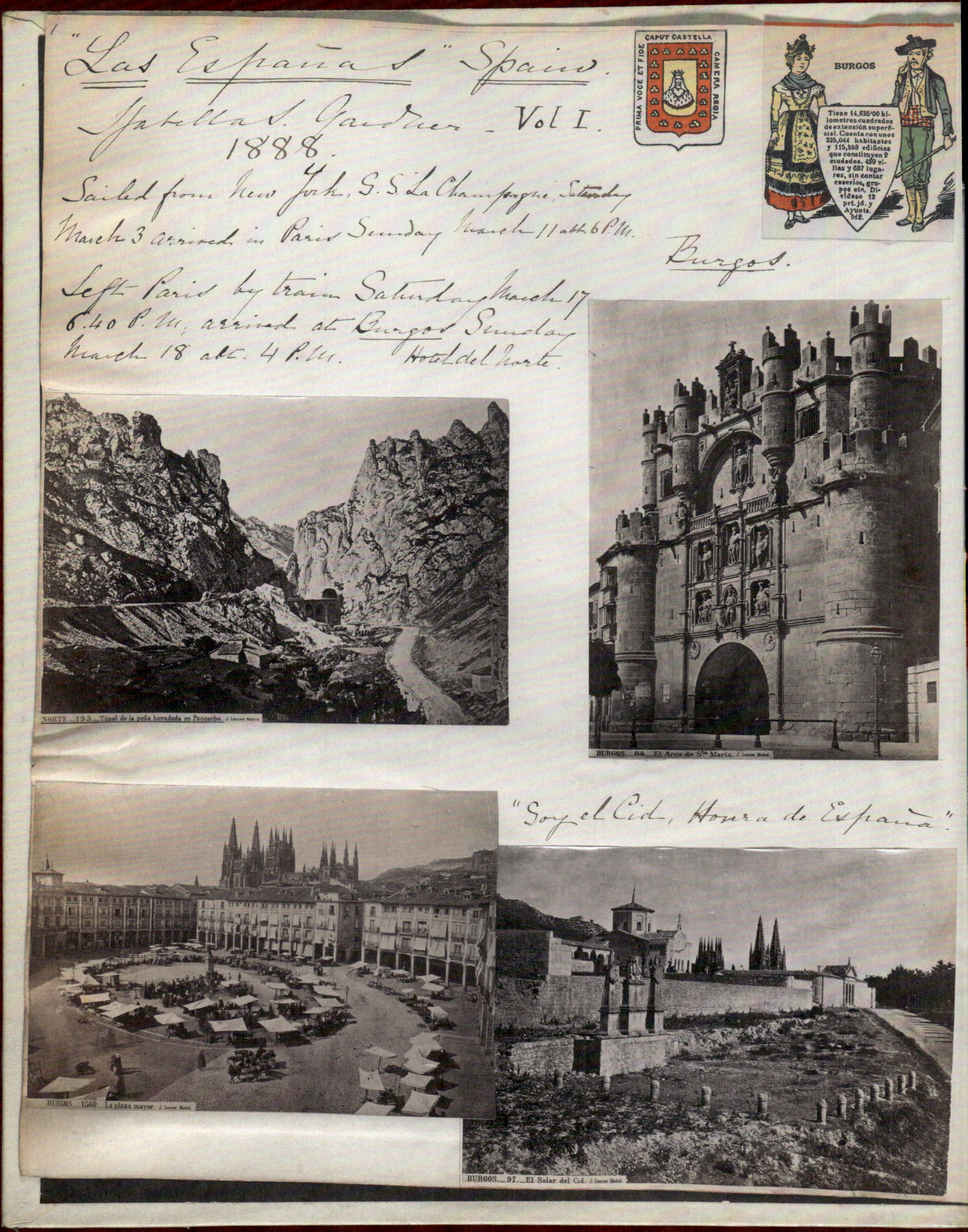
"Las Españas" Spain.
Isabella S. Gardner – Vol I.
1888.
Sailed from New York, S. S. La Champagne, Saturday
March 3 arrived in Paris Sunday March 11 abt. 6 P.M.
Left Paris by train Saturday March 17
6.40 P.M., arrived at Burgos Sunday
March 18 abt. 4 P.M. Hotel del Norte.
CAPUT CASTELLA
PRIMA VOCE ET FIDE
CAMERA REGIA
BURGOS
Tiene 14,635'00 kilometros cuadrados de extensión superficial. Cuenta con unos 335,044 habitantes y 115,258 edificios que constituyen 2 ciudades, 459 villas y 687 lugares, sin contar caserios, grupos etc. Dividese 12 prt. jd. y Ayunts. 512.
Burgos.
NORTE 193 Túnel de la peña horadada en Pancorbo. J. Laurent Madrid
BURGOS 84 El Arco de Sta. Maria. J. Laurent Madrid
"Soy el Cid, Honra de España".
BURGOS. 1560 La plaza mayor. J. Laurent Madrid
BURGOS. 97. El Solar del Cid. J. Laurent Madrid

Burgos – Fortified Eminence –
March 18–19.

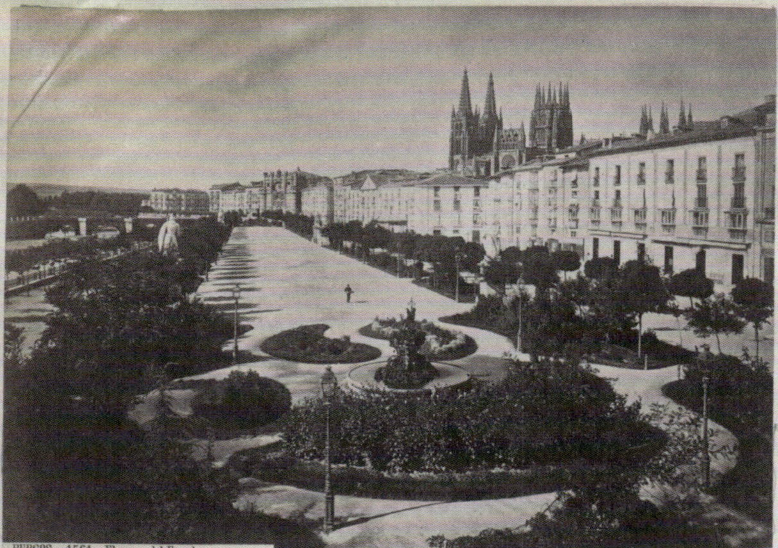

BURGOS. 1561. El paseo del Espolon. J. Laurent Madrid

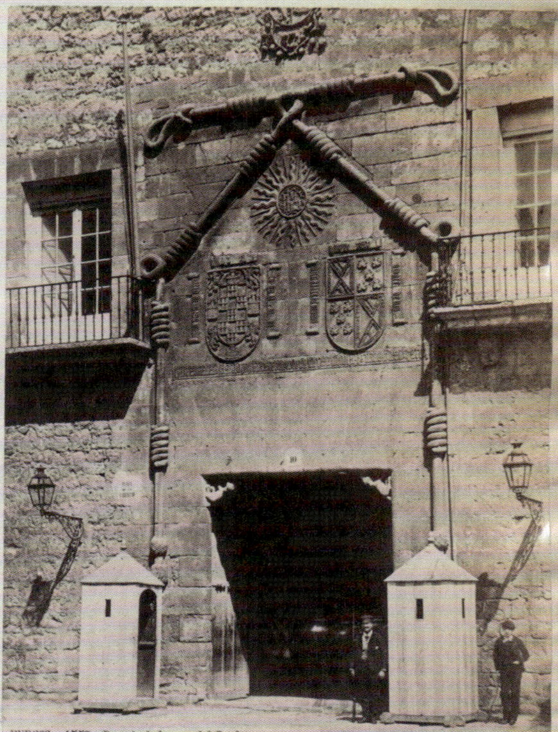

BURGOS. 1559. Portada de la casa del Cordon. J. Laurent Madrid

Cathedral.

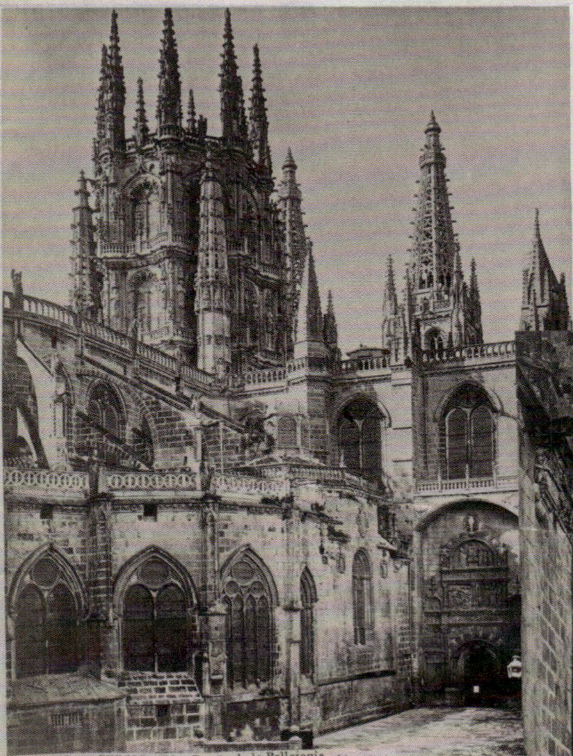

CATEDRAL DE BURGOS. 91. Puerta de la Pellejeria. J. Laurent Madrid

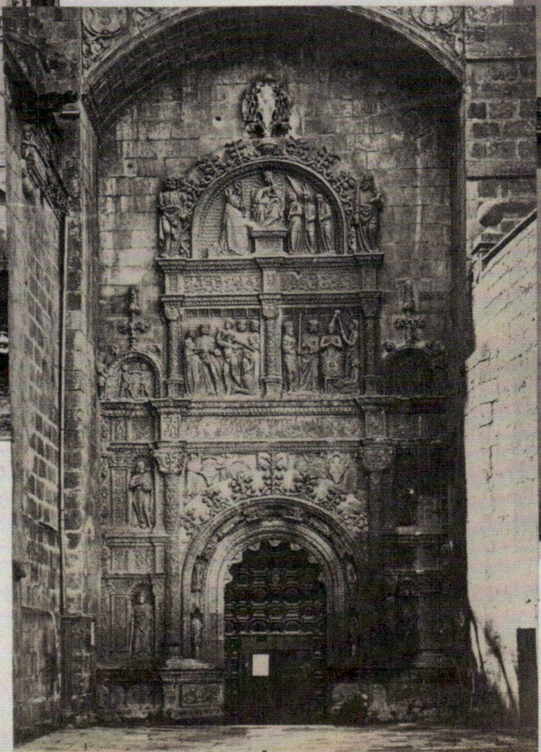

AL DE BURGOS. 94. Portada de la Pellejeria. J. Laurent Madrid

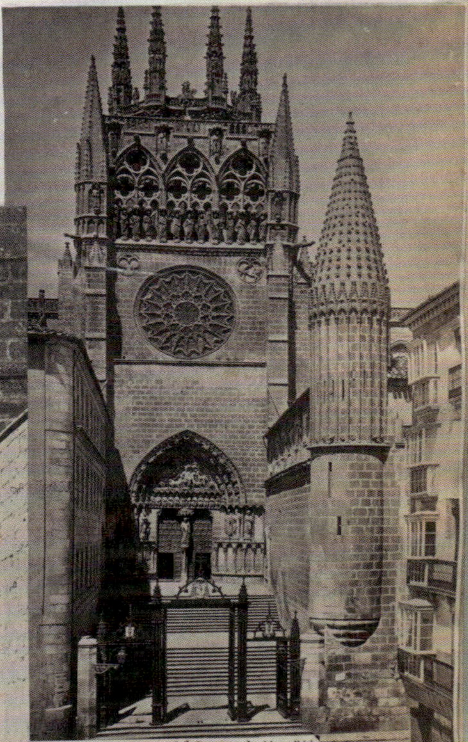

BURGOS. 90. Puerta del Sarmental. J. Laurent Madrid

SPAIN
Volume I, pages 23–24, March 1888

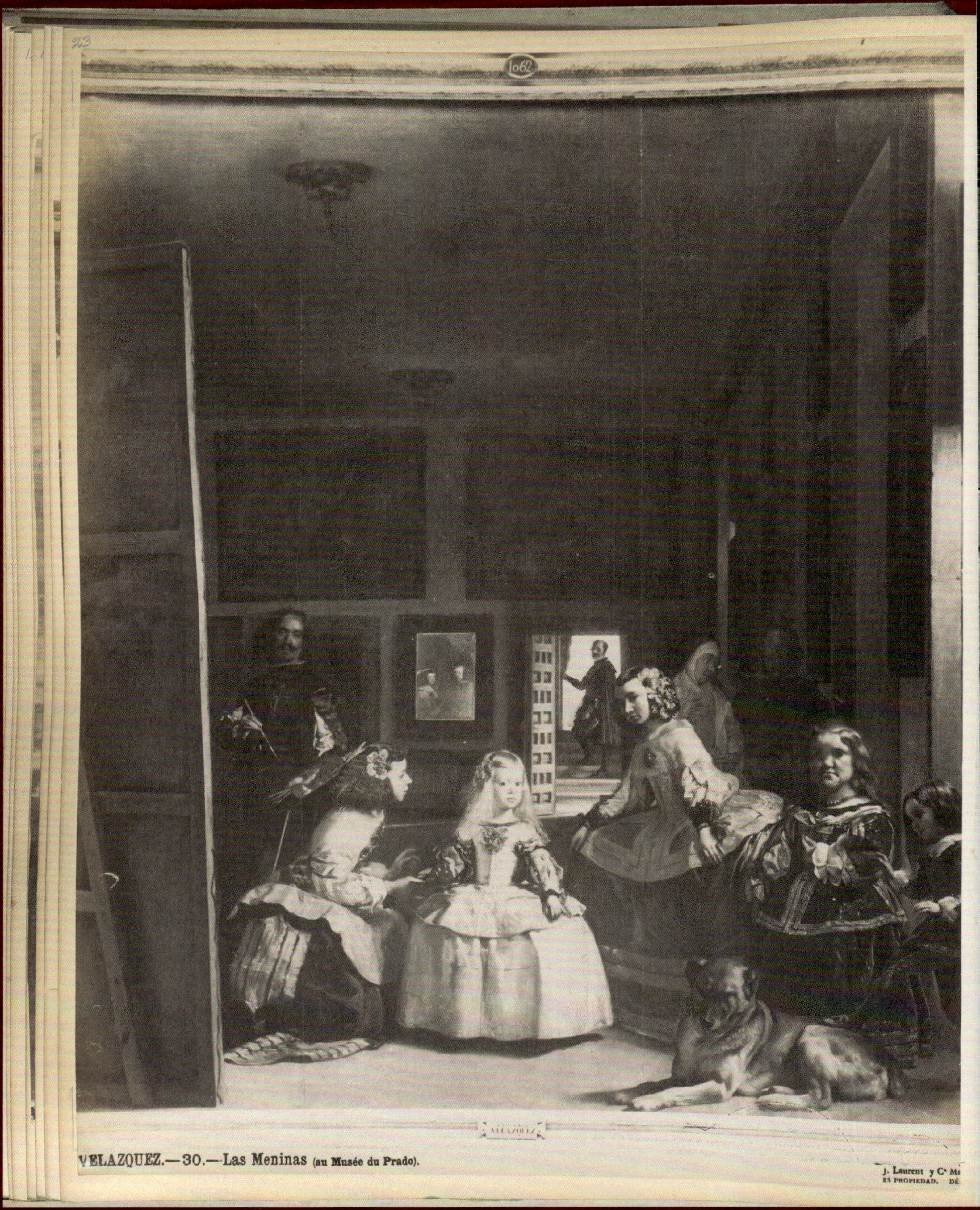

Madrid.

24

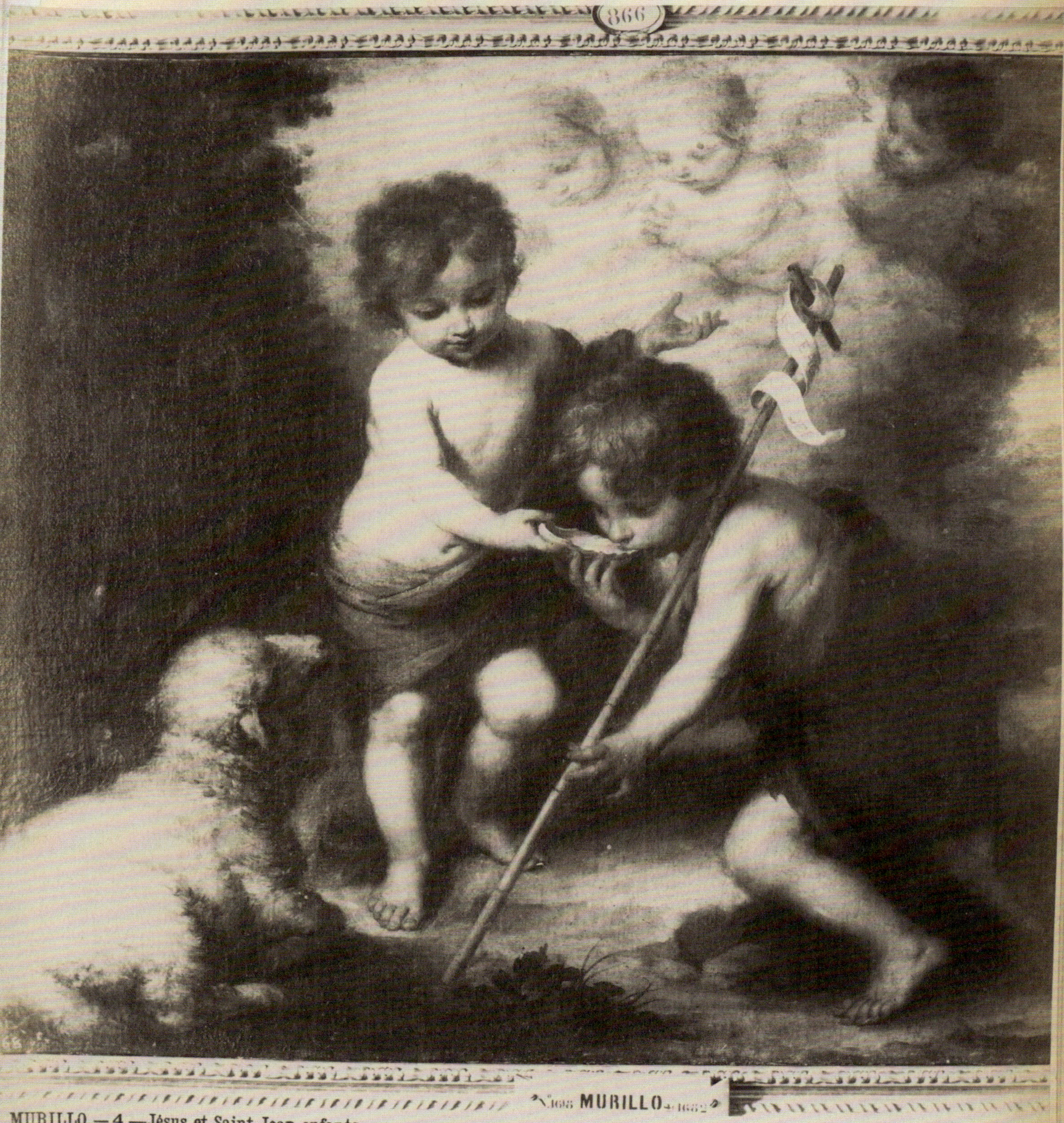

MURILLO.—4.—Jésus et Saint Jean enfants (LES ENFANTS A LA COQUILLE). (au Musée du Prado.)

55

Madrid.

The Academy.
Academia –

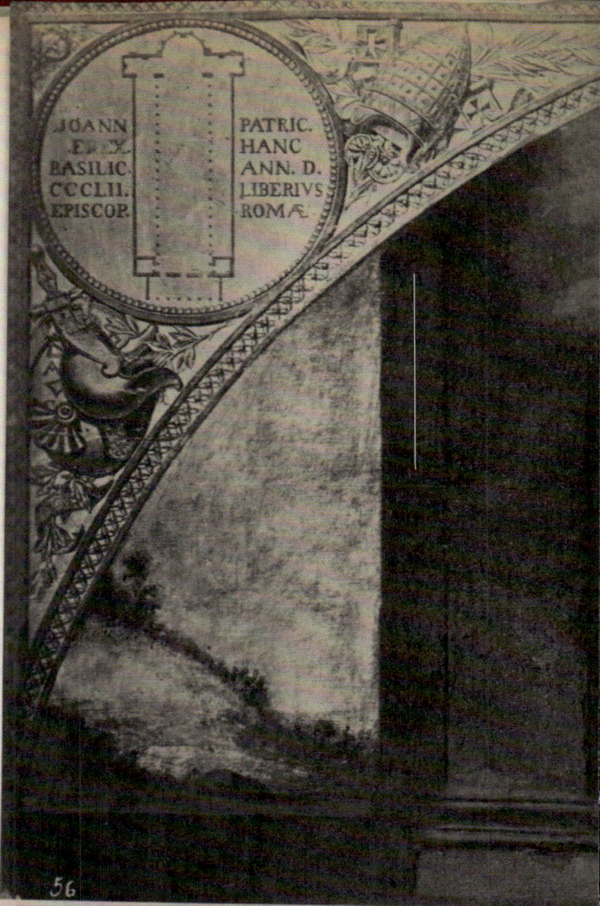

MURILLO._488._La vision du Patricien ron

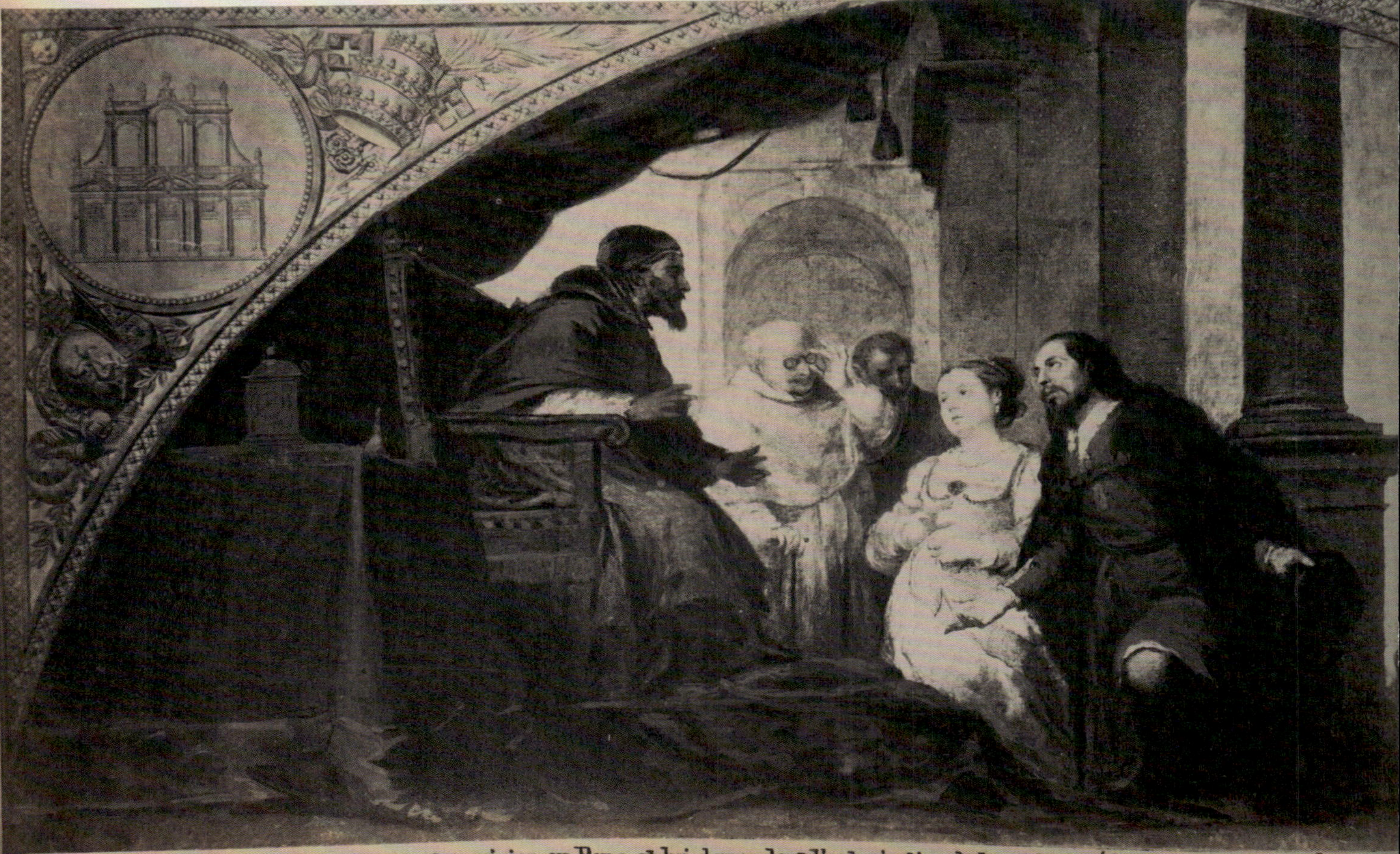

MURILLO._489._Le Patricien racontant sa vision au Pape et lui demandant l'autorisation de la réaliser. (à l'Academie de St Ferdi

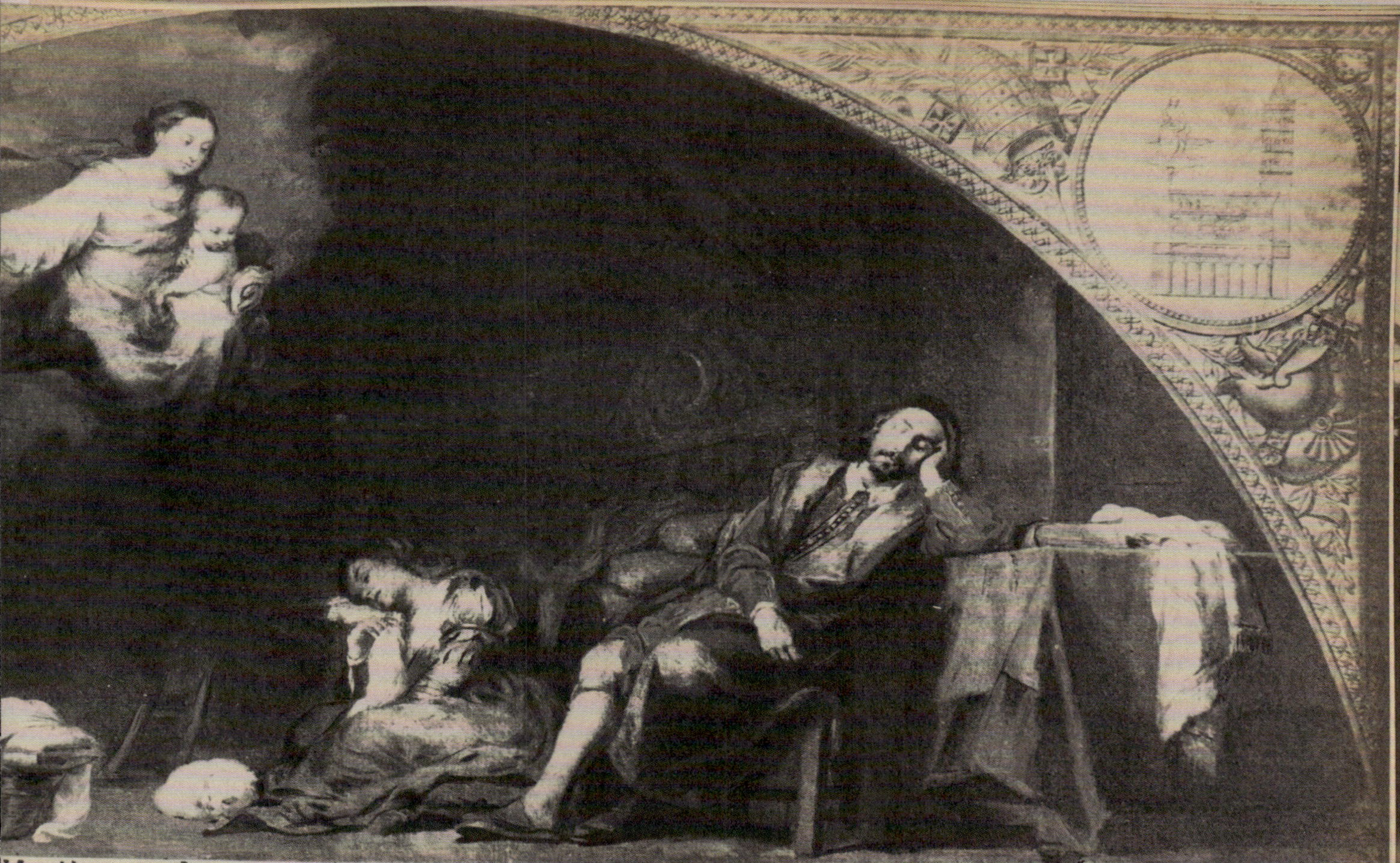

'Académie de St. Ferdinand).

J. Laurent y Cia Madrid
Es propiedad Déposé

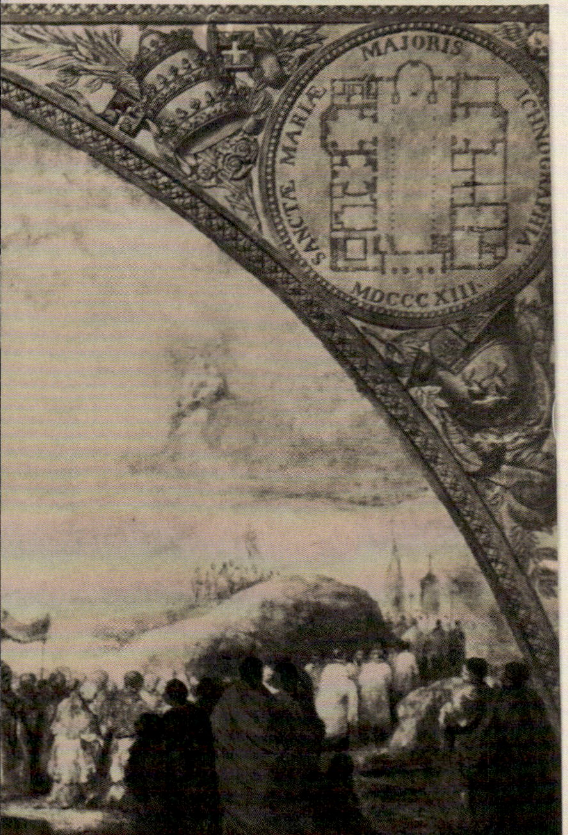

J. Laurent y Cia Madrid
Es propiedad Déposé

56

Left Madrid March 23,
at 6. 20 P. M. Friday.
Arrived at Seville
Saturday March 24
at 9.20 a. m.
La Fonda de Madrid.

SPAIN
Volume I, pages 57–58, March 1888

Madrid

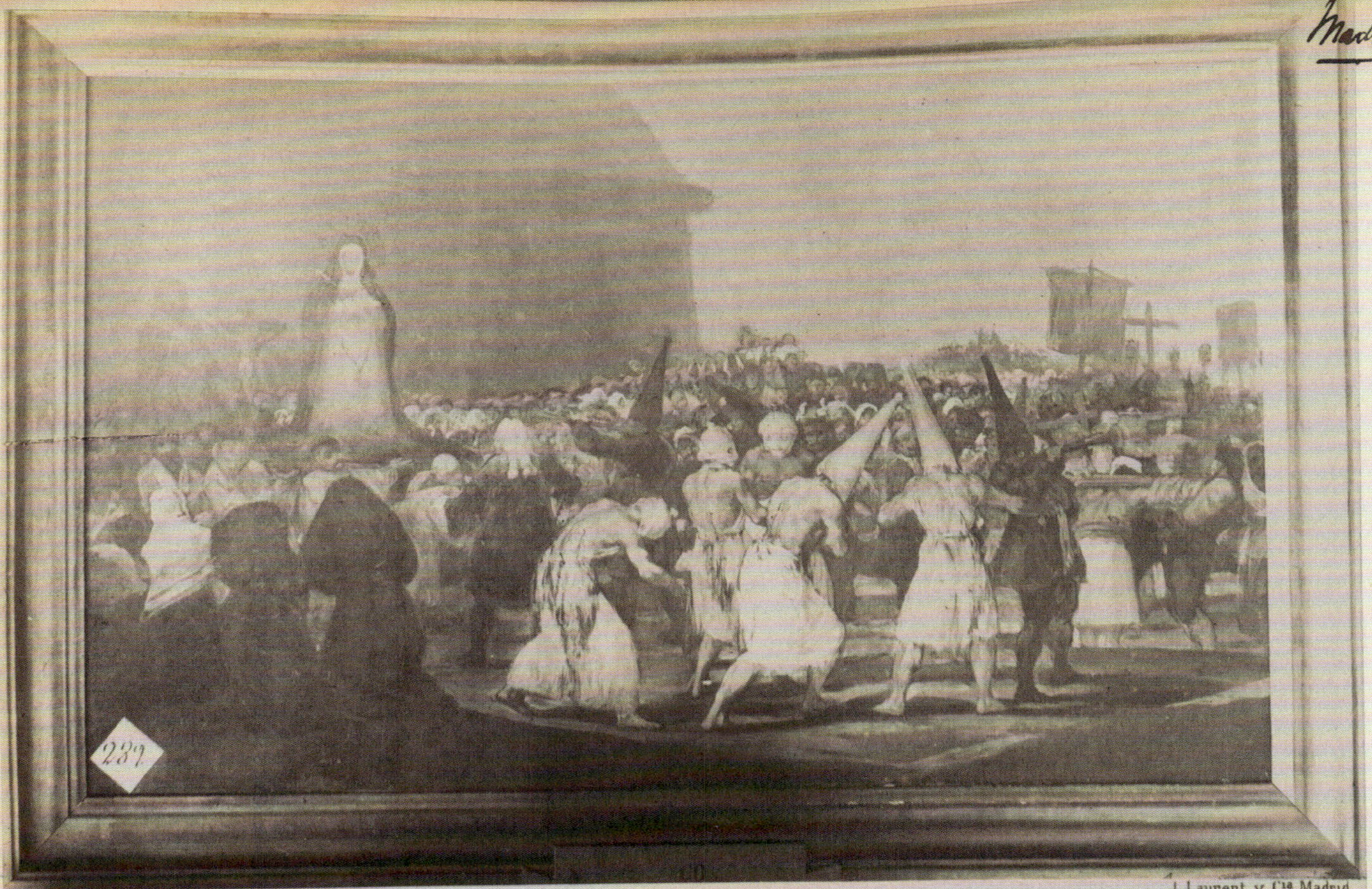

GOYA. 501. La procession du Vendredi Saint. (à l'Académie de St Ferdinand). J. Laurent y Cie Madrid Es propiedad. Déposé

Sevilla

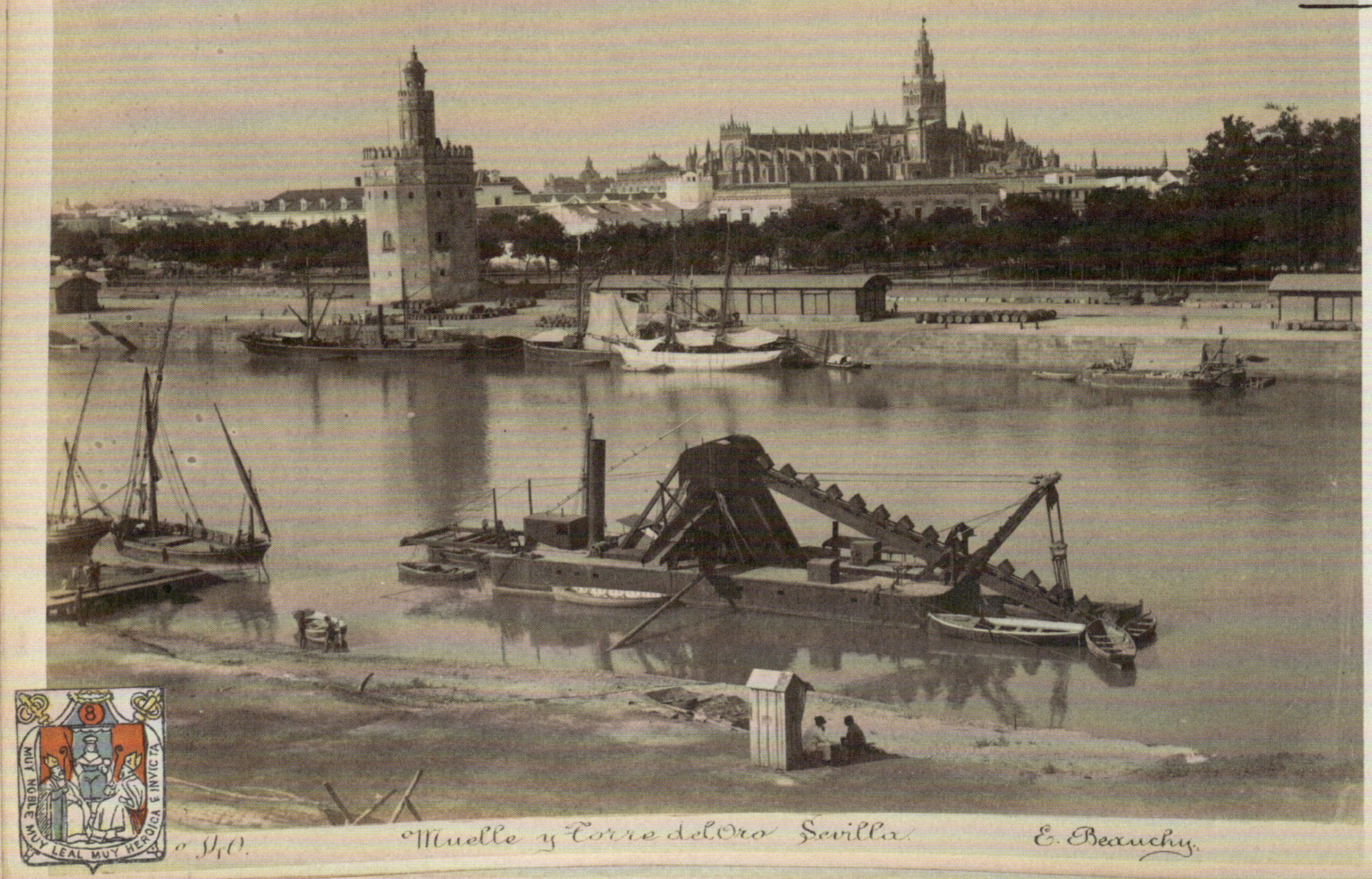

58

Sevilla

Hispal

Ishbiliah

Sibilia

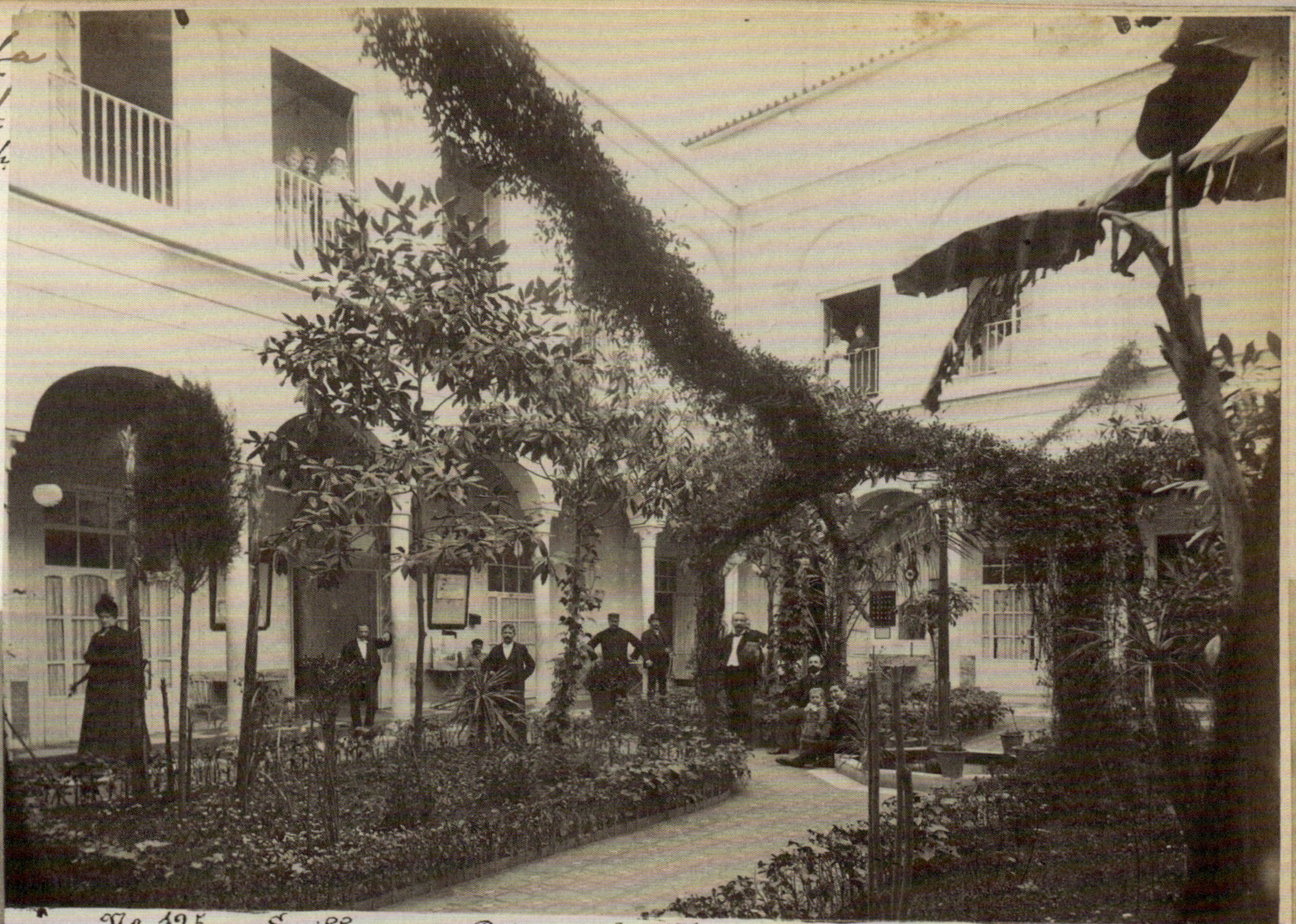

No. 325. Sevilla Patio de la Fonda de Madrid. E. Beauchy

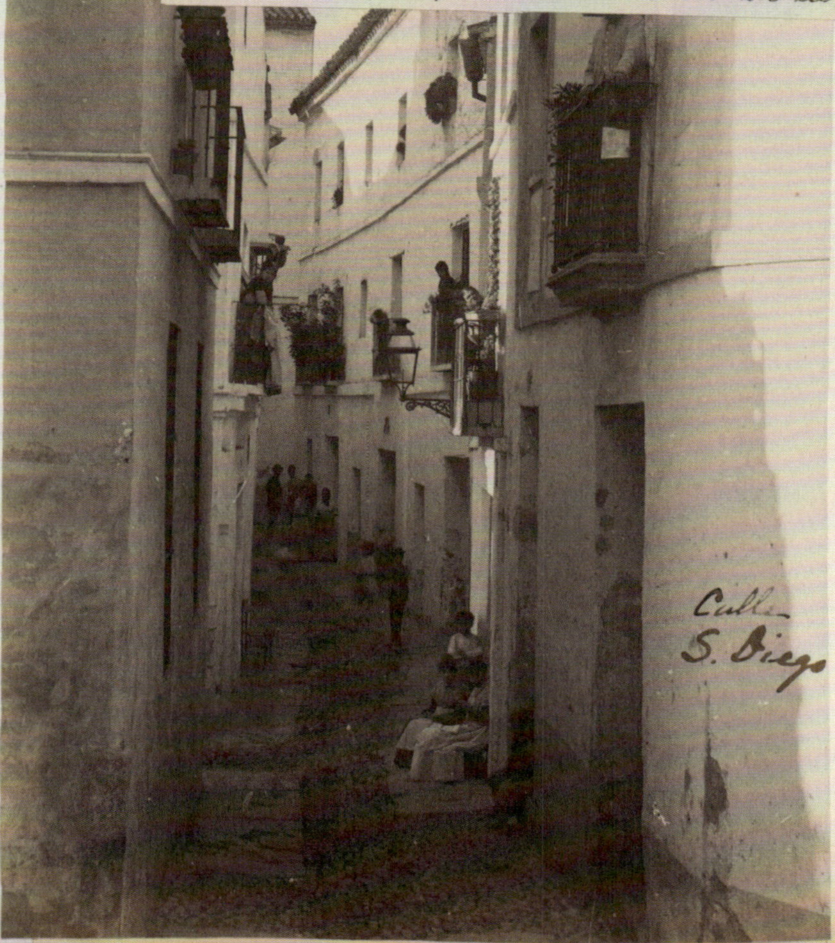

Calle S. Diego

Sevilla March 24 – May 8.

"Quien no ha visto Sevilla
No ha visto maravilla" –

"Here one feels existence"

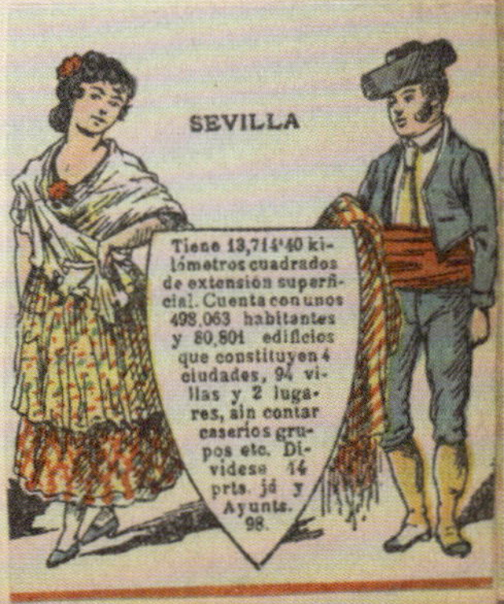

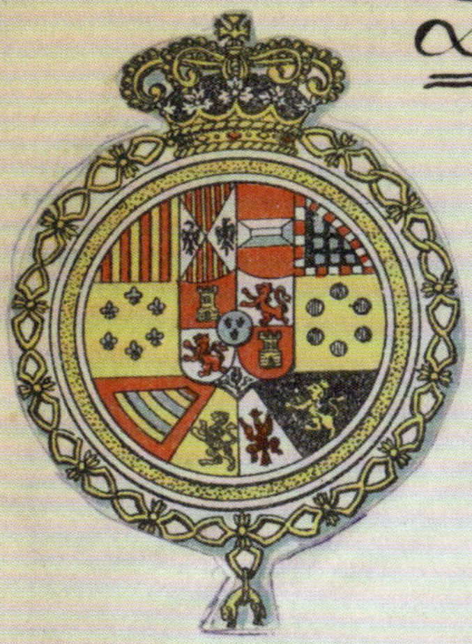

Las Españas Spain.

Isabella S. Gardner Vol. II.

1888.

"Quien dice España dice todo."

Sevilla — muy leal y noble —

Ferdinand VII added, muy heroica — Lopez in 1843 added, invicta

March 24 — May 8.

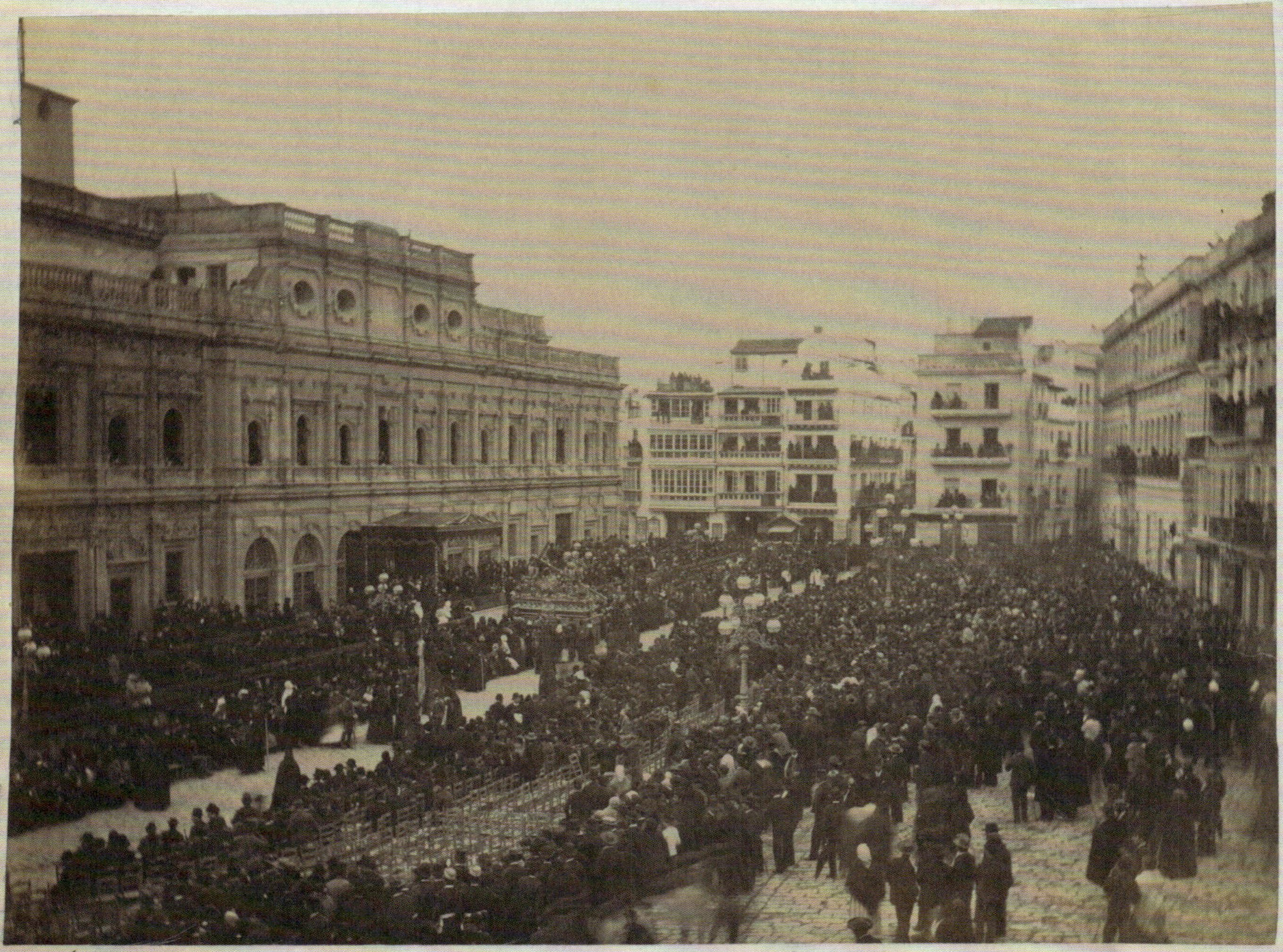

March 25 (Palm Sunday) March 29. March 30 (Good Friday).

Sevilla.

"Condidit Alcides —
renovavit Julius urbem,
Restituit Christo Fernandus
tertius heros."

Inscribed on the Puerta de la Carne.

Paraphrased over the Puerta
de Xerez —
"Hercules me edificó,
Julio Cesar me cercó
De muros y torres altas;
Un Rey Gotho me perdió,
El Rey Santo me ganó
Con Garci Perez de Vargas."

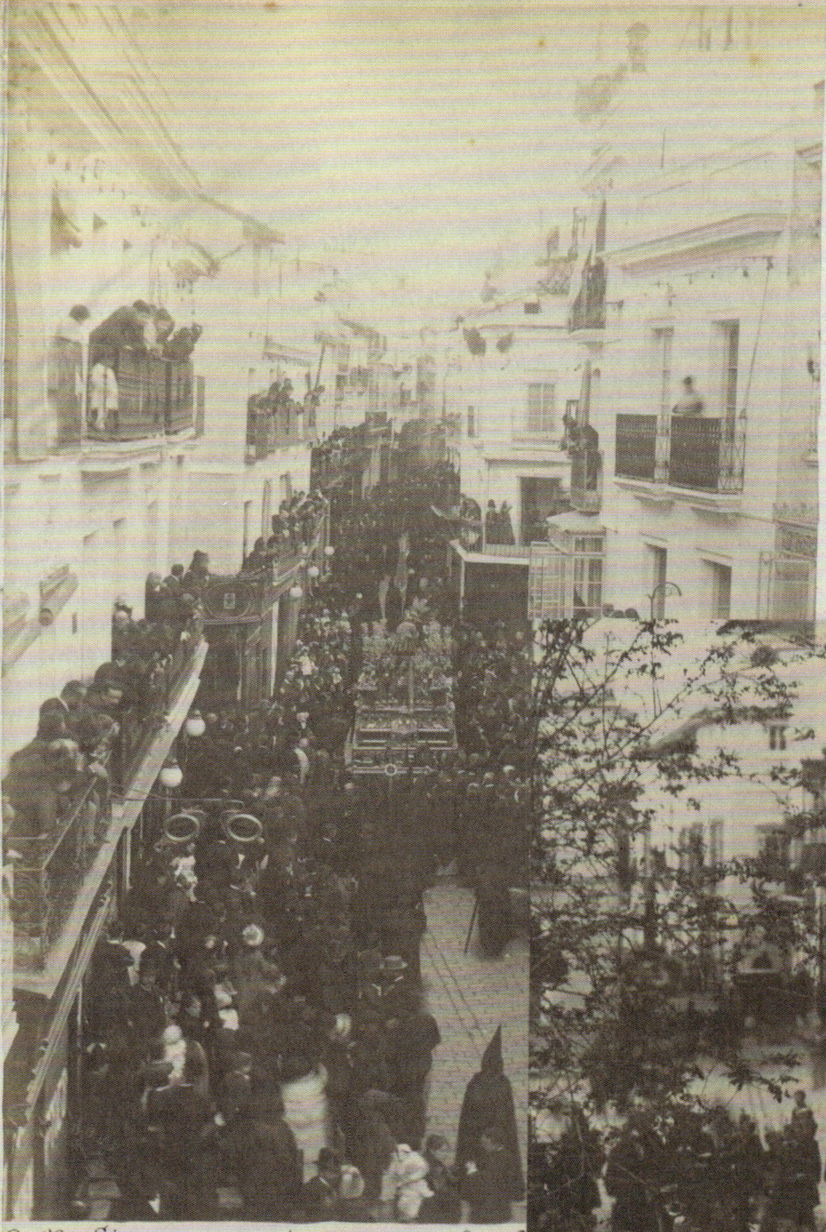

Calle Sierpes en acto de pasar las C

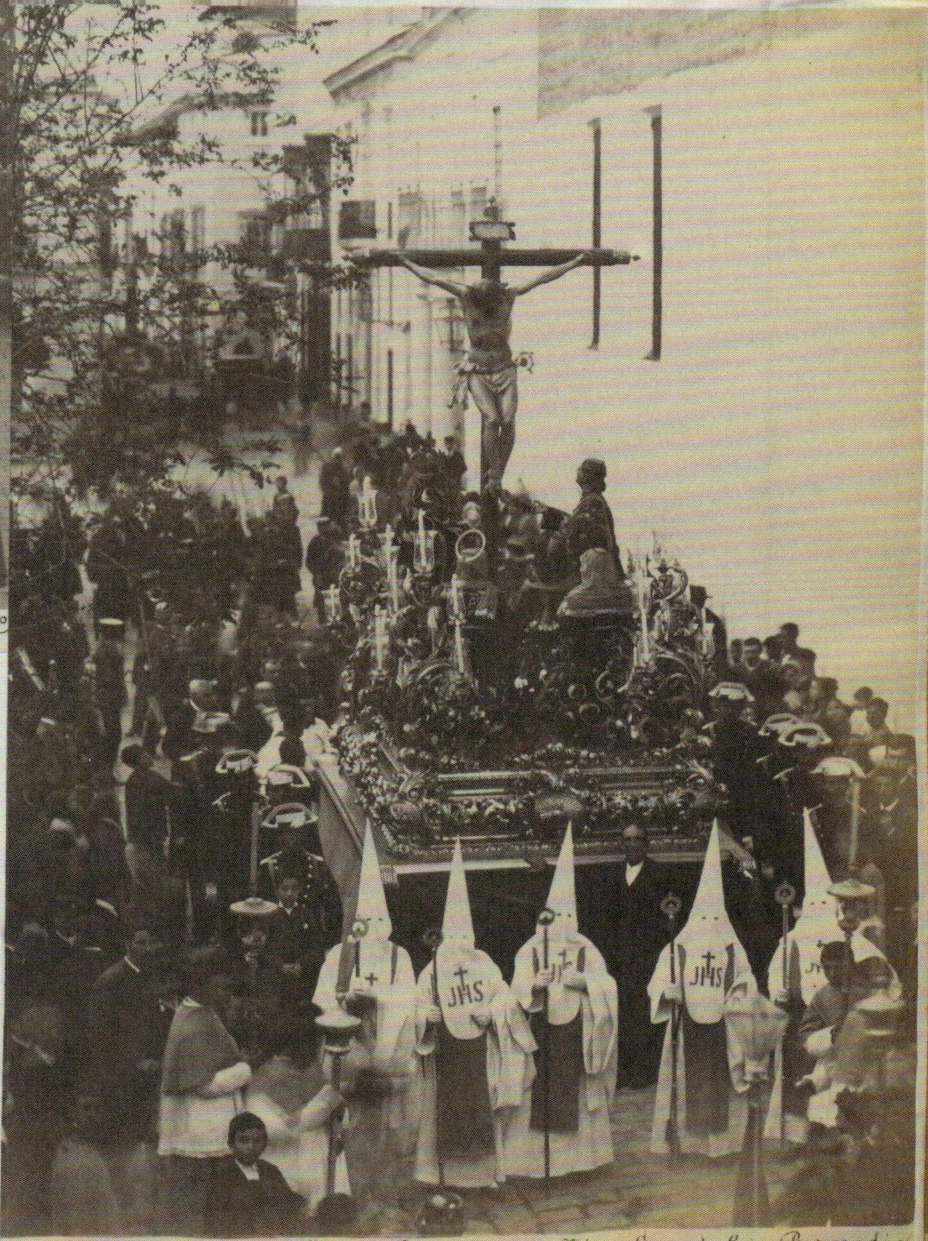

Santo Cristo de las Siete Palabra y Ntra. Sra. de los Remedio.

ORTUGAL
es 3–4, March 1888

3

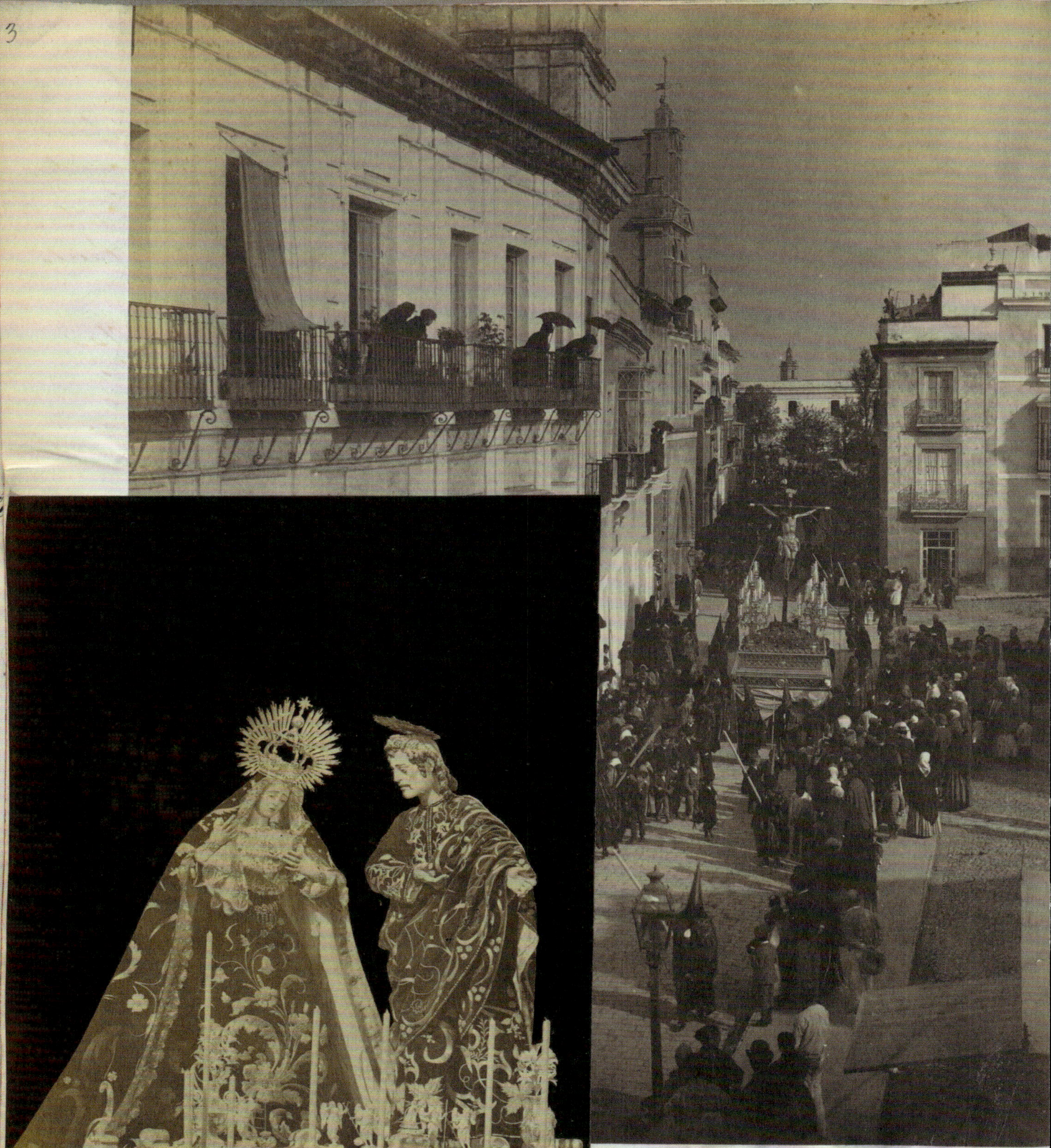

Sevilla.

Sevilla

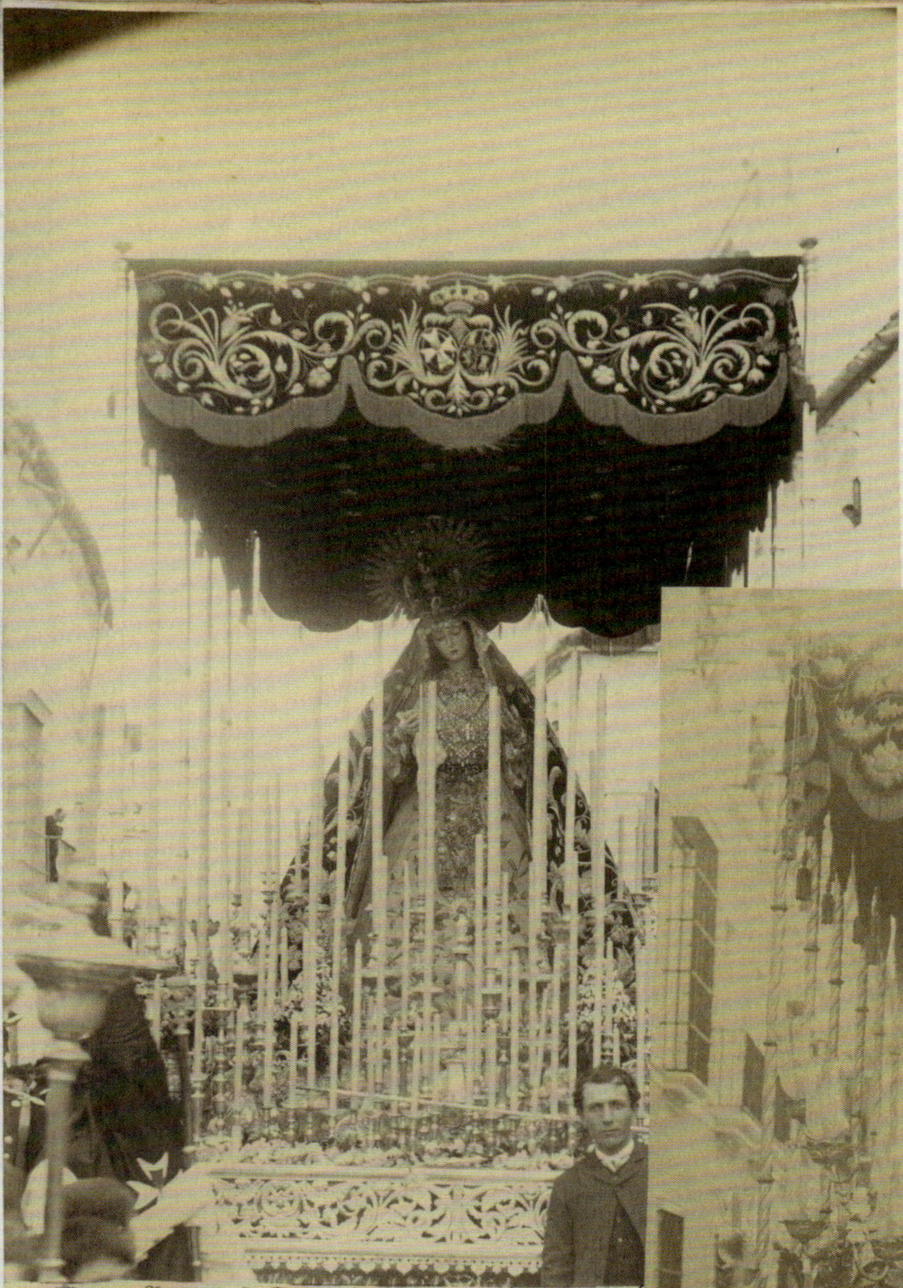

Ntra Sra del Rosario en sus misterios dol

Processions in Holy Week March 25-29-30.

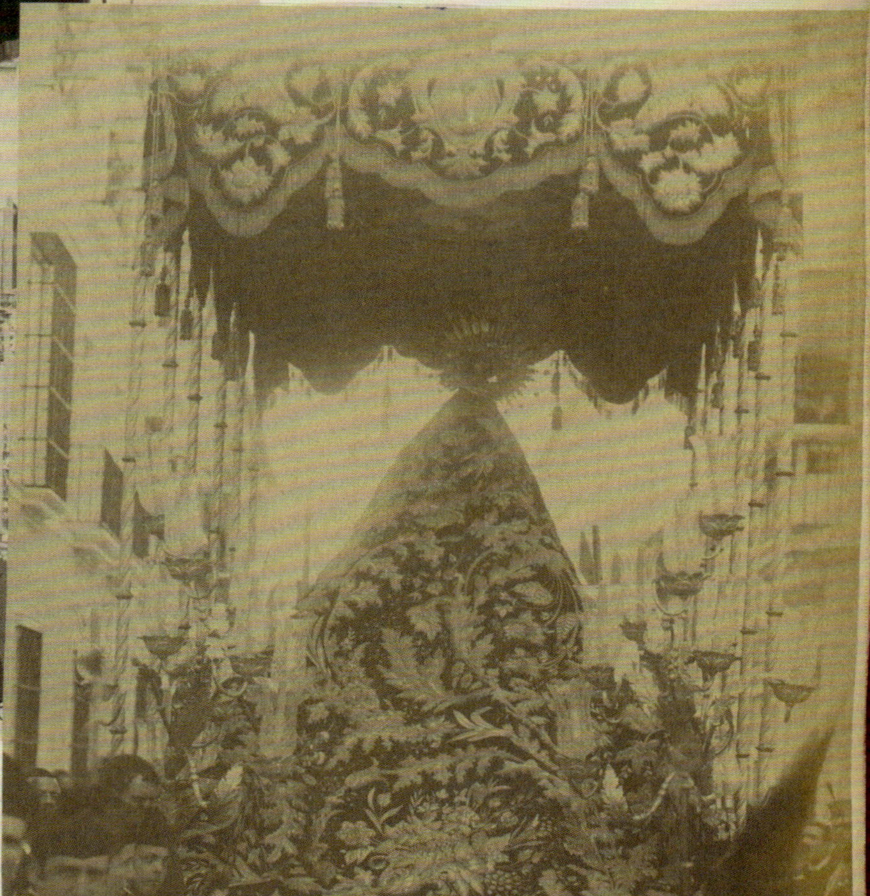

Manto de la Virgen de la Carreteria

SPAIN AND PORTUGAL
Volume II, pages 5–6, March 1888

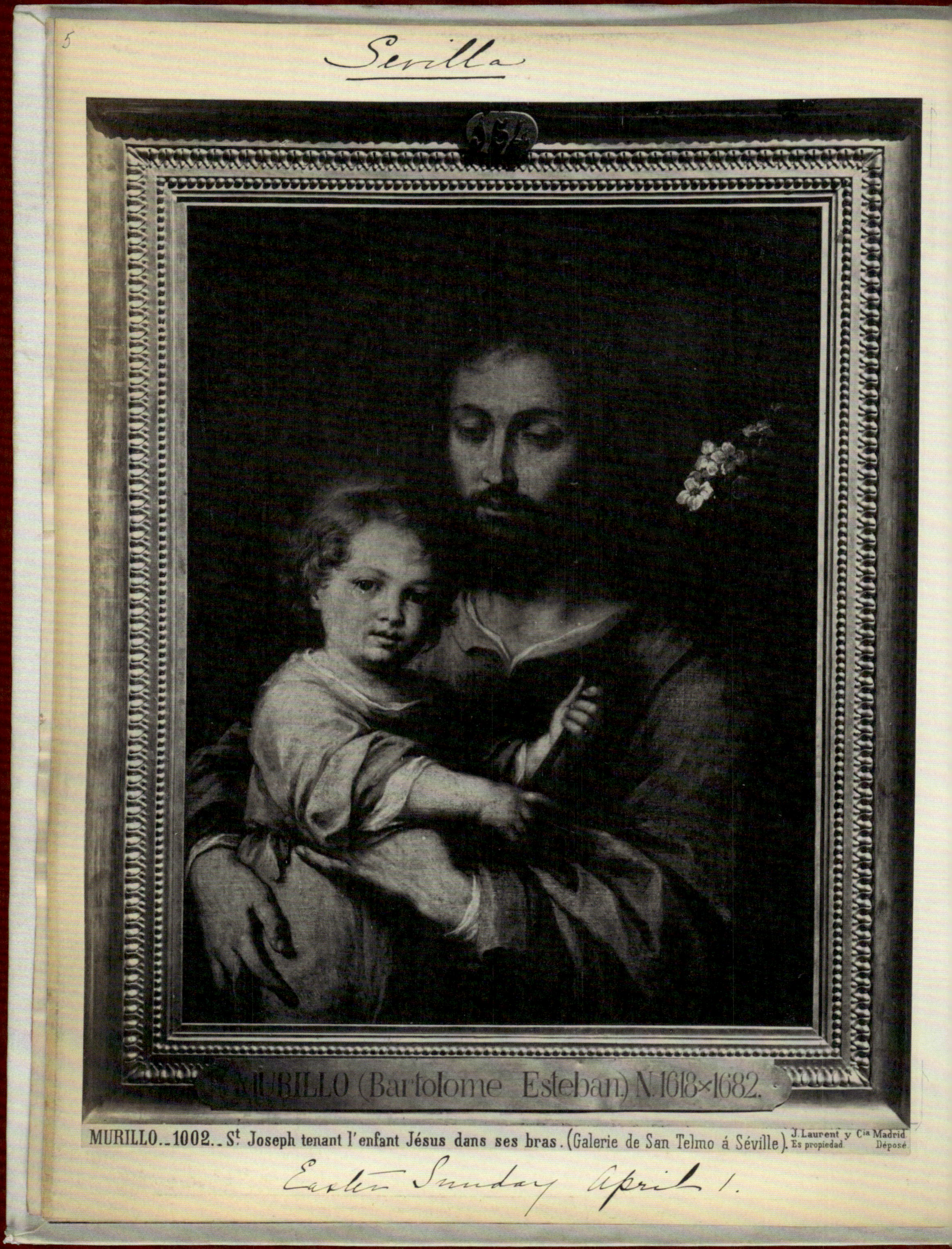

Seville

Easter Sunday April 1.

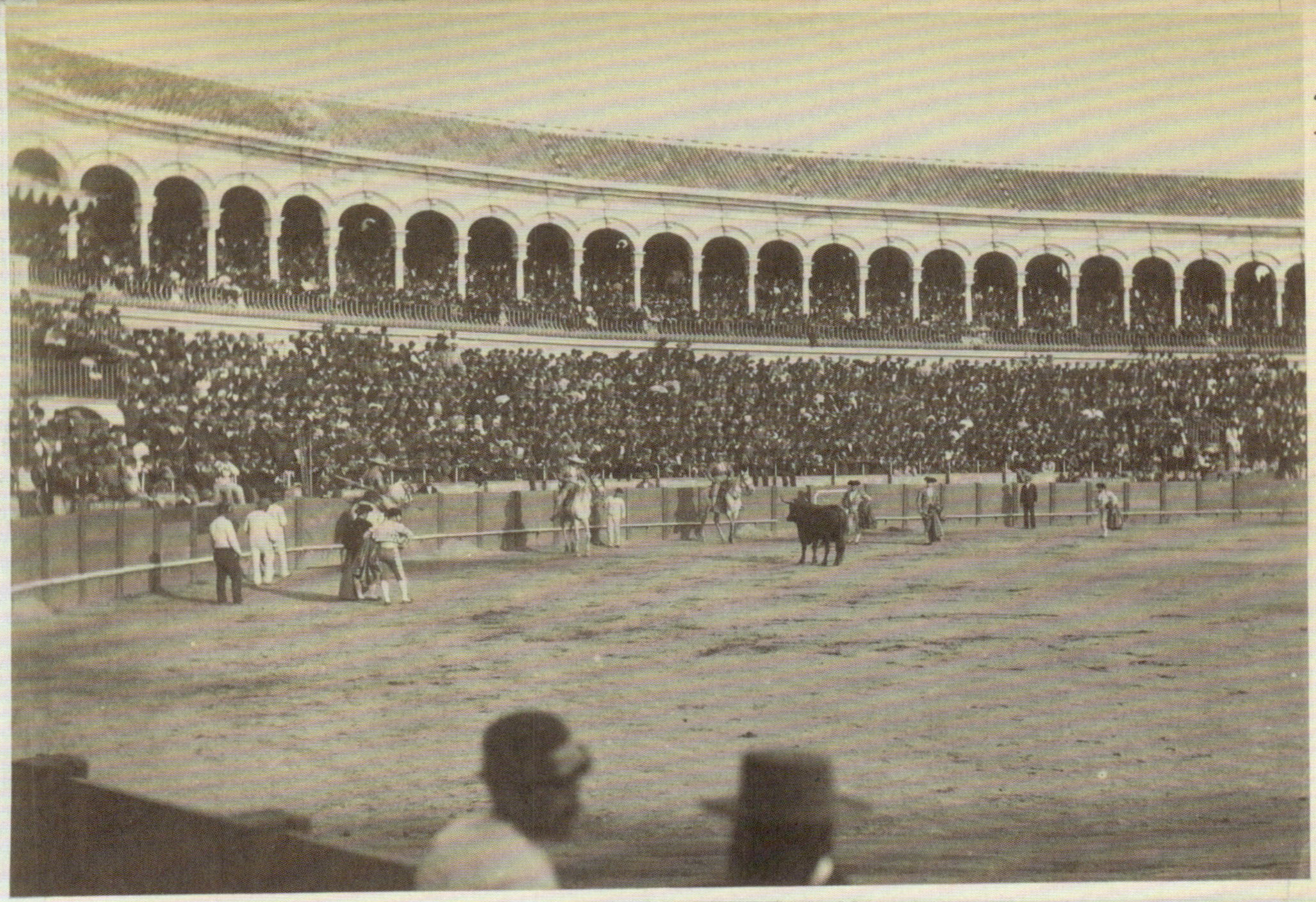

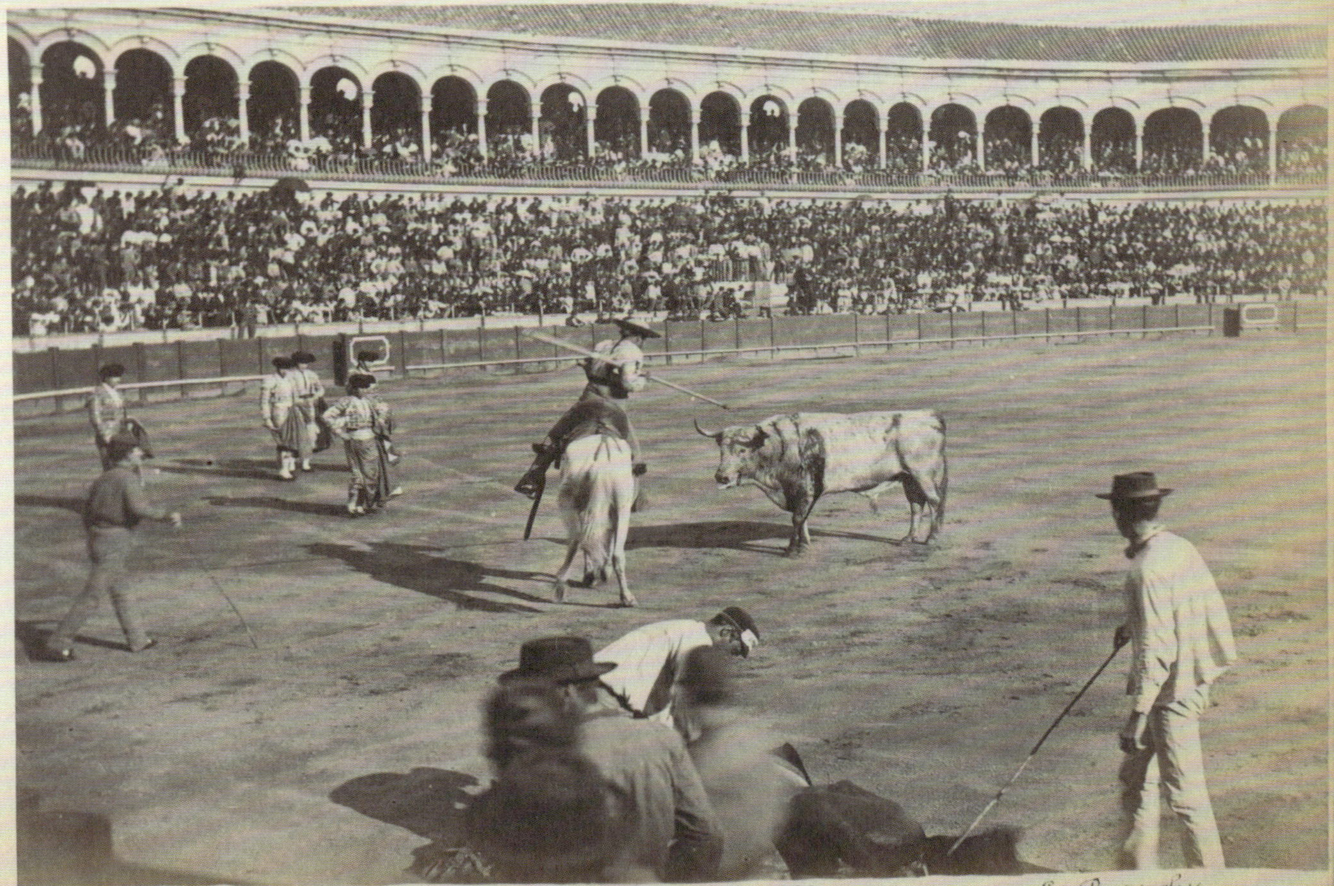

No 186. Corrida de Toros. Sevilla. E. Beauchy.

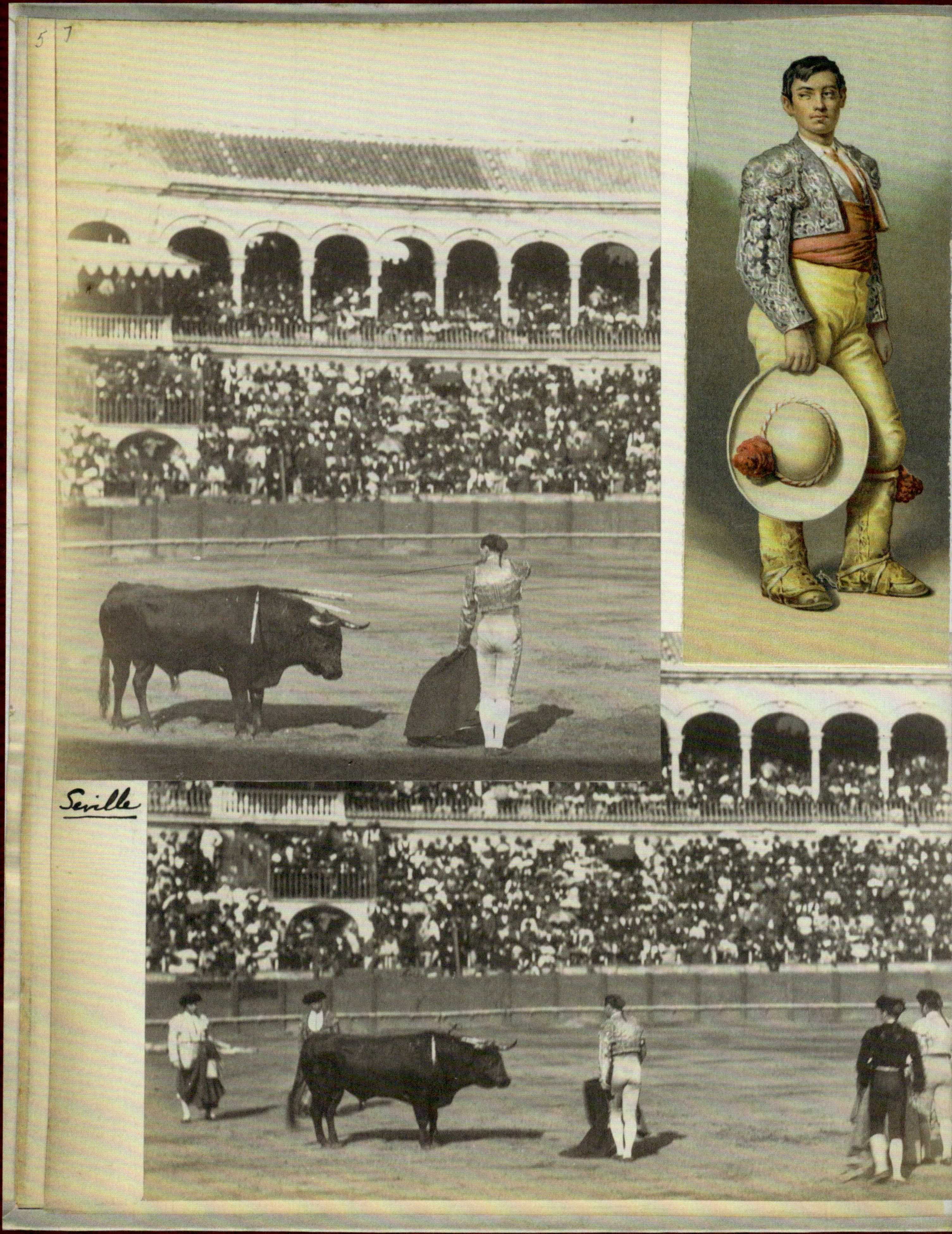
Seville

8

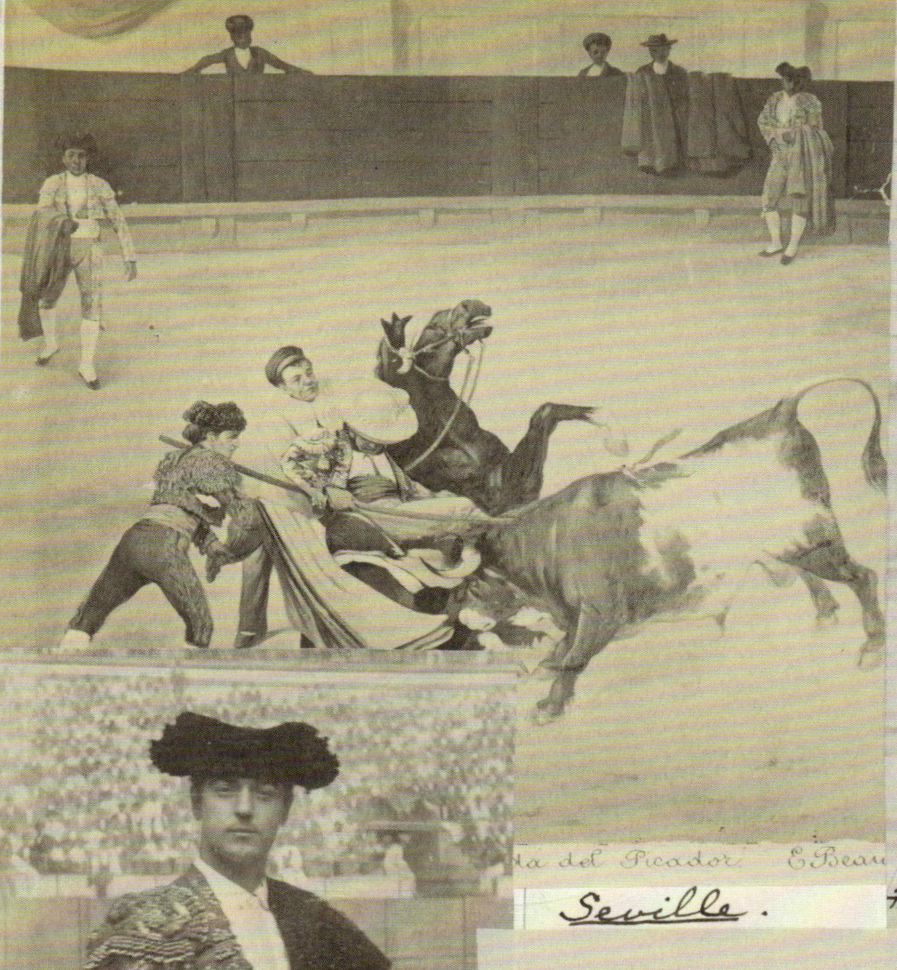

da del Picador E. Beau

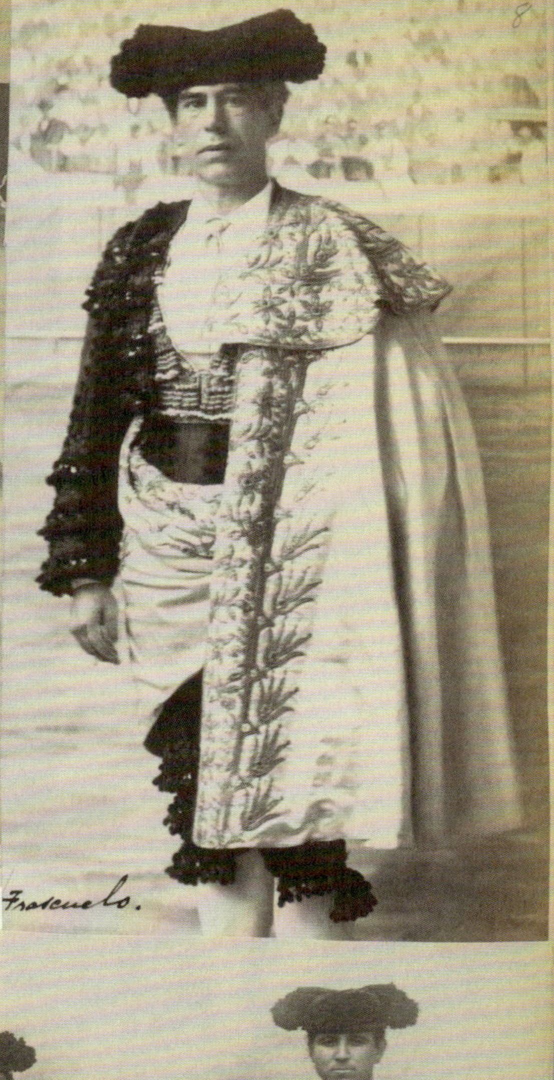

Frascuelo.

Seville.

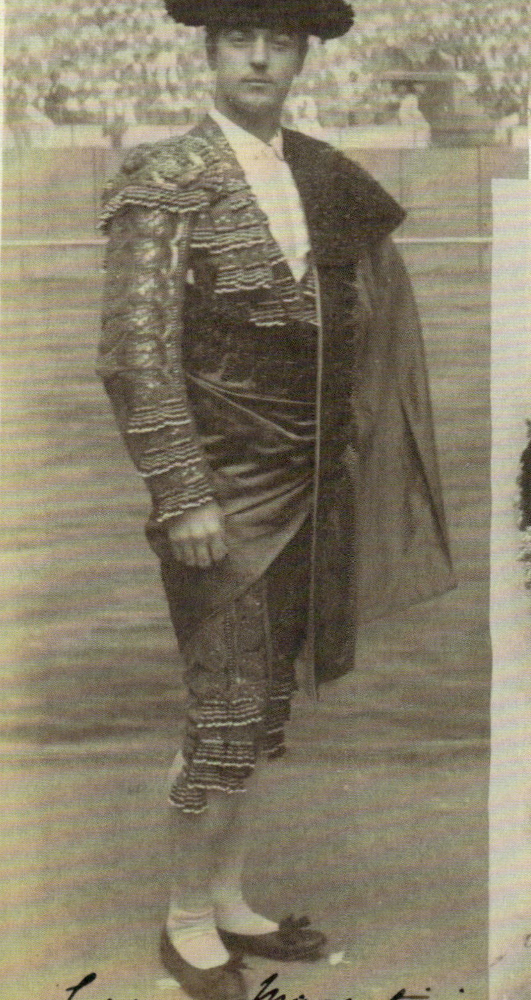

Luis Mazzantini.

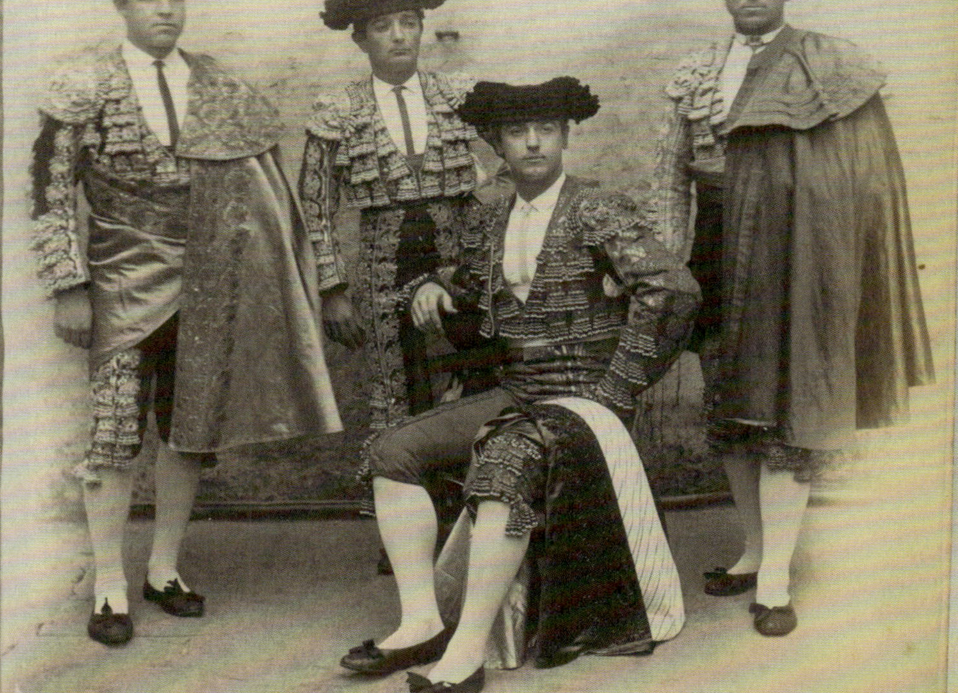

Luis Mazzantini No 114 Luis Mazzantini y Cuadrilla

13

Seville

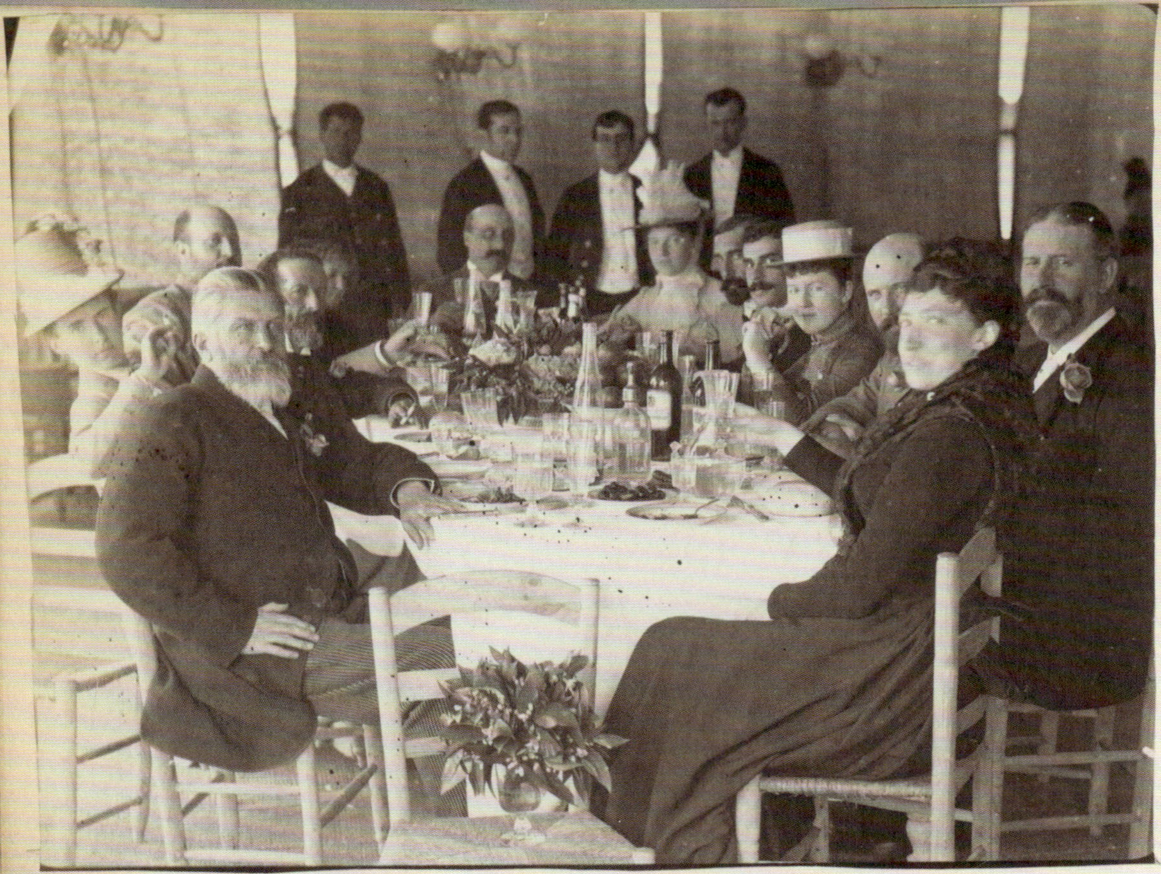

Had lunch party in Club tent at Feria Grounds. Thursday April 19.

Alva Garden.

Photographed there Friday April 20.

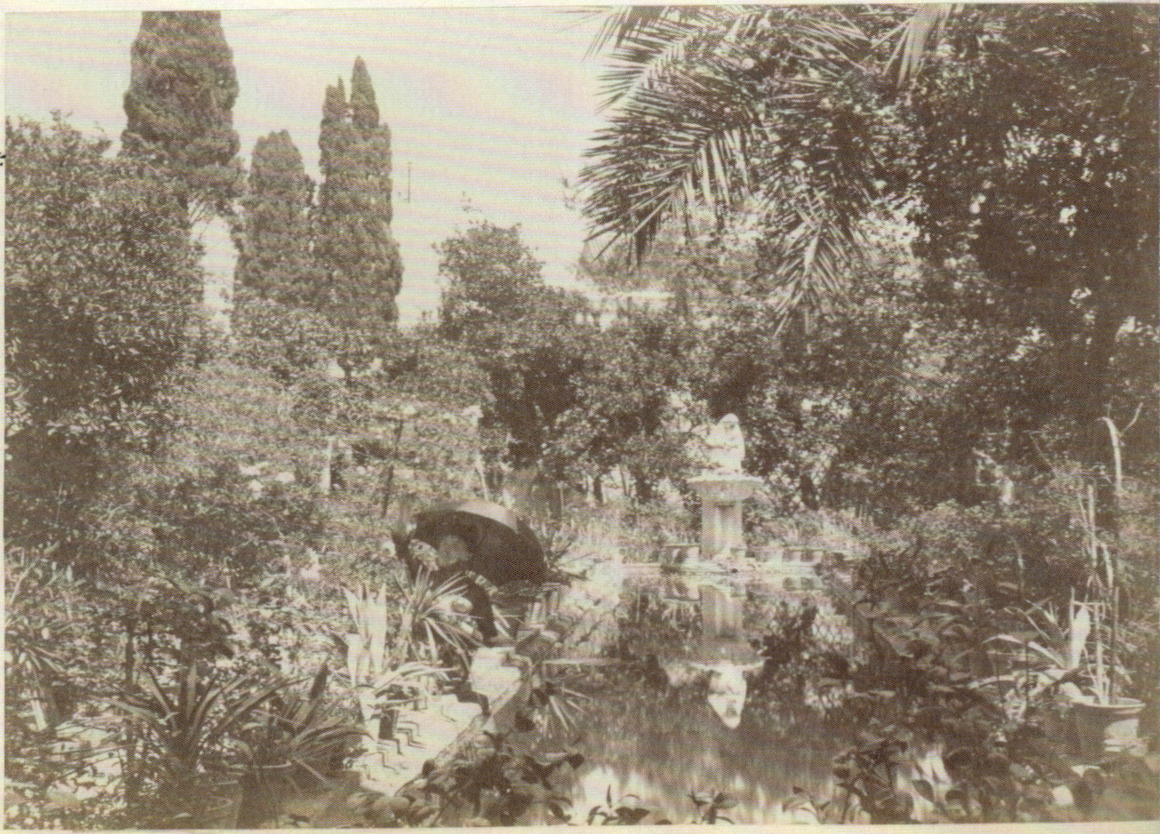

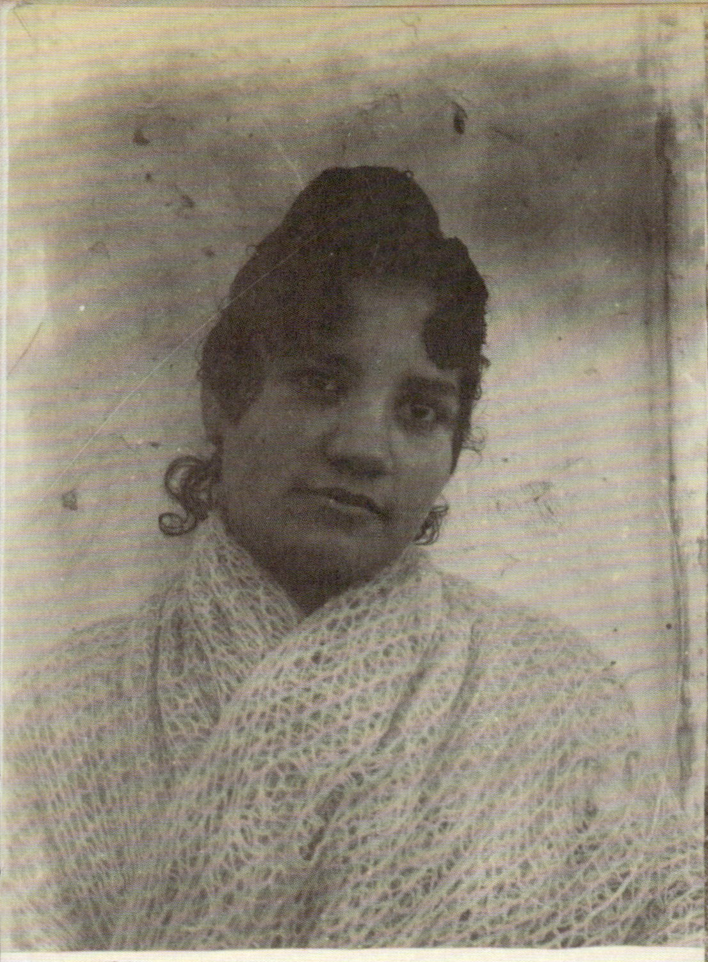

Cigareras in
Tabacco Factory.

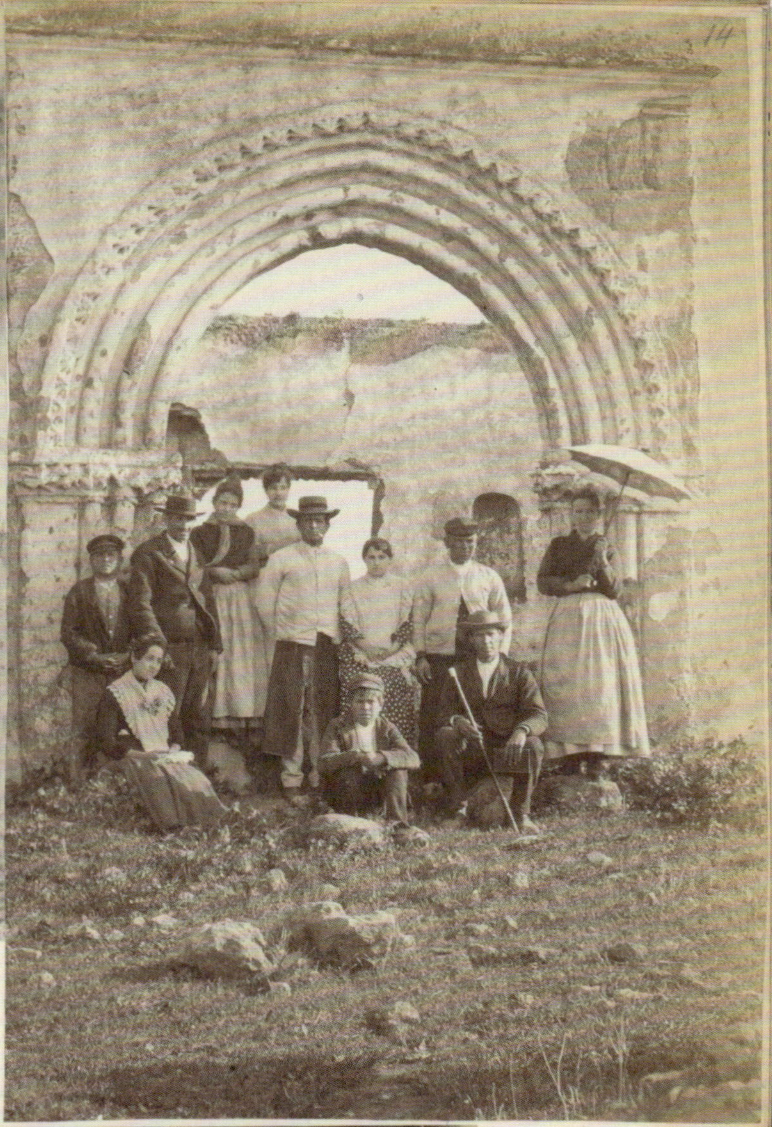

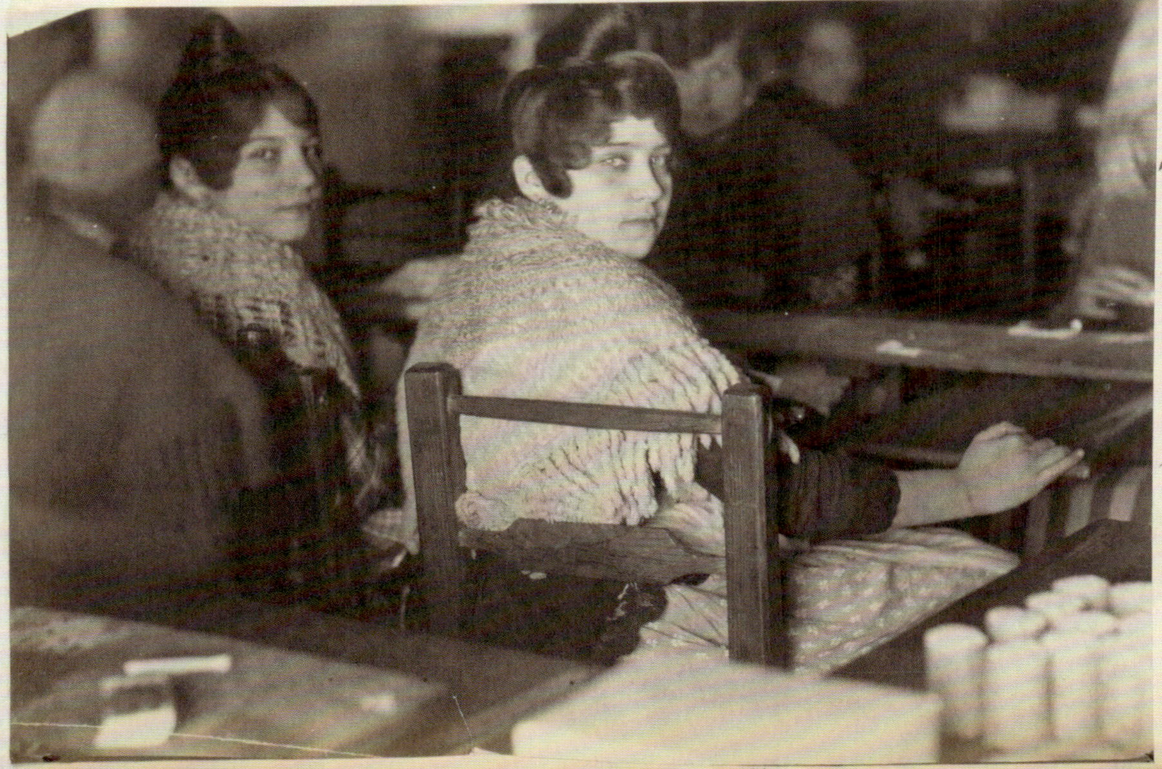

Alcalá de
Guadaira
"Castle of the
river Aira" —
Hiempa,
"Place of many
Springs" —
Monday April
16.

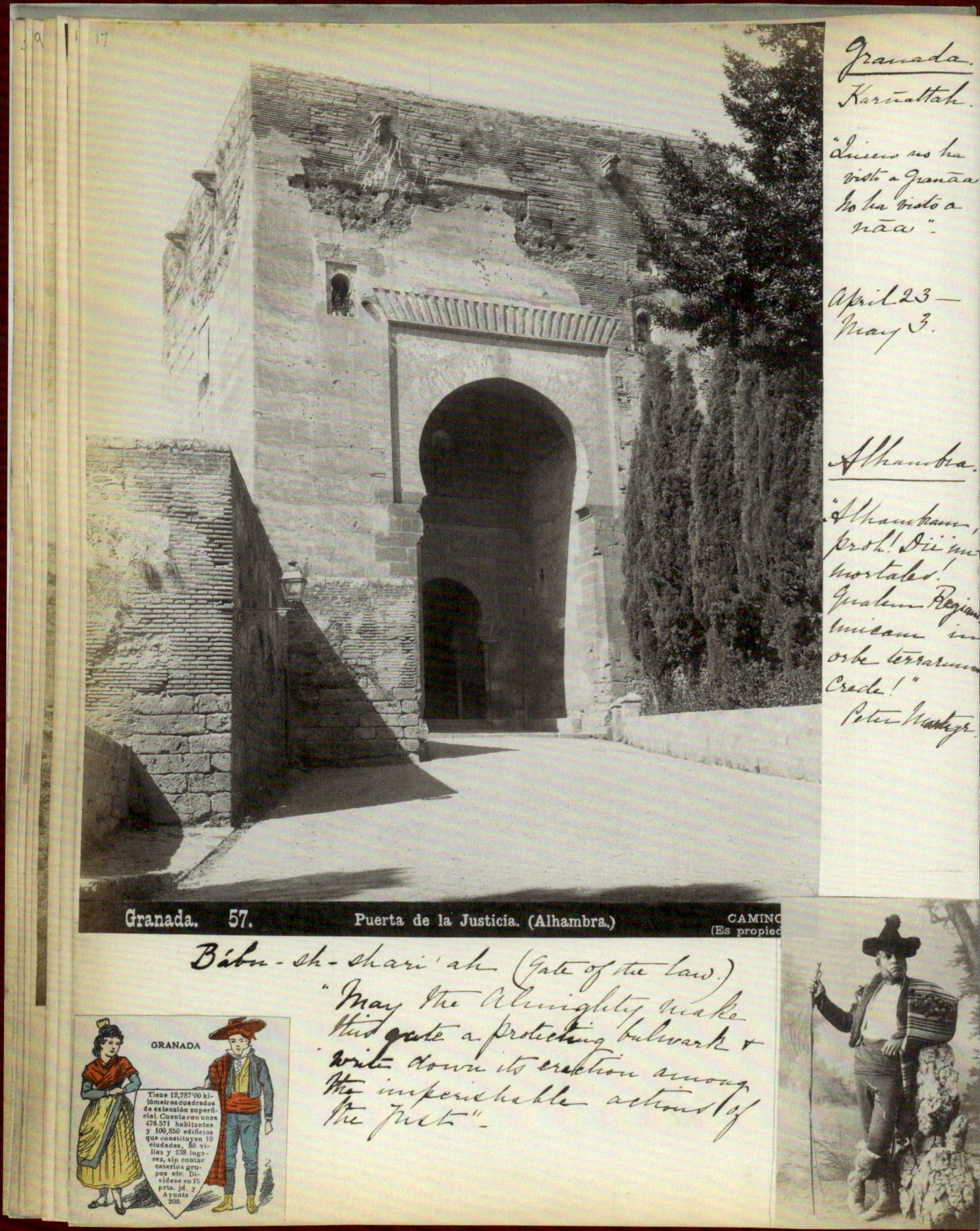
Granada.
Karnattah.
"Quien no ha visto a Granāa No ha visto a nāa".
April 23 –
May 3.
Alhambra.
"Alhambram, proh! Dii immortales! Qualem Regiam unicam in orbe terrarum Crede!"
Peter Martyr
Granada. 57. Puerta de la Justicia. (Alhambra.)
Bábu-sh-shari'ah (Gate of the law.)
"May the Almighty make this gate a protecting bulwark & write down its erection among the imperishable actions of the Just".
GRANADA

Granada. 18
April 23–May 3.

Torres Bermejas
Kal'at Al-hamra

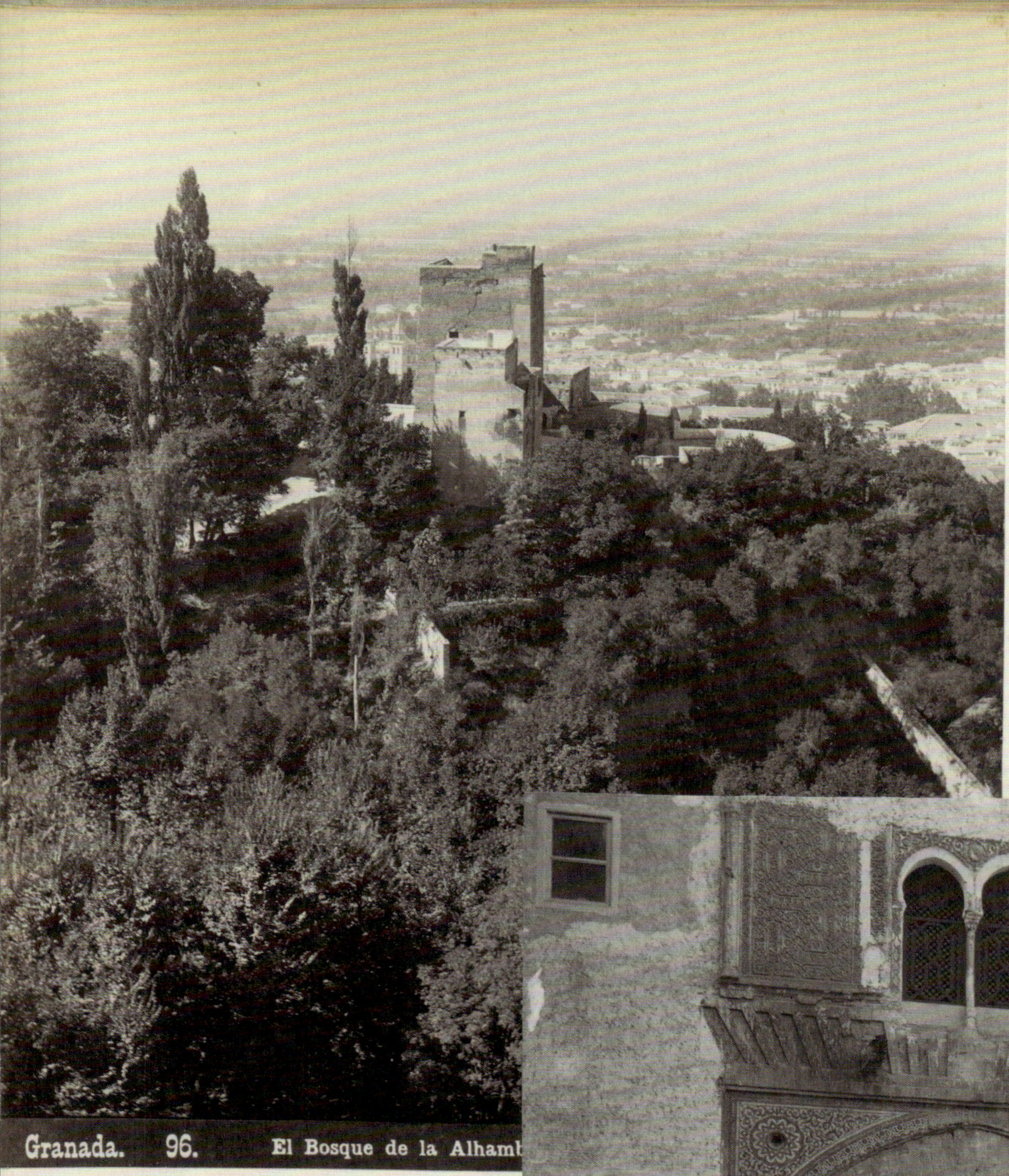

La Torre del Vino

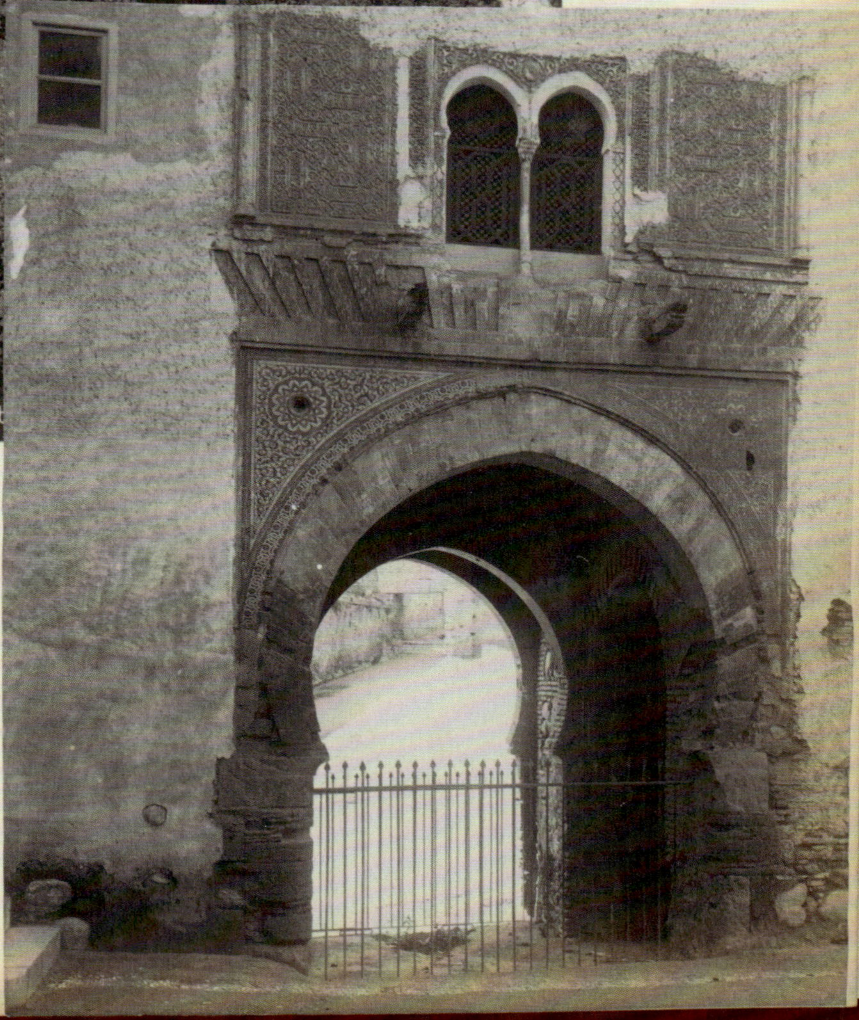

SPAIN AND PORTUGAL
Volume II, pages 21–22, April 1888

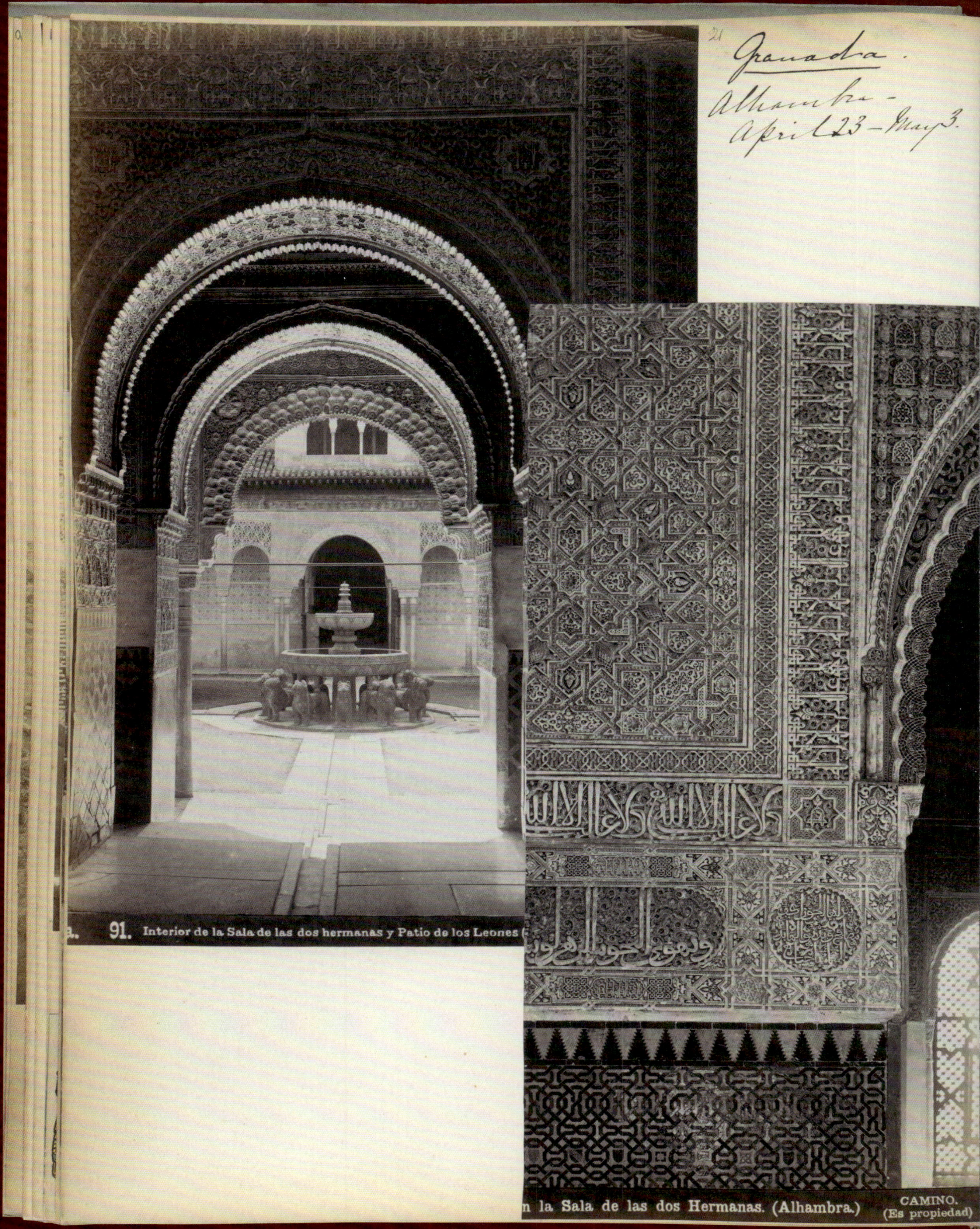

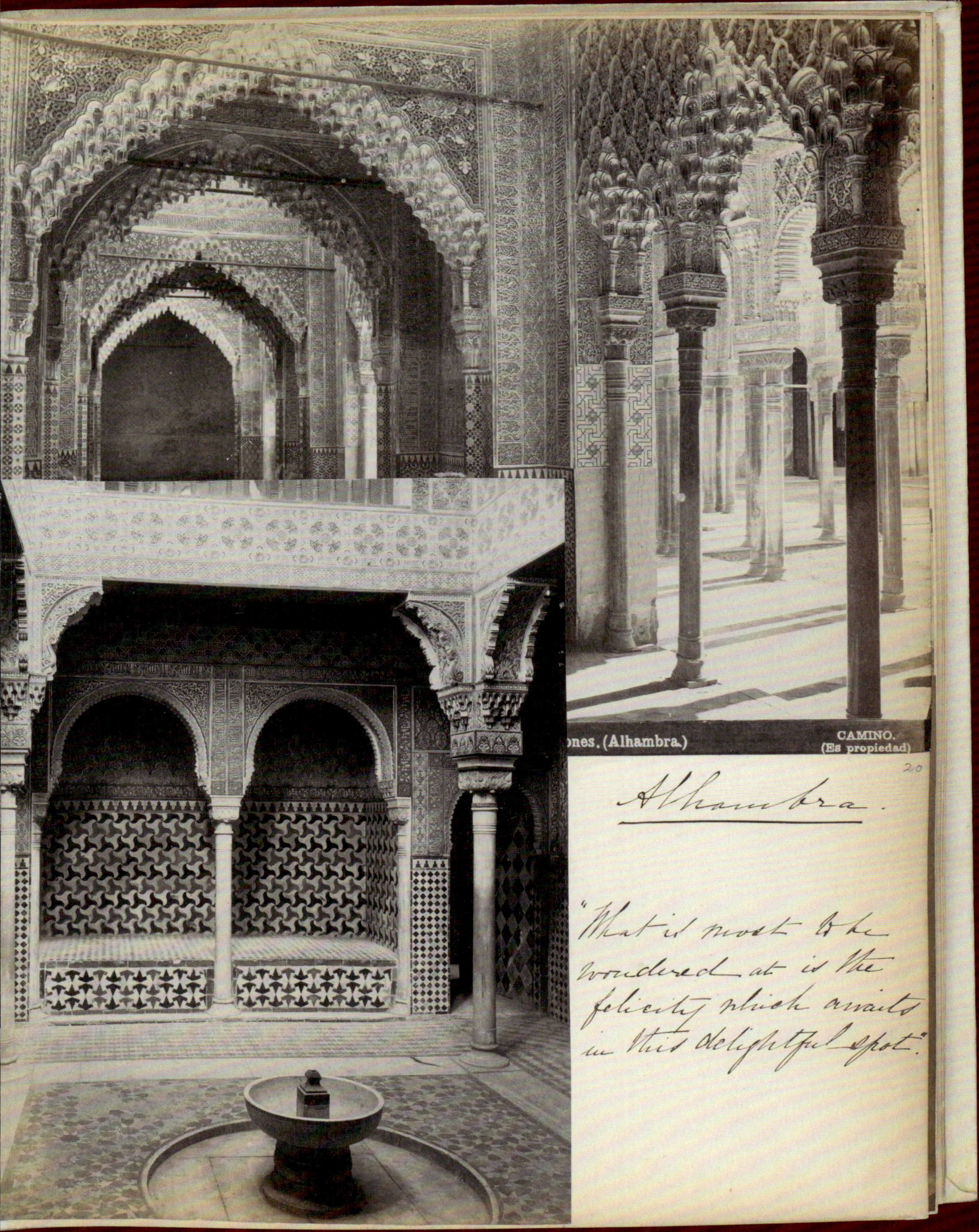
ones. (Alhambra.)
CAMINO.
(Es propiedad)
20
Alhambra.
"What is most to be wondered at is the felicity which awaits in this delightful spot."

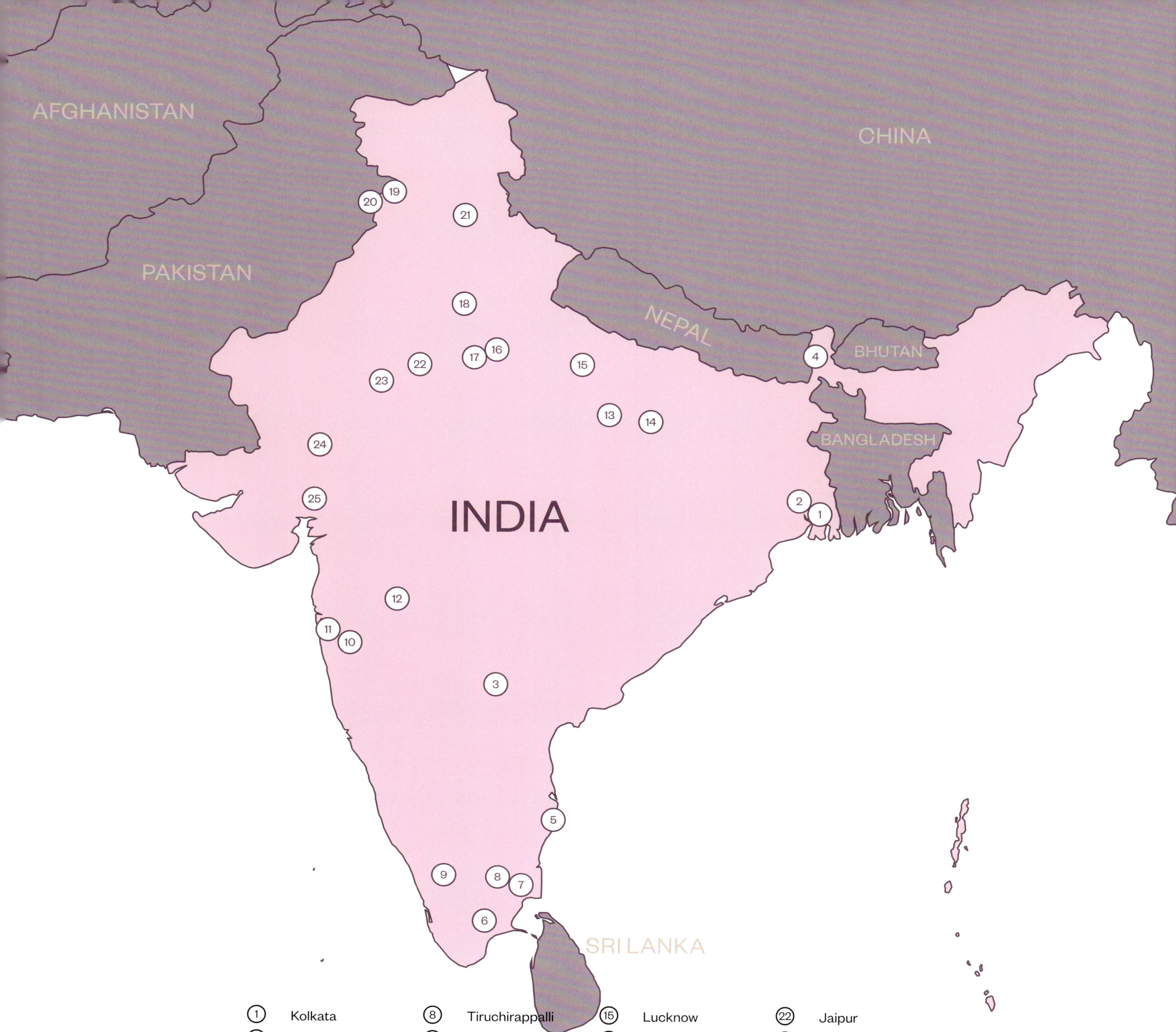

1. Kolkata
2. Howrah
3. Hyperbad
4. Darjeeling
5. Chennai
6. Madurai
7. Thanjavur
8. Tiruchirappalli
9. Coonoor
10. Lonalva
11. Mumbai
12. Ellora
13. Prayagraj
14. Varanasi
15. Lucknow
16. Agra
17. Fatehpur Sikri
18. Delhi
19. Amritsar
20. Lahore
21. Shimla
22. Jaipur
23. Ajmer
24. Mount Abu
25. Ahmedabad

India

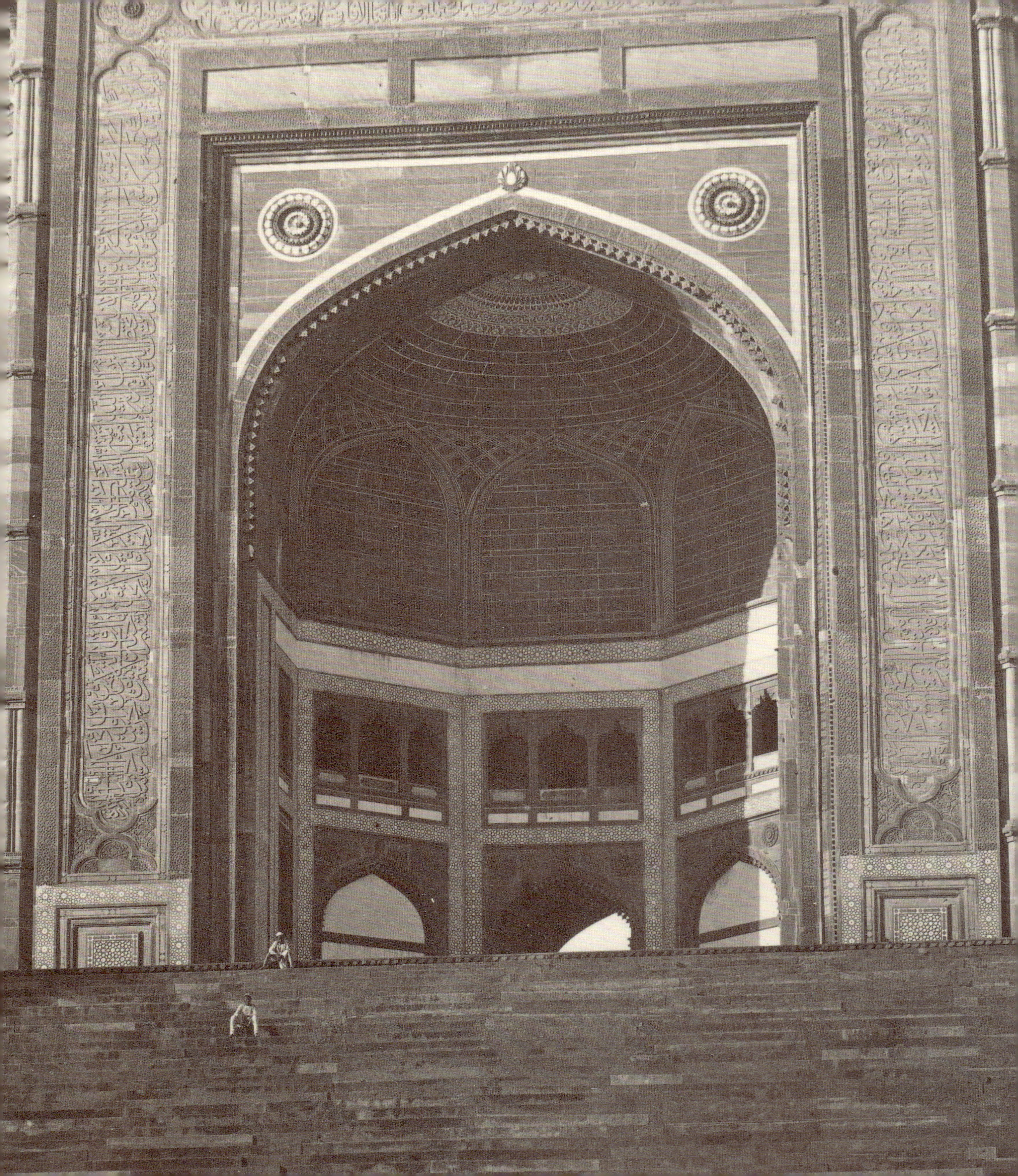

Isabella Stewart Gardner's Travel Albums through India

Pujan Gandhi

For a woman destined to build and distinctly house for the public one of the most remarkable art collections in American history, it would seem that Isabella Stewart Gardner's visit to the Indian subcontinent could not have been timed better. She and her husband, John "Jack" Gardner, disembarked at Calcutta (Kolkata), then seat of the British Crown rule (1858–1947), on 21 January 1884 during the Calcutta International Exhibition. Encompassing the grounds of the Indian Museum, an institution founded by the Asiatic Society of Bengal in 1814—making it the earliest "encyclopedic" museum in all of Asia—as well as the adjacent maidan (park), the nearly twenty-two-acre exhibition featured elaborate displays of the finest arts, crafts, natural specimens, and manufacture from regions throughout India and the Empire, attracting more than one million visitors during its four-month run. Such international exhibitions were part of a sweeping global current: between 1866 and 1914 these were hosted by different cities nearly every two years (fig. 1).[1] A bombastic yet efficient gathering of the latest in global industrial technology and design (the Eiffel Tower was built for the 1889 Paris Exposition Universelle), the phenomenon was akin to, if not more broadly impactful than, the art fairs or biennales of today—curiosity, commerce, and power commingled in a format heralding the benefits and possibilities of a truly cosmopolitan exchange.

The exhibition formats were based on the phenomenal success of the 1851 Great Exhibition of the Works of Industry of All Nations held at Hyde Park, London, in a purpose-built, nineteen-acre steel and glass structure nicknamed the Crystal Palace. There, thirty-four countries, along with thirty nations within the British Empire, presented booths boasting their latest innovations in manufactures, machinery, raw materials, and the fine arts. *The Art Journal*'s illustrated catalogue of the 1851 fair begins with Queen Victoria's proclamation: "It is my anxious desire to promote among all nations the cultivation of all those arts which are fostered by peace, and which in turn contribute to maintain peace in the world."[2] While this declaration of "Pax Britannica" was unquestionably a tool to substantiate and maintain the status quo of

Fatehpur Sikri, March 1884 (detail) from Isabella Stewart Gardner's *Travel Album: India, Pakistan, Yemen, and Egypt, Volume VI*, page 24

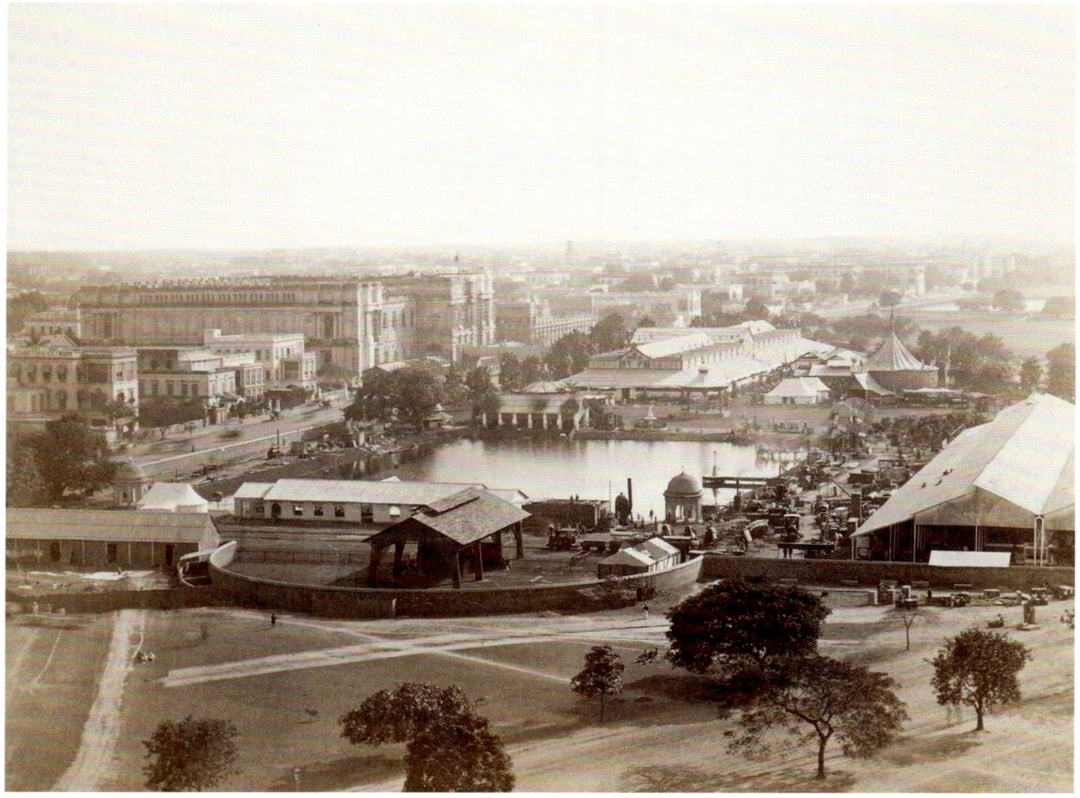

Fig. 1 Shivshanker Narayan, Aerial view of the 1883–84 Calcutta International Exhibition, India between December 1883 and July 1884. Albumen silver print 21.5 x 28.3 cm. PH1982:0432:003 Canadian Centre for Architecture

empire, for India—arguably the Crown's most significant recent acquisition, notably given the largest "booth"—the educative premise proved instrumental in bringing to public attention the skills of its artisans. Many examples of Indian textiles, metalwork, jewelry, ivories, jades, and lacquers acquired at the Great Exhibition laid the foundation for the South Kensington Museum, now the Victoria & Albert Museum, London. In a similar manner, many of the works showcased at the Calcutta International Exhibition would enrich the holdings of the Indian Museum, still on display in the present day.

Despite these broader trends, it's highly unlikely that Isabella Stewart Gardner acquired anything at the fair: in her daily diary, Gardner notes visiting the "Exhibition" three days in a row but fails to comment on the establishment of the Indian Museum; she makes brief mention of "native jewels" but in all succinctly describes the experience as a "Great disappointment." This is a rare negative report from a nearly three-month journey throughout the subcontinent.[3] Most, however, embraced the wonders of the novel, immersive displays of the International Exhibition format. One visitor expressed of his experience at the Great Exhibition: "philosopher and poet alike are agitated, and silent; gaze whithersoever they may, all is marvelous and affecting; stirring new thoughts and emotions, and awakening oldest memories and associations."[4] Politically, the exhibition was especially significant in emphasizing the interconnective possibilities of the Empire (the Australian colony of Victoria proudly reporting that "out of 1,100 packages dispatched, only 45 reshipped back to Melbourne"[5]), and proved to be a source of national pride. In India, various courts of its regional vassal states—

many of which Gardner would later visit in her journey—were encouraged to curate lavish and diverse displays, with princes proudly present at the opening ceremonies. To ensure a robust attendance, executive commissioner Jules Joubert (1824–1907) led an extensive marketing campaign, promoting the event with lectures at "Hindu and Mohammedan colleges,"[6] while the government sponsored special train services for South Asian workers living outside Calcutta. From colonist to colonized, the exhibition successfully inspired a broad audience (save Isabella), and furthered Queen Victoria's quest for the arts fostering a great "Pax Britannica."

Whereas the catalogue of the first Great Exhibition in London was illustrated in traditional print methods, the Calcutta International Exhibition proved fertile ground for the rapid developments in the medium of photography, which arrived in India in 1840—just one year after it was introduced in Europe—and was simultaneously fueled by the interests of Crown and commerce. A. W. Turner, under the direction of James Waterhouse (1842–1922), head of the photographic section of the Survey of India in Calcutta, produced some two hundred images that closely detailed the wares of each regional court, from the jewels of Bengal to the silver of the Shan state in Burma (fig. 2).[7] His fastidious and frank approach reflected that of his colonial commission, whose primary intention was to document, categorize, and report back to the powers that be. Meanwhile, Shivshankar Narayan (active 1860s–1890s), an entrepreneur photographer who established his studio and business in Bombay (Mumbai), arguably approached the exhibition with greater agency. He made sure to capture the epic sweep and opulence of the exhibition's regional displays (fig. 3), in turn, evidencing ways in which colonial subjects found opportunities to materially contribute to formulations of Indian identity.[8] For an organizer, the Calcutta International Exhibition would have been an ideal constellation of elements: beautiful objects, innovative methodologies, (mostly) superlative visitor experiences, record-breaking and diverse attendance, and not least, lasting significance. Today, these remain enviable benchmarks for any exhibition, and one would think the experience would have been a point of personal inspiration for a future founder of a museum. The question remains: why was Gardner not impressed?

First, it's helpful to remember that Gardner's collecting interests as we know them today were largely nascent during this period—she would begin this endeavor in earnest after receiving her inheritance in 1891, building on connections and experiences she'd accumulated over previous decades to enlist the guidance of agents, such as Ralph Wormeley Curtis (1854–1922), and art historian-cum-dealer Bernard Berenson (1865–1959) to build her collection and lay ground for her museum, born, as she later explained, out of the need for beauty and art in "this very young country."[9] Gardner, however, was collecting rare books and manuscripts during this period, demonstrating a voracious interest in all aspects of arts and letters, including non-Western traditions. In 1882, prior to their journey to Asia, Gardner befriended the writer F. Marion Crawford (1854–1909), who had recently studied Sanskrit in India; that same year, she and Jack hosted a series of private lectures by Edward Sylvester Morse (1838–1925), on the spiritual and aesthetic life of Buddhism in Japan.[10] Travel

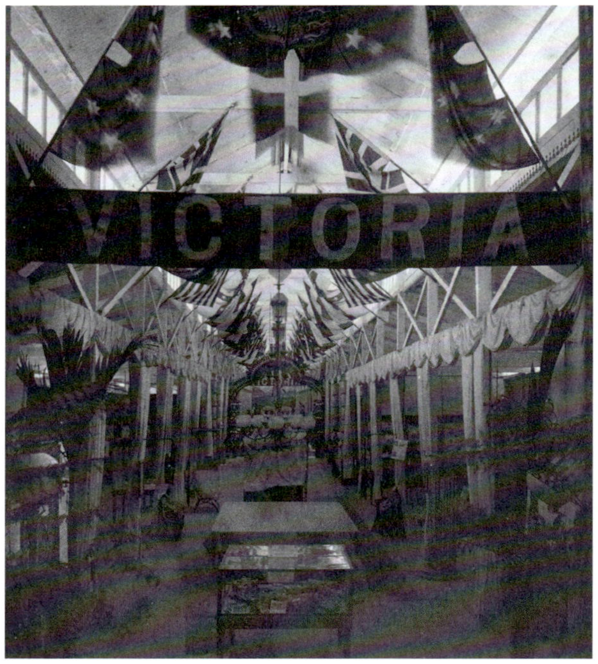

Fig. 2 A. W. Turner, "Victorian Court" in *Photographs of the Calcutta International Exhibition*, 1883–84, plate. 183. Central Secretariat Library, New Delhi

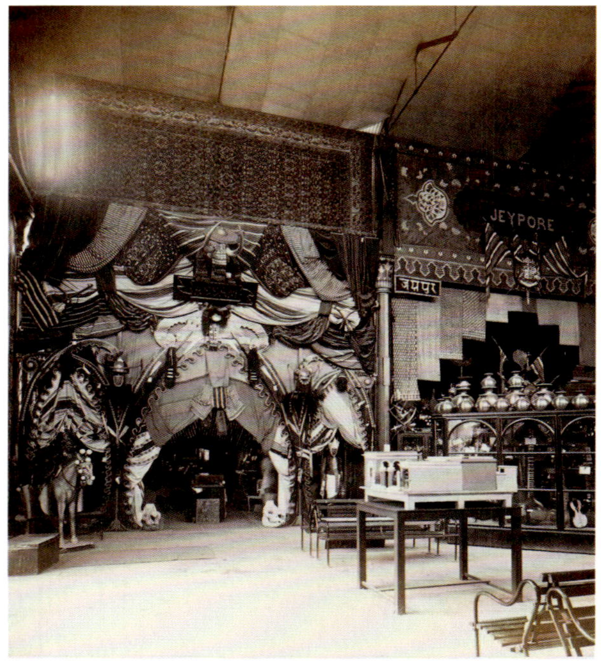

Fig. 3 Shivshanker Narayan, Interior view of a pavilion showing the Assam and Jeypore exhibits, Calcutta International Exhibition of 1883–84, India between December 1883 and July 1884. Albumen silver print 22.7 x 27 cm. PH1982:0432:009, Canadian Centre for Architecture

was equally, if not more, significant to Gardner's personal and intellectual development, and instrumental in bringing Gardner out of depression following the death of her only son in 1865. As she ably journeyed on her expansive, well-researched itineraries, she began recording her impressions via letters, journals, souvenirs, and small collectibles, and most tellingly, assembled over twenty-eight travel albums between 1867 and 1897. These albums were composed primarily of professional and amateur commercial photographs collected at various sites, which were then pasted, occasionally annotated, and arranged with other media and mementos, with some pages still bearing the press of dried flowers (Plate 27).

While in India, Gardner created two such albums, revealing intimate glimpses of her personal proclivities—and thus her lack of interest in the Calcutta International Exhibition—as well her positionality as an American following in colonial footsteps. In Gardner's records of criss-crossing the subcontinent, we see ways in which her visit relates to—and diverges from—the British colonial project: she embraces the aesthetics of the picturesque while rejecting the mechanization of industrialism; she shares an interest in taxonomy (people, places, flora, and fauna) but appears less interested in exploitative riches; and while she enjoys the company of her colonial hosts, she's simultaneously seeking beauty beyond those confines, and the vitality of precolonial life. Gardner's albums offer a materially significant—and beautiful—opportunity to consider the history of late-nineteenth-century Indian photography beyond the paradigm of colony and metropole, through a highly personal lens.

Like those of many privileged enough to travel during this period, Gardner's albums were created to preserve personal memory. In particular, the process of delighting in "picturesque" vistas of India had already taken hold, often emerging from codependent and entrenched commercial or colonial interests. Famous voyages to India included those of painters William Hodges (1744–1797), between 1781 and 1783, and Thomas (1749–1840) and William Daniell (1769–1837), between 1785 and 1794, who sensed the opportunities provided by the "East," and upon their return to England published several volumes of etchings, aquatints, alongside reimagined watercolors and paintings that took the nation by storm. This view presented India as the great, wild, crumbling, and unknown nation prime for capture, "betterment," and exploit.[11] This spirit would also inspire Samuel Bourne (1834–1912), a British banker turned landscape photographer who set sail for India in 1863. Armed with supplies of chemicals, glass plates, and a mobile darkroom, in a little less than a decade Bourne would produce over 2,500 photographs of the landscapes and architectural wonders of the growing British Empire, and was especially gifted in capturing vistas of scale and detail, seventeen of which are included in Gardner's album (Plates 28 and 29). With associate Charles Shepherd (active 1858–1878), Bourne established the Bourne & Shepherd Studio in Calcutta in 1865, which continued to thrive after his return to England, employing photographers such as Colin Murray (1840–1884) (fig. 4 and Plate 27). Gardner includes fifteen of Murray's images of Jaipur, Mount Abu, and Ahmedabad in her album and records visiting Bourne and Shepherd's studio in Bombay—from which the highest proportion of her attributed photographic acquisitions came—on 2 April 1884.[12] She, Bourne, and the studio shared an aesthetic affinity for romantic visions of India's architectural past: Bourne was likewise disaffected by the grand colonial edifices in the capital of Calcutta, once describing it as a "place totally devoid of architectural beauty."[13]

Alongside a growth in commercial photography, the growing ambitions of empire resulted in successive government-sponsored photographic surveys beginning in the early 1850s, which mapped territory with volumes focusing on architectural history, naturalia, and ethnography, such as the eight-volume *People of India* (1868–75), comprising 468 images that served as a "guide" to help distinguish India's various tribes and ethnic groups.[14] As Maria Antonella Pelizzari explains, it was the "cultural texts of British Colonialism that lead to the photography archives of India."[15] Gardner's album, and the landscapes, peoples, and architectural sites she chooses to include in it, cannot be removed from this context; this was the source of visual and material culture that would inflect any such assemblage of this time (fig. 5 and Plate 20). However, it is worth noting that as an American woman, Gardner was not necessarily the intended audience of such output. This is made rather specific by an assertion in photographer John Murray's 1891 *Handbook for Travellers in India and Ceylon*, that "a trip to India is no longer a formidable journey or one that requires special preparation … the Englishman who undertakes it merely passes from one portion of the British Empire to another."[16] Isabella and John Gardner's itinerary followed the well-trodden colonial

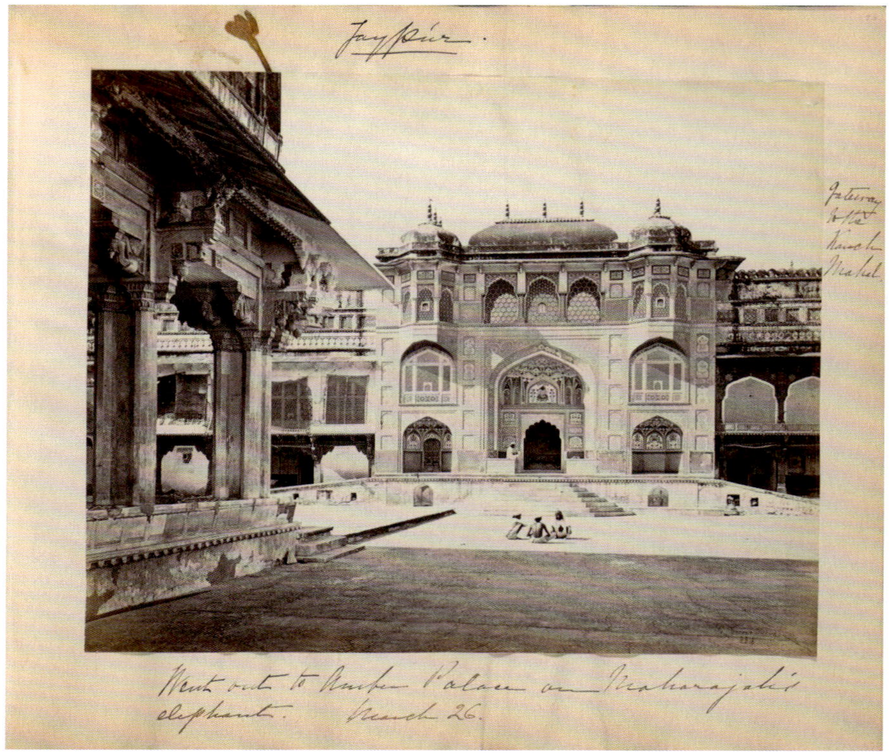

Fig. 4 Jaipur, India, March 1884 in Isabella Stewart Gardner's *Travel Album: India, Pakistan, Yemen, and Egypt, Volume VI*, page 70

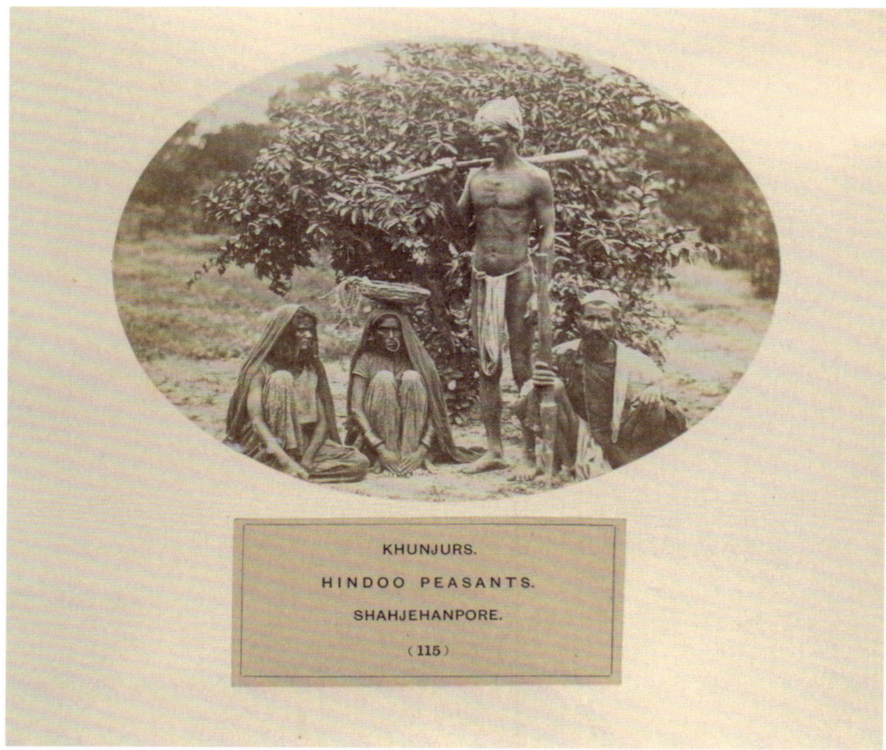

Fig. 5 "Khunjurs, Hindoo peasants, Shajehanpore," from the India Museum's *The People of India: A Series of Photographic Illustrations*, 1868–75. The Miriam and Ira D. Wallach Division of Art, Prints and Photographs: Photography Collection, The New York Public Library

path, but their visit to India predated the publication by more than five years—and it is difficult to know if, as Americans, they swelled with the same sense of colonial pride.

Situating Gardner's travel albums in the context of Anglo-Indian history reveals some compelling omissions. One simple observation is that she does not include a depiction of India's vast railway system, which from 1853 quickly developed into the most advanced in Asia, and even today remains among the most frequently cited "contributions" of British colonial rule. This veers from the tendencies of most British travelers who also created travel albums, including the Hobson family's *Le tour du monde I: 1899-1900-1901*, which includes a mixture of bought photographs and those they self-produced with the newly available portable camera box, proudly recording themselves in the great colonial landscape, riding the train to Darjeeling, and presumably picnicking among the ruins of Agra (figs. 6 and 7). Elizabeth Heath considers the sporting of privilege and gentility in these early "selfies" a product of a colonial ideal, whereby "actual lineage and social status mattered little; rather, all Britons could reimagine themselves as members of an aristocracy."[17]

While Gardner's positionality within an "American aristocracy" in her albums is debatable, emphasis is placed less on self-situating and more on capturing the essence of her personal experience, observations, and interests. In the album page devoted to her 14–16 February stay in Nilgiri Hills of southern India, she includes a typical image of white travelers enjoying the grounds of the Glen View Hotel but then pierces its bottom frame with one of Indian men—presumably outside the hotel enclosure—with an oxcart, conveying hay (Plate 23). This is just one of several instances in which Gardner inserts unexpected, associative dimensions in her albums, slightly outside the bounds of contemporary colonial decorum. She's more explicit in seeking out and including in her albums images of palanquins—which by the 1860s had all but disappeared from the subcontinent—somehow managing to imagine one by a quiet tank (temple reservoir) in the bustling, colonial city of Bombay (Plate 24). She then devotes a page to crudely rendered tourist paintings, characterizing the region's historical transition to cart and buggy (Plate 25). These images are still products serving an "othering," Orientalizing gaze, but their inclusion in Gardner's album continues to suggest her greater interest in India's precolonial past.

In this endeavor, Gardner's ability to arrange images in highly personal, multisensory ways would have proved far more rewarding than the contrived, hyperconsumerist, sterile spectacle she would have seen at the Calcutta International Exhibition. There, exhibits featured cast models of historic architecture, with the "magnum opus" being a specially conceived portal, known as the "Gwalior Gate," that was sponsored by the Maharaja of Scandia; despite being overseen by J. B. Keith, a curator with the Archaeological Survey of India (ASI), it had no bearing in built reality, intended more to amalgamate and illustrate the carvings of many periods and places.[18] This would likely have jarred anyone's quest for an "authentic" travel experience, and for Gardner was a decided contrast to her experience in Burma (Myanmar), which she visited just before Calcutta. She wrote giddily of visiting the Shwedagon Pagoda (the

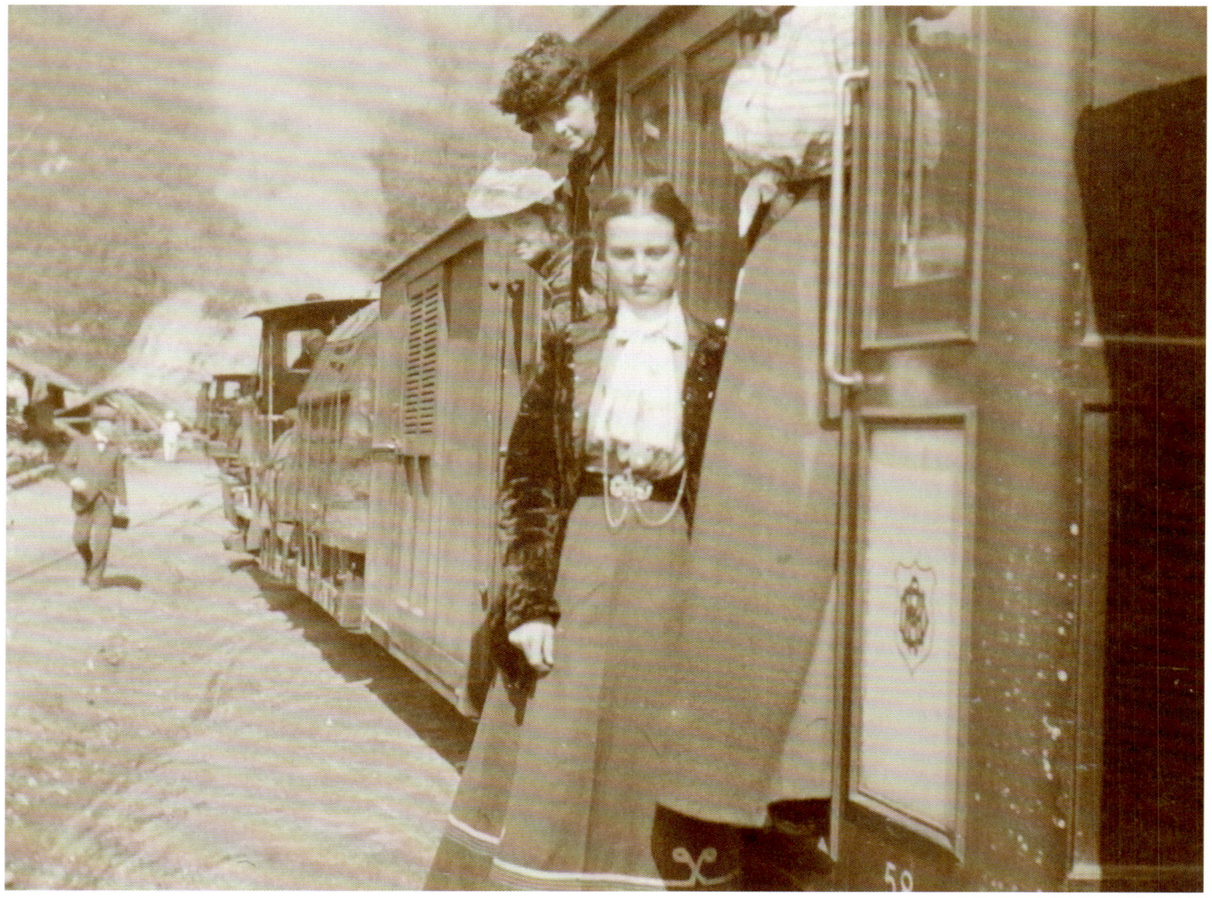

Fig. 6 “En route to Darjeeling,” 1899–1901, detail from Hobson Family Album. The Wolfsonian

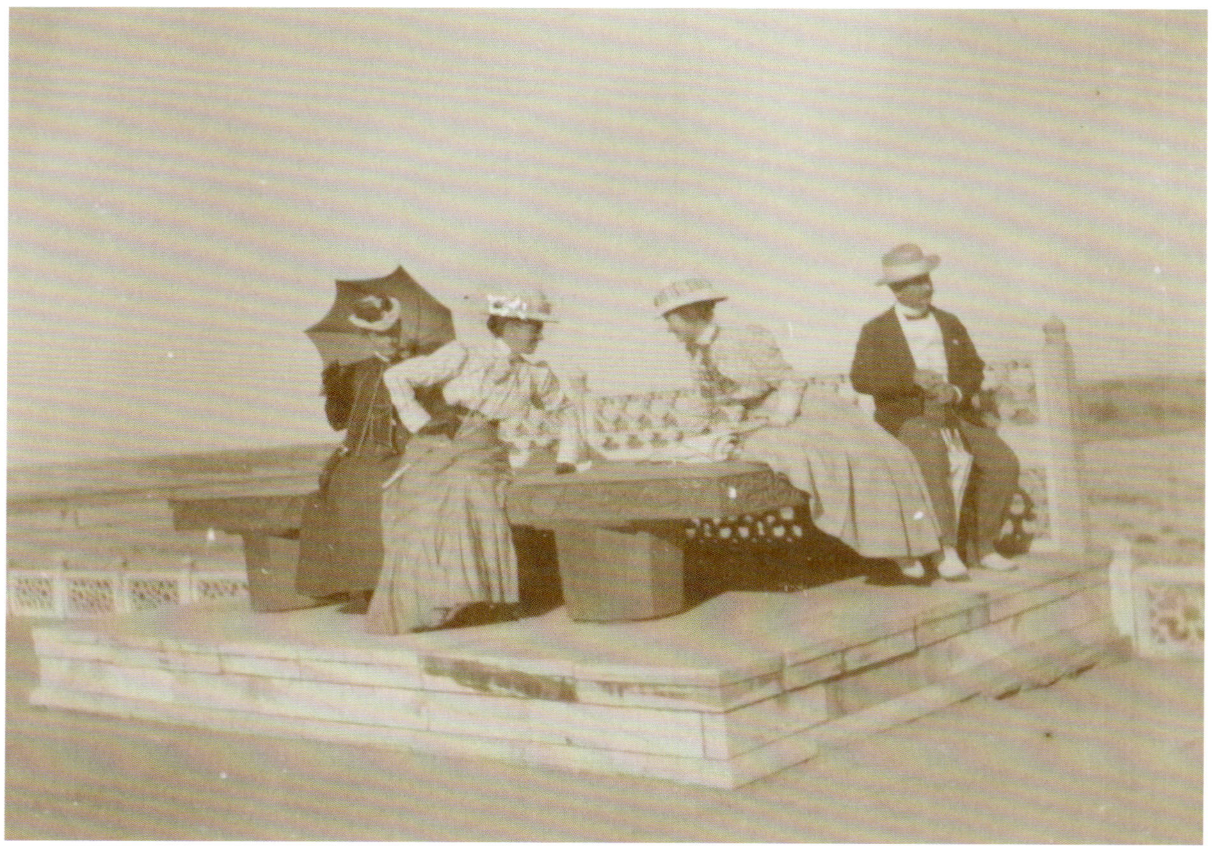

Fig. 7 “Agra, Black Throne,” 1899–1901, detail from Hobson Family Album. The Wolfsonian

Great Pagoda) on the Festival of the Full Moon: "The temple a perfect wonder with 1000s of little bells jingling—Quite the most Eastern city we have seen ... every possible colour and some gorgeous rich stuffs.... Every one lighting candles, carrying flowers, flags, and paper umbrellas."[19] It's fitting then, that Gardner would later campaign for the Museum of Fine Arts, Boston to abandon collecting plaster casts, and when she opened her museum in 1903, she insisted there be music throughout.[20]

From "Parsee Theatre" to street jugglers, Gardner's diary reveals consistent delight in what she perceives as "most Eastern," which, coupled with an exploratory interest in "Eastern" religions, would impact the types of objects she would acquire on her journey (fig. 8). In Calcutta, she would skip the multicarat gems on offer from the regional courts—many of which were created for the purposes of International Exhibitions and aligned in design and ostentation with those gifted to the prince of Wales on his 1875–76 tour of the subcontinent (fig. 9).[21] Instead, on the advice of Marion Crawford, her Sanskritist friend from Boston, Isabella and Jack visited the famous jeweler and antique dealer Alexander N. Jacob (real name, Iskandar Meliki bin Ya'qub al Birri, 1849–1921), from whom she acquired two bazubands (armbands), "certified to have belonged to the Queen of Delhi!"[22] Here, she was referring to Zinat Mahal, the favorite wife of the last Mughal emperor, Bahadur Shah Zafar II (r. 1838–58), who was sent into exile after the failed Indian Mutiny (also called the Great Uprising) of 1857, whereby authority fully shifted to Crown rule; according to Jacob's provenance records, the "armlets were taken from King's palace Delhi during the time of the Mutiny by the British soldiers, and were auctioned by Colonel Innes the prize agent."[23] Gardner was also excited that the armlets included the "nine sacred gems," referencing the *navaratna* tradition of traditional Indian jewelry, with each gem (pearl, ruby, emerald, diamond, coral, blue sapphire, yellow sapphire, hessonite, and chrysoberyl) corresponding to the nine celestial objects (seven planets and ascending and descending modes of the moon), which in turn provide talismanic qualities to the wearer.[24] However, Gardner's armbands include turquoise, which is not one the traditional stones, but was subtly added to the array under the influence of Victorian taste.[25] Again, we see Gardner fully imbricated in the by-products of the British colonial project, but as much as she is able, she attempts to eschew prevailing conventions in search of something more alternative, and more "authentic." And when she finally visits one of the key sites of the Indian Mutiny at the Residency in Lucknow, she stays true to spirit, and pastes into her album an image of Indian musicians playing beneath the rubble (Plate 26).

Given the backdrop of Crown rule, and despite Gardner's objections, it's impossible to negate the impact of the Calcutta International Exhibition on the journey recorded in her travel albums. Specifically, she would likely have seen a suite of twenty photographs produced by Messrs. Nicholas and Co., called "Photographic Views of Southern India," which would prefigure her tour of temples in Tamil Nadu—a highlight of the itinerary. At least one image from the series, a "great tank out of the city" (Vandiyur Mariamman Teppakulam Tank, Madurai), would be included in her album, appearing above an

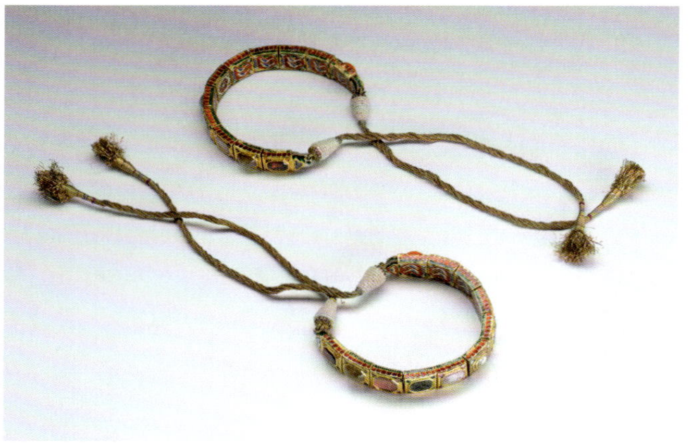

Fig. 8 Indian, Jaipur, Pair of Armlets (Bazubands), about 19th century. Gold on a lac core, with precious and semi-precious stones, enamel decoration, and cords of silk and gold with seed pearls, 22.7 cm long. Isabella Stewart Gardner Museum, Boston

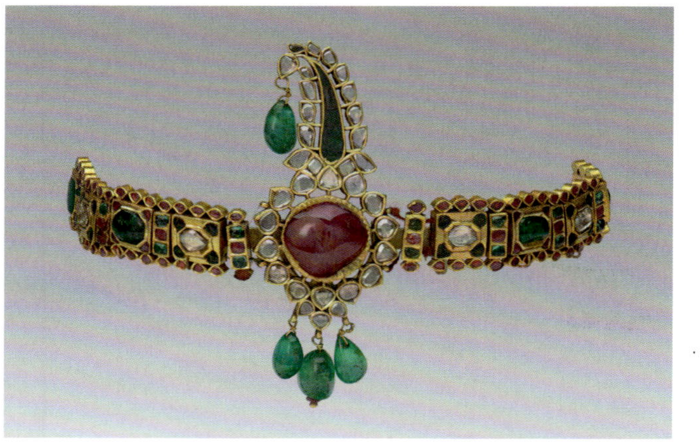

Fig. 9 South Indian, Turban ornament, mid–19th century. Gold, diamonds, emerald, rubies, and red cord, 12.0 × 24.6 × 0.8 cm. Royal Collection Trust.

image of manual laborers (Plate 22). Gardner notes in her diary: "The country dotted all over with small temples and images of gods and animals—An odd method of raising water; men going from one end of a pole to the other." That she is able to transform these observations to the visible and tactile in her album is a testament to the medium of photography itself, which in turn was patronized and popularized via the prevailing Exhibition model of the period. In fact, before Shivshankar Narayan captured the sites of the Calcutta International Exhibition, he made a name for himself displaying thirty-three images of his ASI-sponsored survey of Ambarnath temple at the 1873 Welt-Ausstellung in Vienna, and would later exhibit at the 1886 Colonial and Indian Exhibition in London.[26] Similarly, while Gardner makes annotative, often spiritual flourishes in her albums (Plate 21), her overweening emphasis on architecture was in keeping with architectural historian James Fergusson's (1808–1886) curation of the Indian display at the 1867 Paris Exhibition, which included over "500 representations of Indian Buildings," and is credited with spurring increased funding for architectural preservation throughout the subcontinent. The photographic medium in India fostered a sense of discovery and promise, as put by Samuel Bourne: "it is here classified as one of the fine arts. Are we more enlightened, or simply more just and unprejudiced in this land of rising British enterprise than the would-be patrons of art in professedly free by someone clique-ridden England?"[27] His proclamation elucidates the double-edged sword that characterizes so much of art production during the British period, that of innovation and "progress" on one side and violence and uninhibited exploitation on the other.

Embracing many forms of beauty and interspersing her point of view, Gardner's albums remain a result of a long-established colonial impulse: the need to document and record the natural specimens, peoples, and places of a conquered land was justified by enlightenment logic, and employed as a cultural, economic, and political device. In India, this process took form with the rise of the British East India Company (1599–1858), which grew from a small British trading company seeking access to lucrative

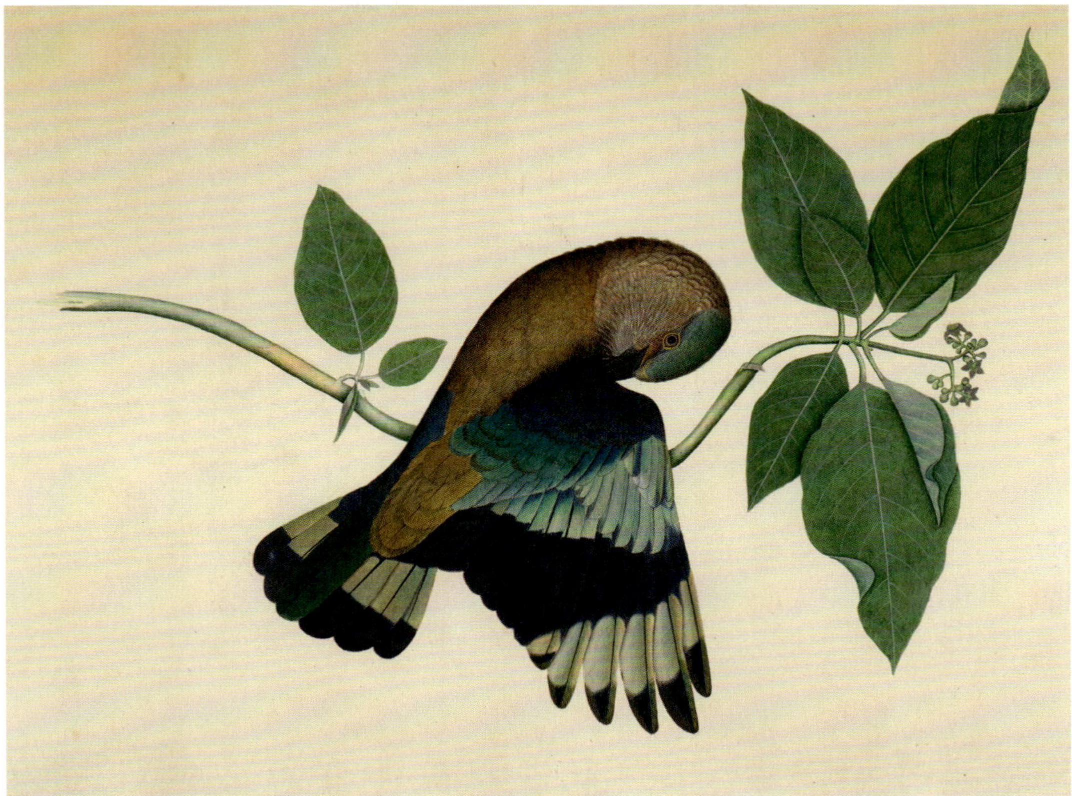

Fig. 10 Sheikh Zain al-Din, *Indian Roller on Sandalwood Branch*, 1779. Opaque colors and ink on paper, 76.2 × 96.52 × 2.54 cm. Minneapolis Institute of Art (2018.53.9)

spice routes into a military and administrative power that by the second half of the eighteenth century effectively ruled over the Indian subcontinent, generating nearly half of Britain's trade.[28] Among the most storied patrons during this period was Lady Mary Impey (1749–1818), wife of company minister Sir Elijah Impey, whose famed Impey Album (produced in Calcutta between 1777 and 1883) commissioned three Indian painters, Shaikh Zain al-Din, Bhawani Das, and Ram Das, who hailed from the neighboring—and recently defeated and depleted—regional court of Patna, to create 326 natural history paintings to record the newfound flora and fauna of their adopted home.[29] Lady Impey's interest in nature was in keeping with the times; she had likely read Buffon's encyclopedic *Histoire naturelle* and the writings of influential Danish botanist and taxonomer Carl Linnaeus.[30] As a result, she directed the artists' attention toward a popular European subject matter, provided large-scale, imported Whatman paper, and embraced the way in which they imbued their subjects with the tender naturalism of late Mughal painting. Thus, we see the emergence of an Anglo-Indian aesthetic predicated on new systems of patronage, whereby Indian artists adopted their skills at the force of new tastes.

Reading Gardner's description of arriving in Calcutta likewise reveals her sustained interest in horticulture and the natural world, as she notes the "Date Palms … Botanical Gardens.… On the other side the Palace of the King of Amdh (Oude) Flocks of birds flying about it—Being exercised." Her brief observation echoes Shaikh Zain al-Din's *Indian Roller on Sandalwood Branch* (fig. 10) from Lady Impey's menagerie. Of course, by the time of Gardner's arrival, the once prosperous Kingdom of Oude

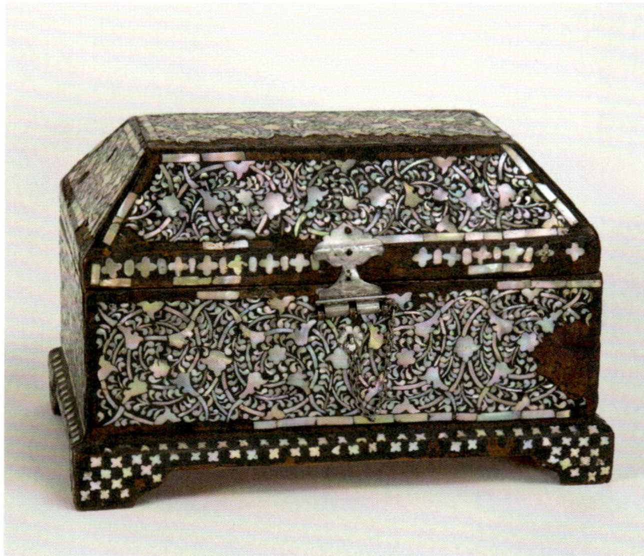

Fig. 11 Indian, Gujarat, Casket, about 1600. Wood coated with colored resin and inlaid with mother-of-pearl, 19.5 × 20 × 19.5 cm. Isabella Stewart Gardner Museum, Boston

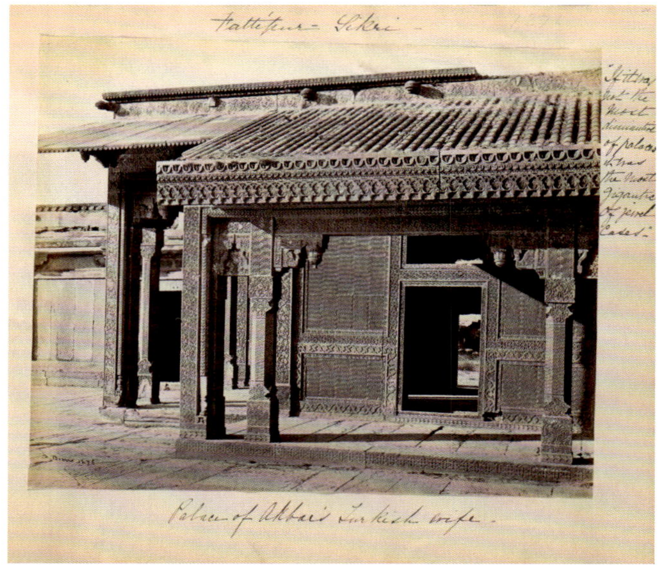

Fig. 12 "Fattepur Sikri, Palace of Akbar's Turkish Wife" from Isabella Stewart Gardner's *Travel Album: India, Pakistan, Yemen, and Egypt, Volume VI*, page 22

(Oudh) had been chipped away by over a century of wars and treaties with the East India Company, and by 1857 was fully annexed under British Crown rule. In fact, Gardner was describing its last king, the flamboyant Wajid Ali Shah (1822–1887), who was bribed into exile and living under constant surveillance while recreating the glories of his former court at his cloistered riverside estate.[31]

Throughout Isabella Gardner's journey of the Indian subcontinent, we see time and time again the complex, Anglo-Indian context from which her travel albums emerge: visits to the Calcutta International Exhibition, stays in colonial hill stations, and even with the photographers whose images she opts to acquire and paste. At the same time, we see her seeking the essences of India's precolonial past, be it in "authentic" Mughal jewelry, the charms of the palanquin, or street musicians. Later, as an art collector, she would acquire works such as a sumptuous, seventeenth-century mother-of-pearl inlaid box (fig. 11), but by and large, we can best assess the influence of her journey through India on her museum via her travel albums: instead of using the grid-by-grid approach of contemporaneous architectural historians, she makes sure to arrange her pages in ways that capture the ornament and grandeur of India's heritage, all the while ensuring each moment elicits personal reflection, and by the very act of an engaged, emotive gaze, lifeforce amid inequality and the balances of progress and decay.

On 18 April 1884, Gardner would set sail from India to Venice—a city that arguably impacted Gardner's collection and museum the most—but looking back on her album pages devoted to Fatehpur Sikri, the abandoned citadel built by the Mughal emperor Akbar (1542–1605), she writes next to a Samuel Bourne photograph of the Hujra-i-Anup Talao, or the Palace of Akbar's Turkish wife (fig. 12): "If it was not the most diminutive of palaces, it was the most gigantic of jewel cases."[32] As Gardner went on to build her own jewel case, she formulated an approach in contrast to the encyclopedic, "expo" model of the museums of her day: "the only way to treat ancient pictures is the one you adopt, to employ them not as a collection but as ornaments, provided they express your sentiments and satisfaction. It is the only way to make them live, to allow them a continued lease on life."[33]

1. Dehejia 2008, 12n15, 12n25.
2. *The Art Journal* 1851, xi.
3. One could easily contrast Isabella Stewart Gardner's accommodating traveling disposition with that of contemporary Bostonian Maturin Murray Ballou (1820–1895), whom John Sr. references in a letter he writes 17 March: "I have read all the books about yr. journeying but George tells me of a new book, by a Boston man, Maturin Ballou." Ballou likewise stayed at the Great Eastern Hotel in Calcutta and published in his 1884 memoir, *Due West; or, Round the World in Ten Months:* "It was rather surprising to find so poor a hotel as the 'Great Eastern' proved to be … [it] does not present the ordinary domestic comforts to be found in an American country tavern. We were told by our banker and others that the complaint in this matter was so general that a company was forming to give to the city a first-class hotel on the American system, a consummation devoutly to be wished." Ballou 1884, 169.
4. Hoffenberg 2003, 177n5.
5. Royal Commission for Victoria 1883–84, 7.
6. Hoffenberg 2003, 179n13.
7. Turner 1884.
8. Hoffenberg 2003, 183.
9. Nielsen, Riley, and Silver 2017, 191.
10. Tweed 2014, 52.
11. Guha-Thakurta 2003, 114.
12. Chong 2009b, 300. The albums include approximately 190 illustrated pages.
13. Sampson 2000, 165n5.
14. Heath 2015, 78n5.
15. Pelizarri 2003, 21.
16. Ibid., 31n19.
17. Heath 2015, 90.
18. Hoffenberg 2003, 182.
19. Isabella Stewart Gardner, entries from Friday, 11 January, and Saturday, 12 January 1884, *Travel Diary: China, Vietnam, Cambodia, Indonesia, India, Singapore, Myanmar, Egypt, and Italy, 1883–1884* (ARC.009111), Isabella Stewart Gardner Museum Archives, Boston.
20. Chong 2007, 215.
21. Those gifts, which included wares ranging from elaborately imagined, enameled, and jewel-encrusted inkwells to gold rosewater sprinkles, formed the basis of a traveling exhibition (1876–83), which was seen by more than 2.5 million British people. The profits from the exhibition were in turn reinvested to acquire even more material for the South Kensington Museum. See https://www.rct.uk/collection/themes/exhibitions/splendours-of-the-subcontinent-a-princes-tour-of-india-1875-6/the-0/the-display-of-gifts-in-britain.
22. Carvalho 2009, 458.
23. Ibid., 458n6.
24. Balakrishnan 2014, 214.
25. Carvalho 2009, 458.
26. Hoffenberg 2003, 187.
27. Dehejia 2000, 16.
28. Dalrymple 2019, xxx.
29. Dalrymple 2020, 15.

30. Topsfield 2020, 42.
31. Llewellyn-Jones 2014.
32. Isabella Stewart Gardner, *Travel Album: India, Pakistan, Yemen, and Egypt, Volume VI*, 1884, 22, Isabella Stewart Gardner Museum, Boston.
33. Chong 2007, 215n8.

Volume V, pages 25–26, February 1884

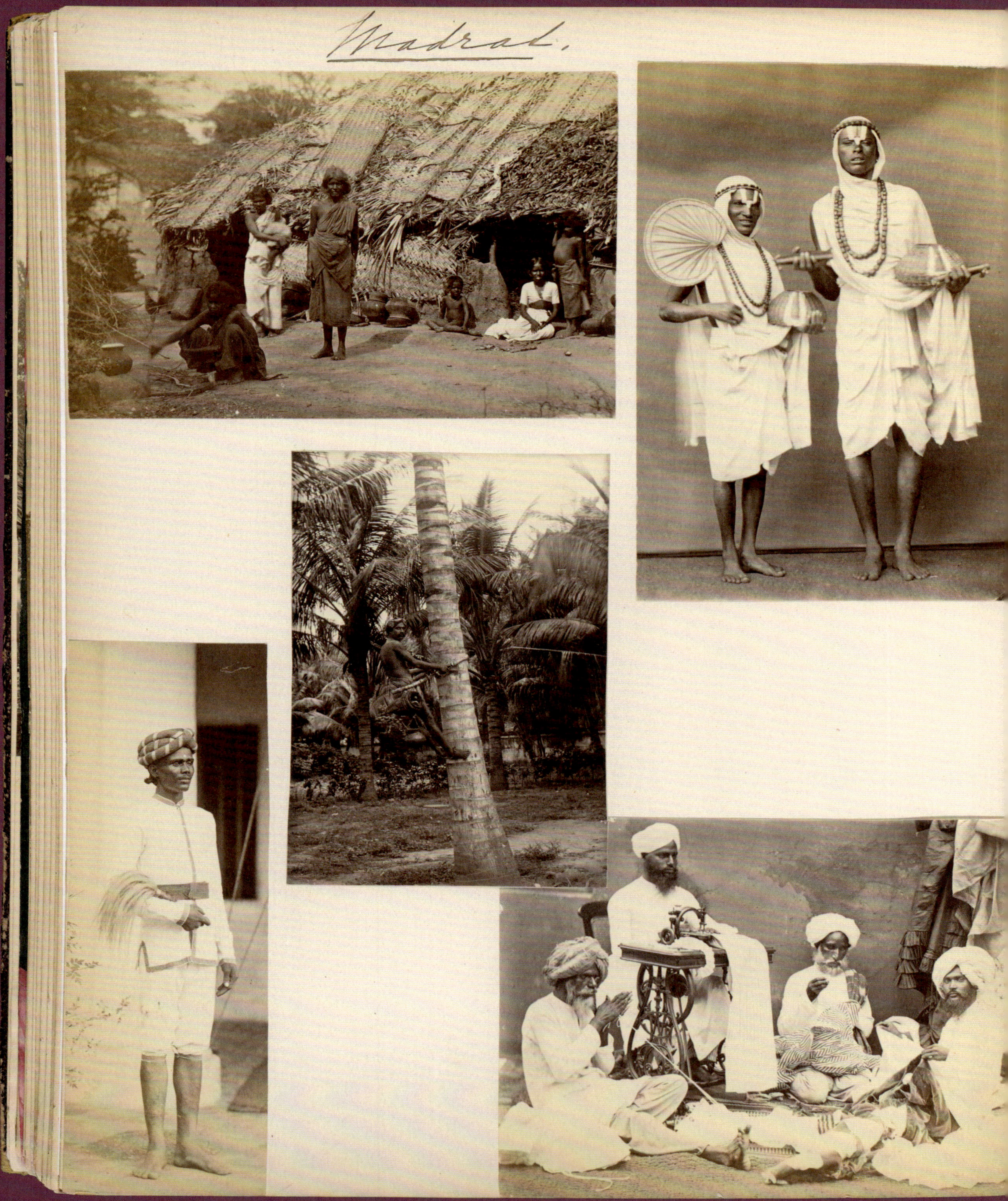

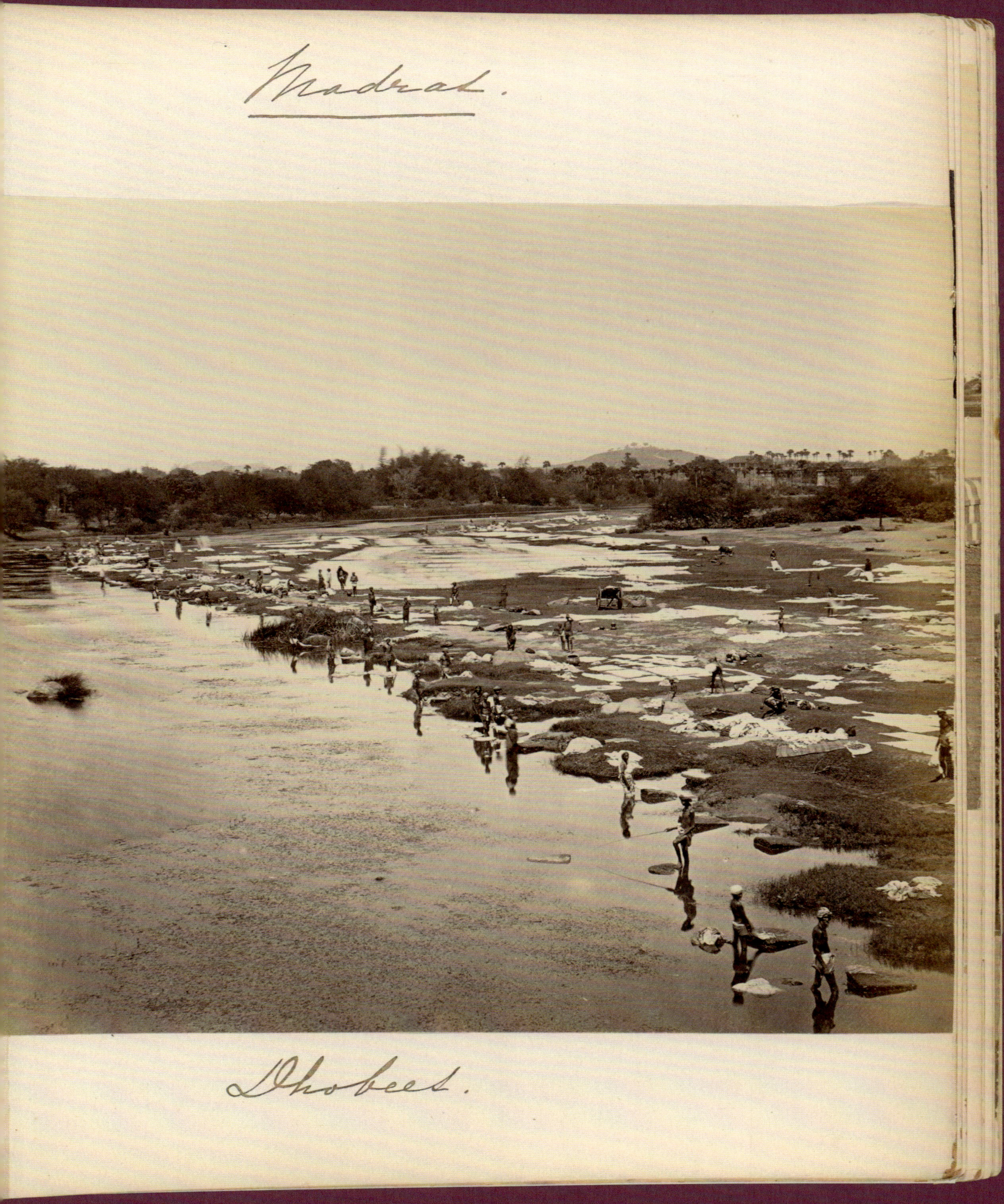
Madras.
Dhobees.

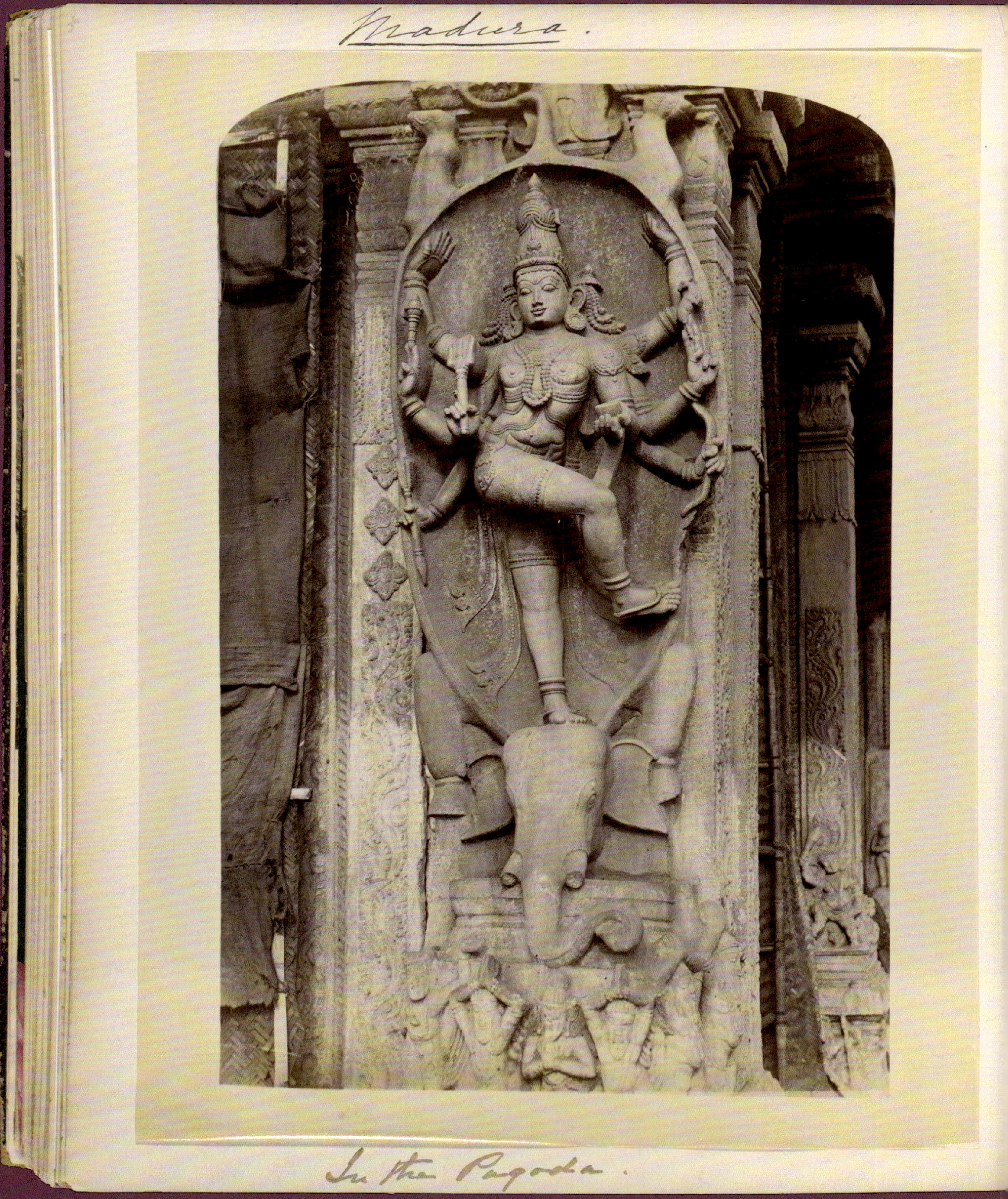
Madura.
In the Pagoda.

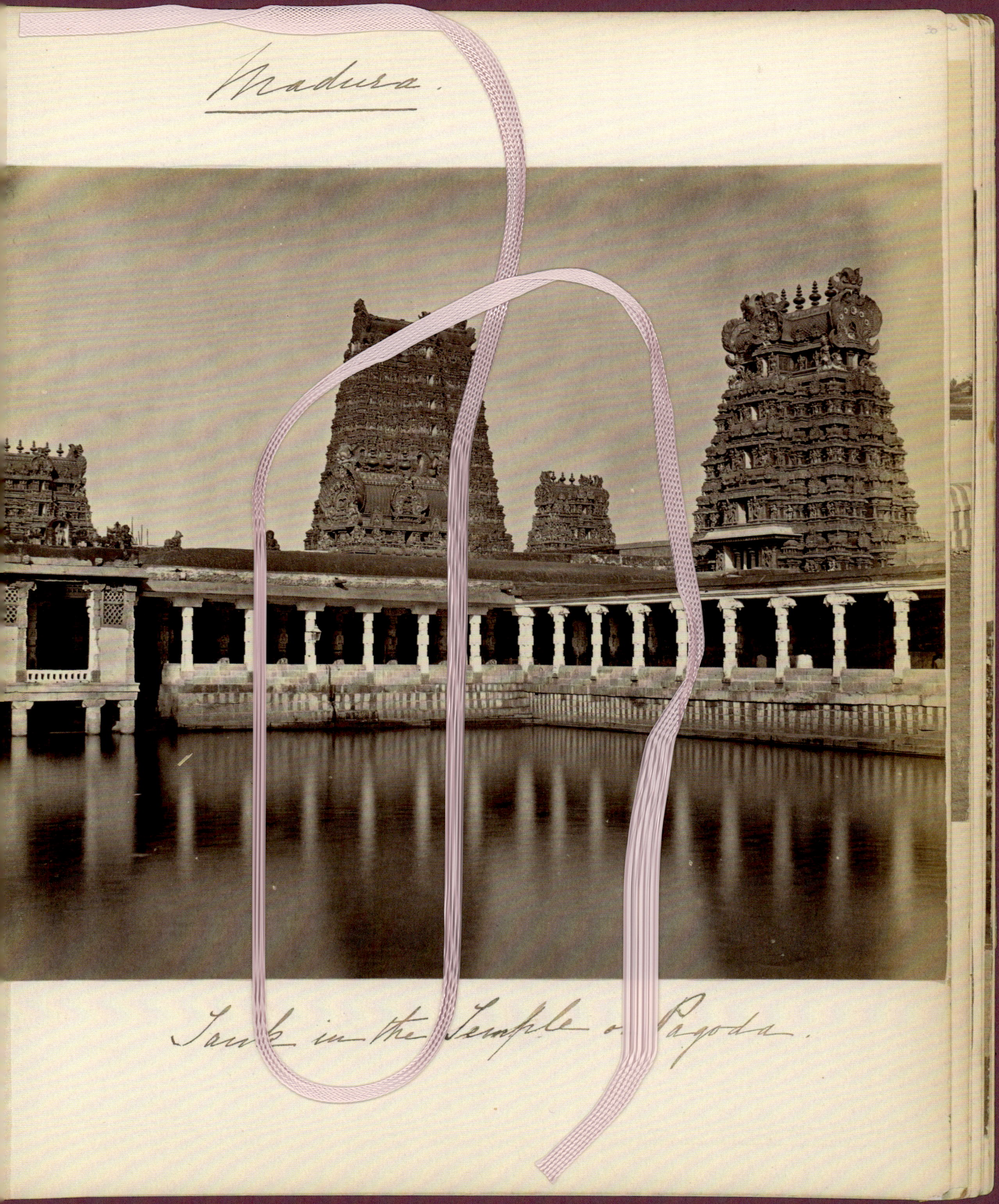
Madura.
in the Temple
Pagoda.

INDIA
Volume V, pages 31–32, February 1884

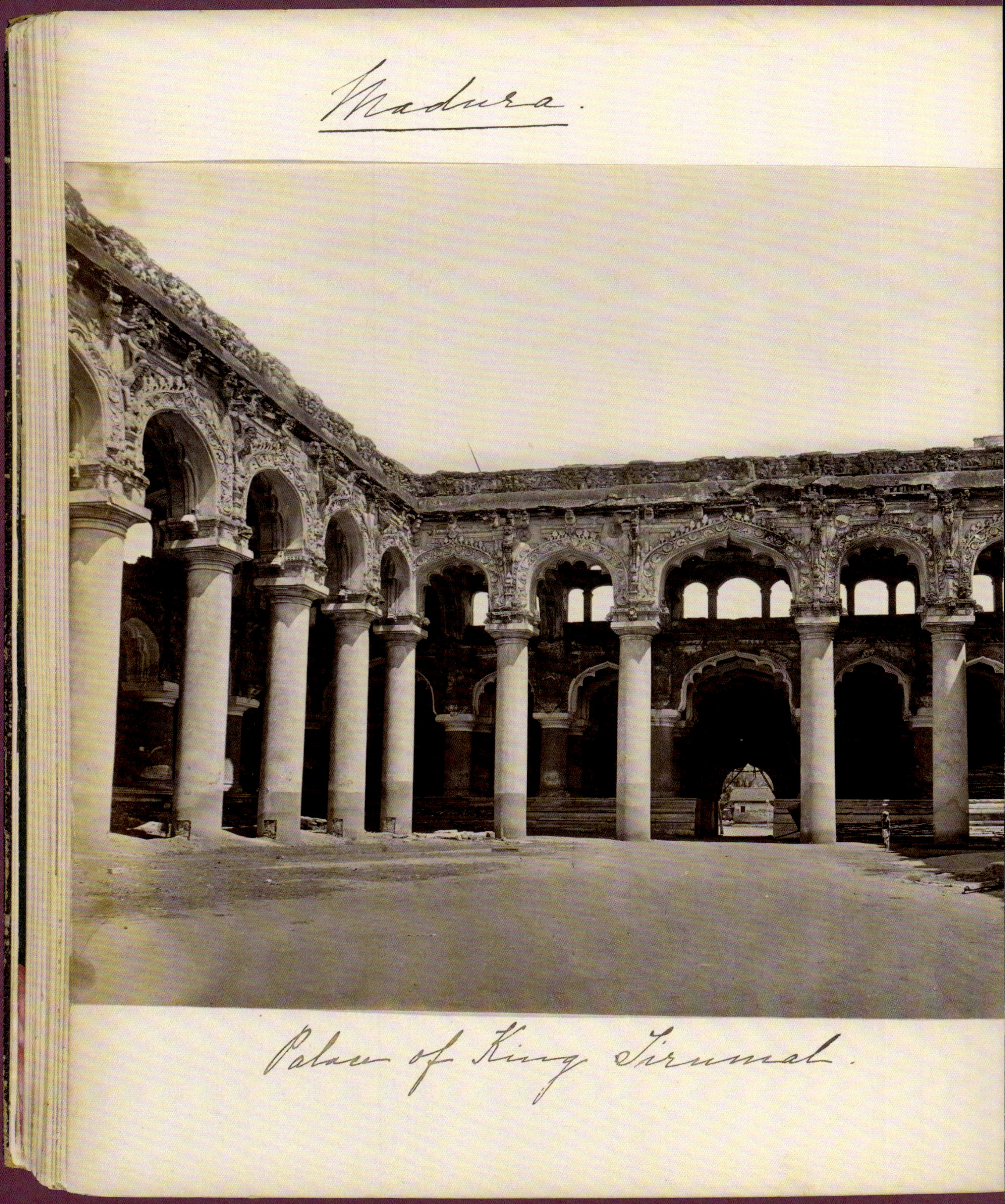

Madura.

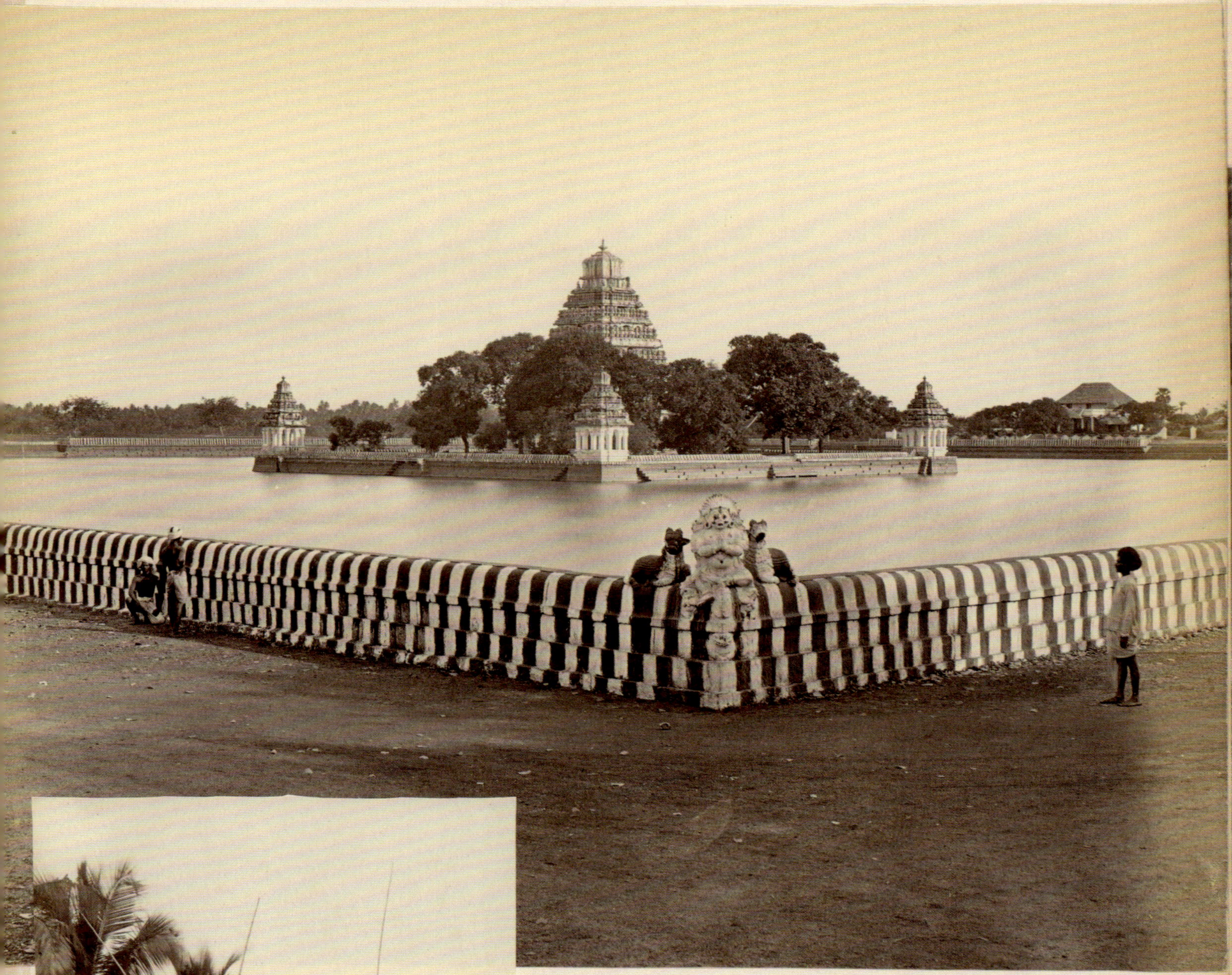

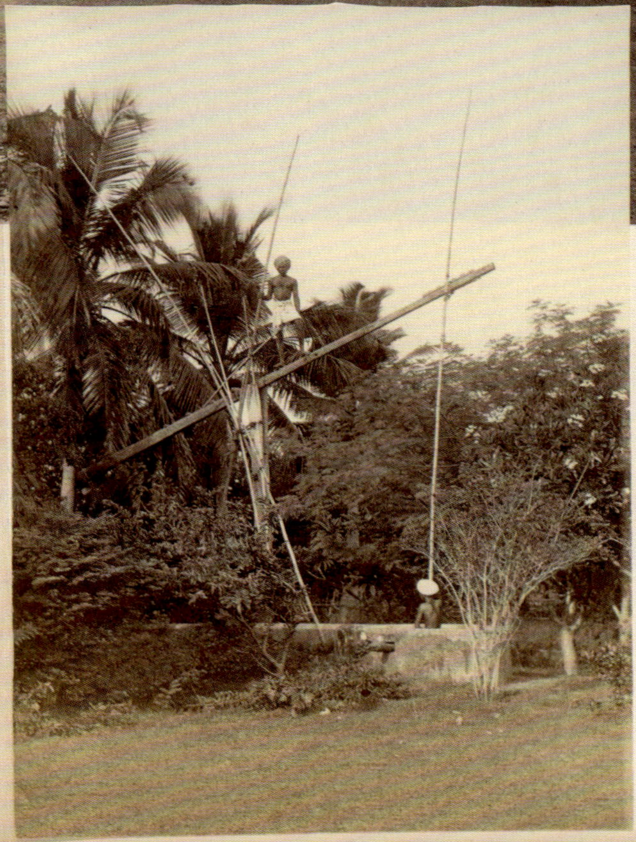

Great Tank out of the City.

INDIA
Volume V, pages 41–42, February 1884

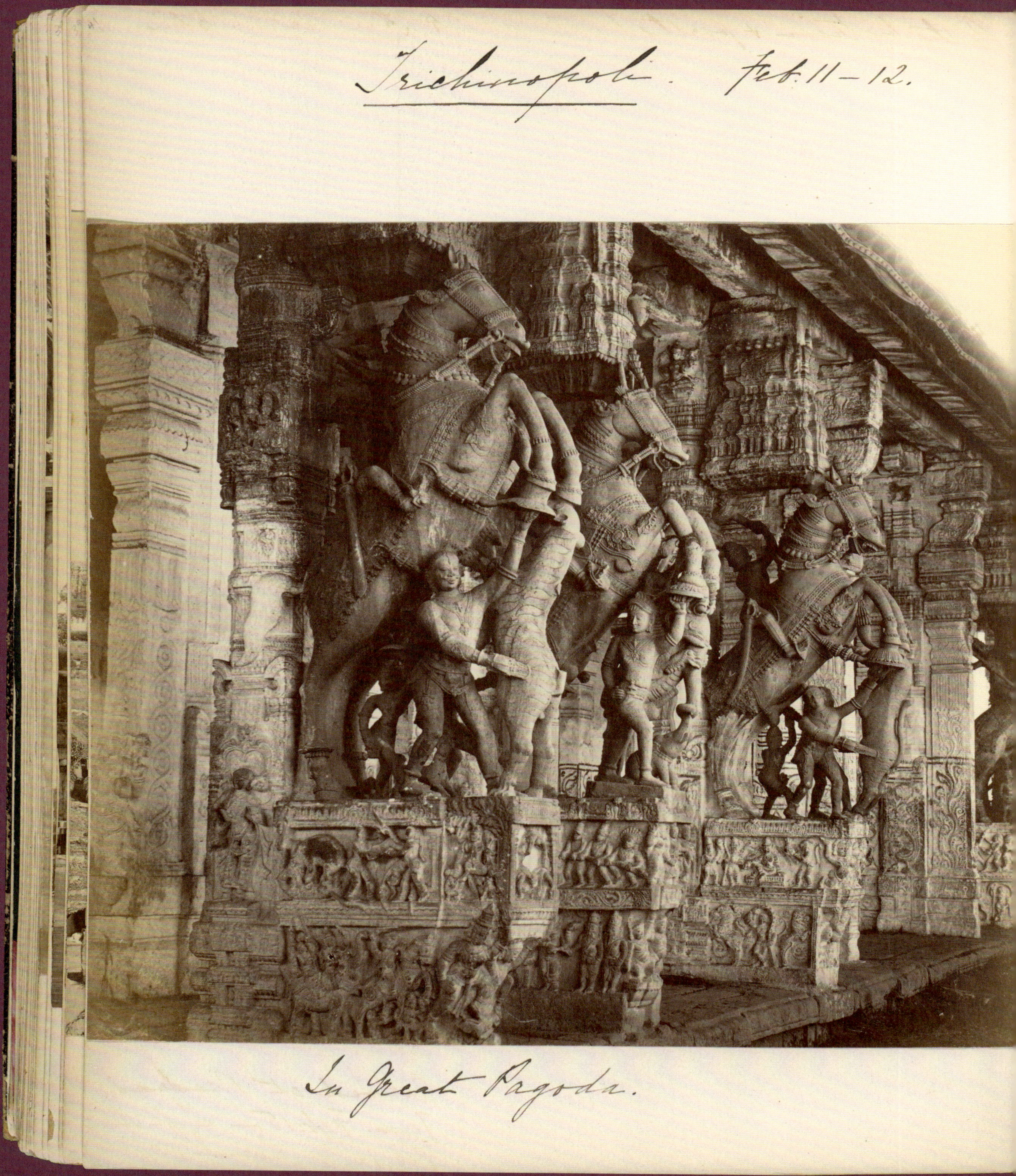

Party by train + party by Tonga to Kunúr. February 13. 42

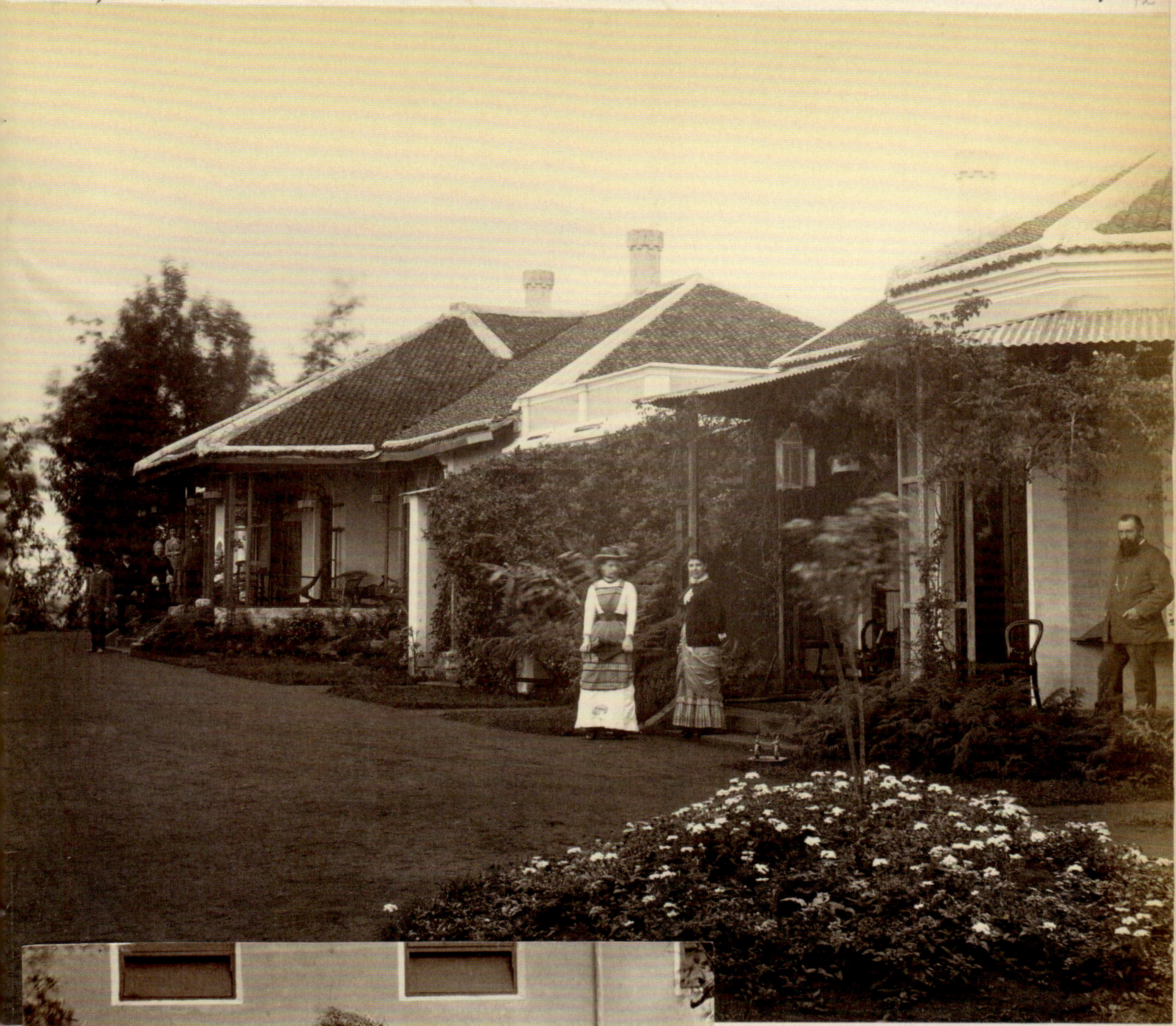

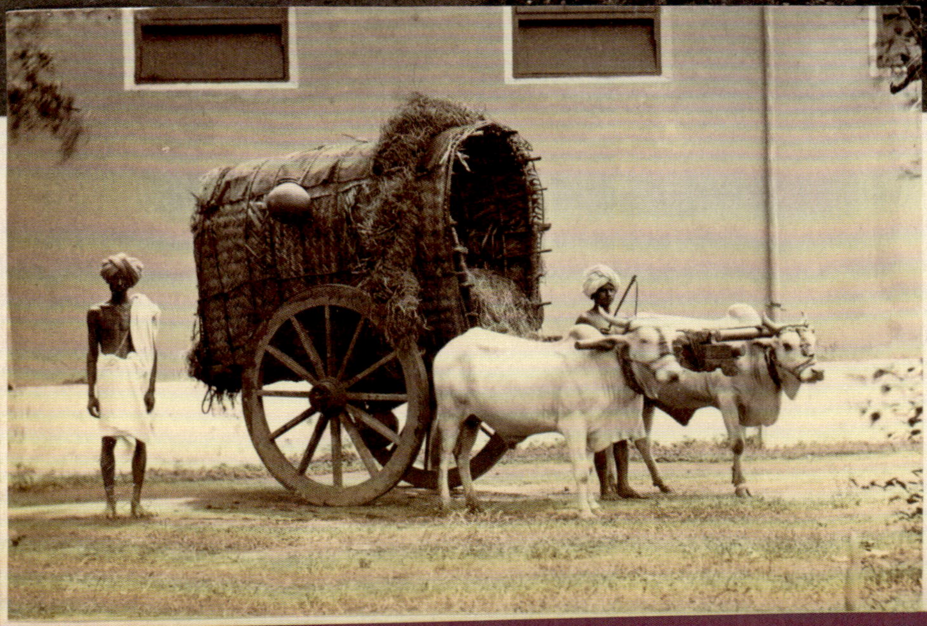

Hotel at Kunúr in the Nilgirris. Left there Thursday February 14, by Tonga. Arrived at Madras 7 a.m. Feb. 15.— Left there Feb. 16. 5 P.M. for Lassali— +

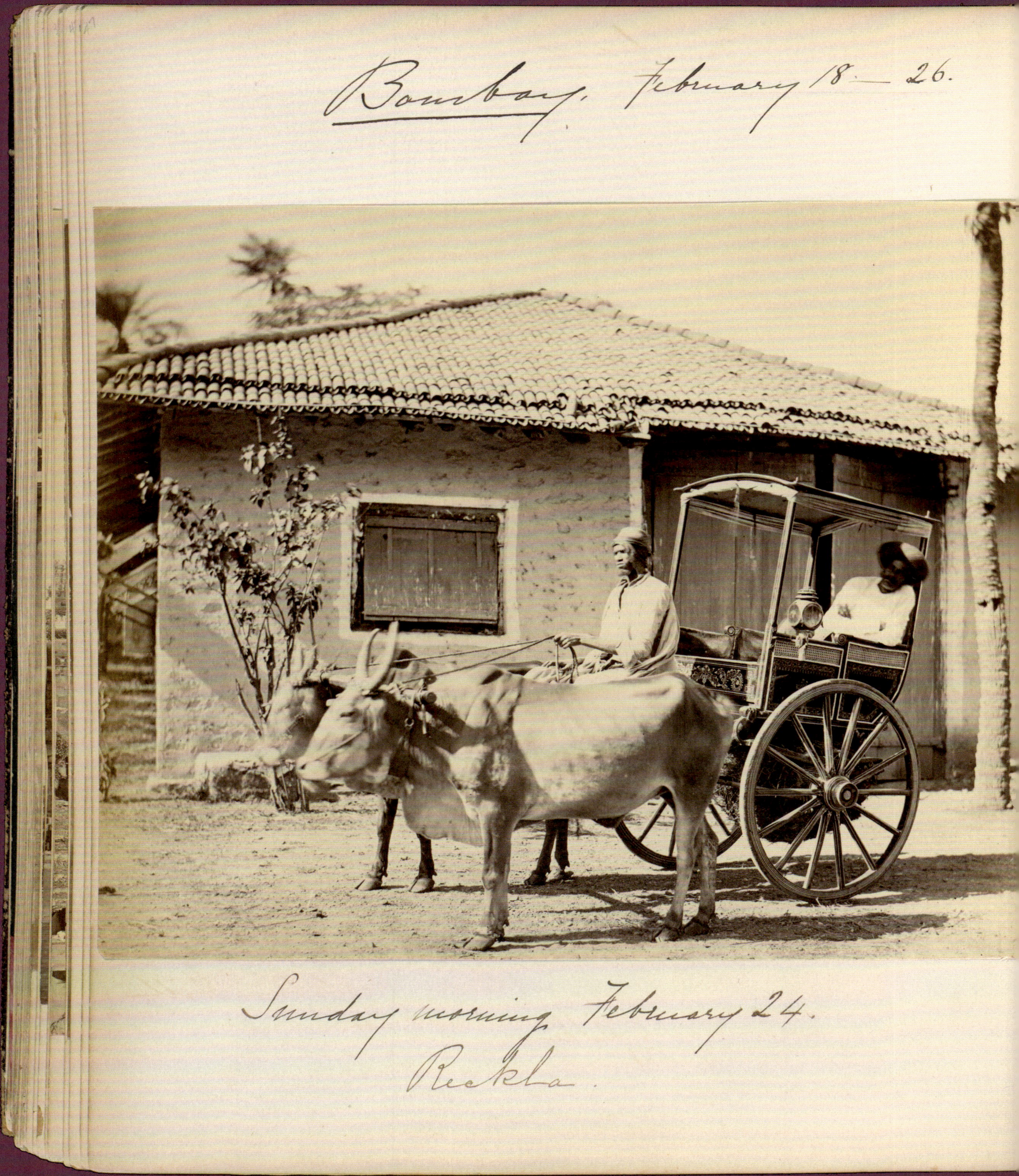
Bombay. February 18 — 26.
Sunday morning February 24.
Reckla.

Bombay.

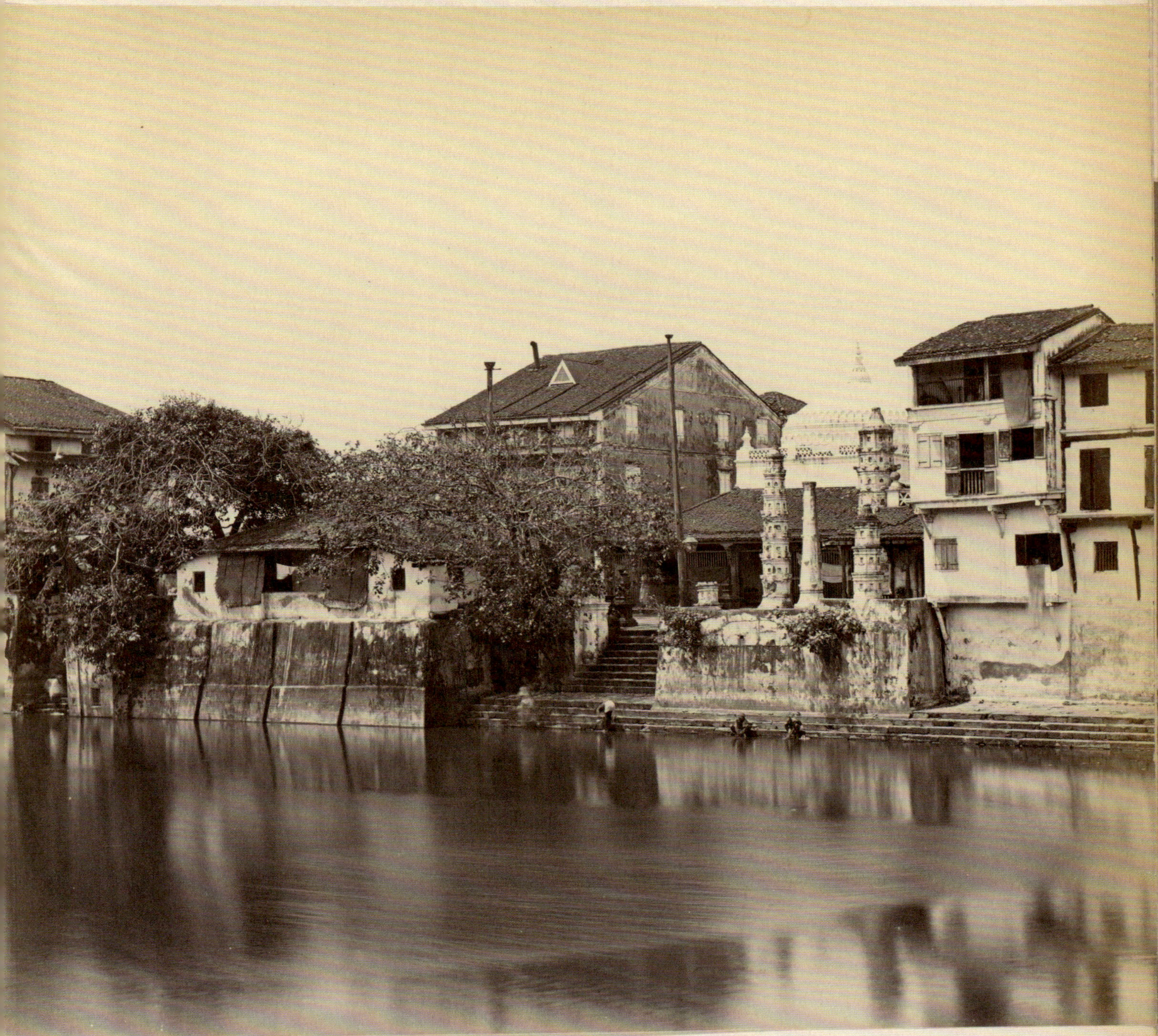

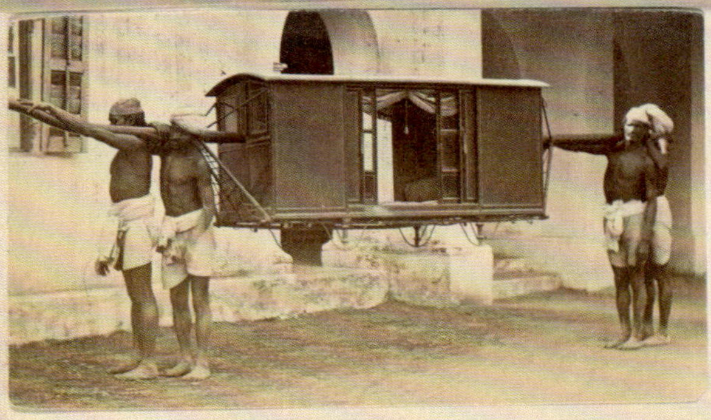

Walkeshwar Tank & Temples.
Holy village Malabar Hill.
Feb. 24.

A palki.

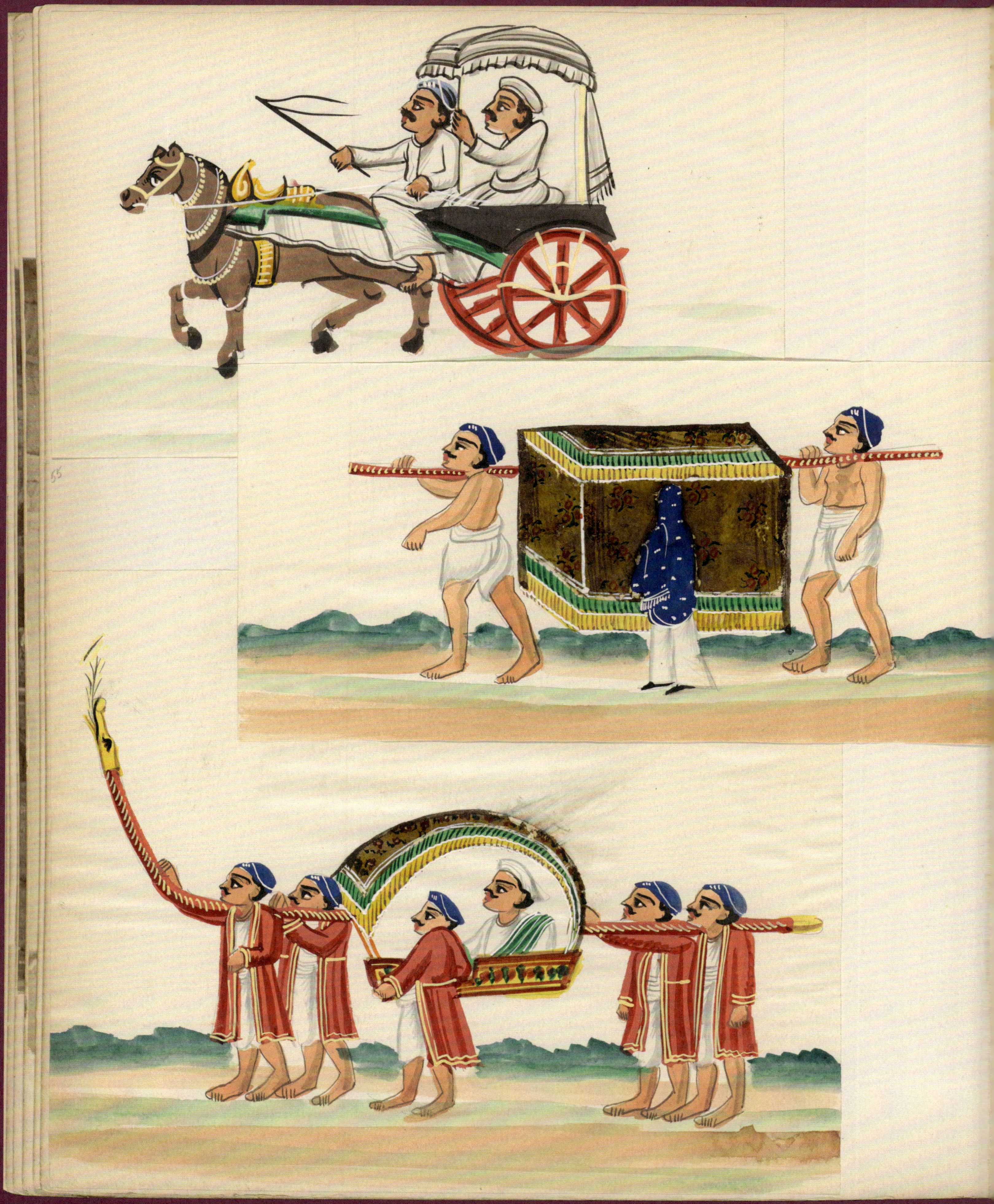

Banáras March 3 — 5.

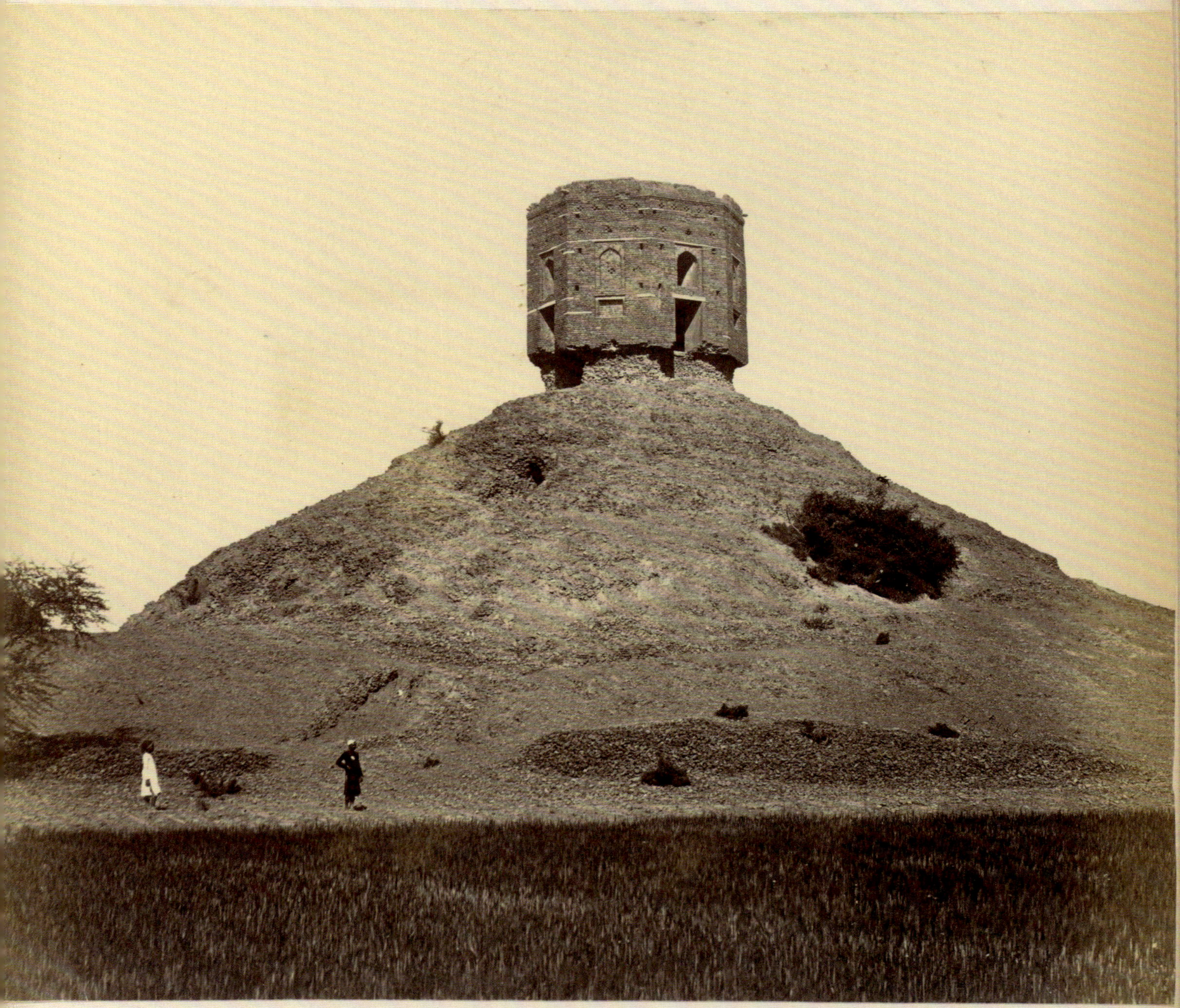

Tope at Sárnáth.
Monday Feb. 3

INDIA
Volume V, pages 69–70, March 1884

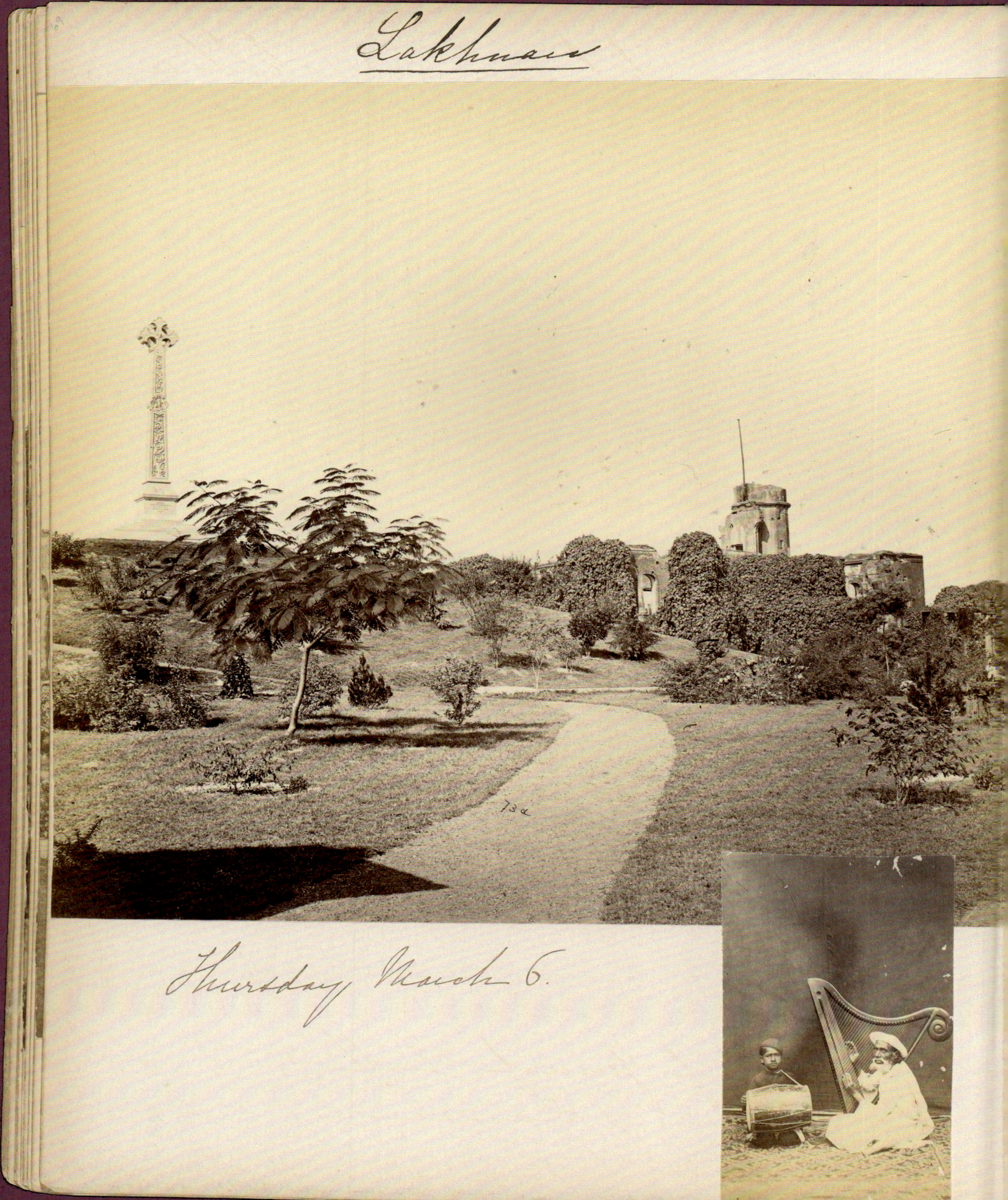

Lakhnau. (Nakhlau.)

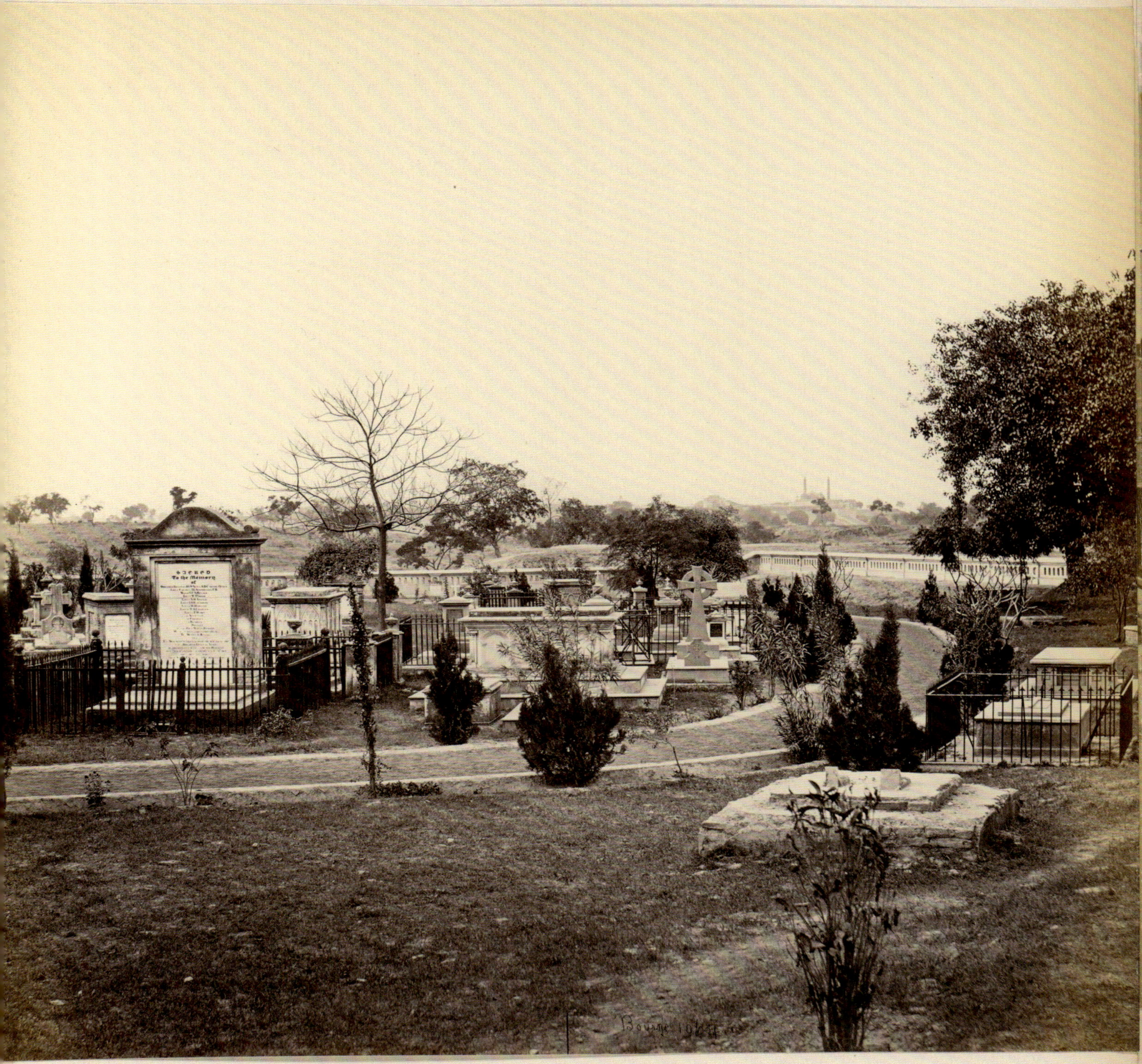

Graves of General Neill & Sir Henry Lawrence.

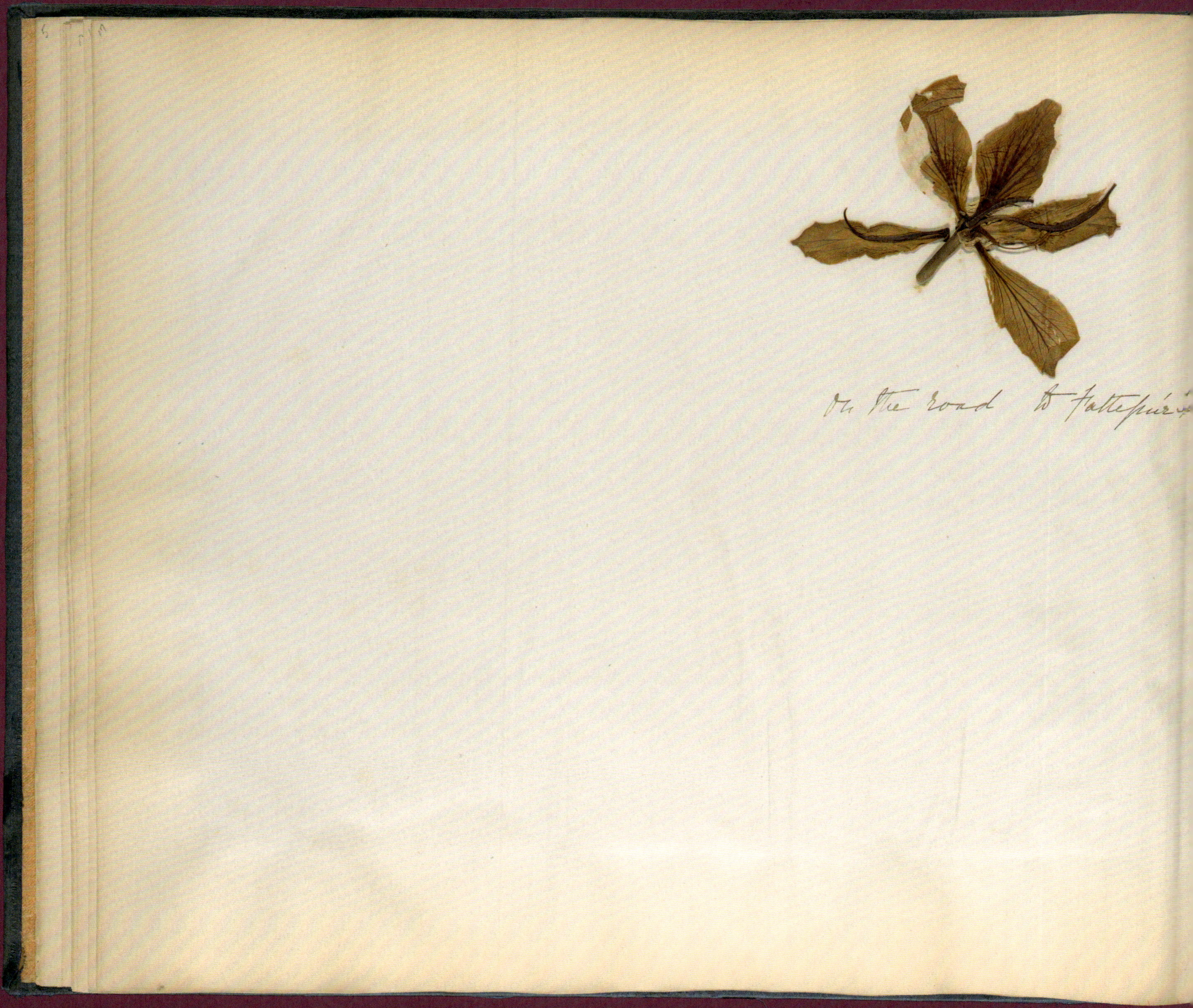
On the road to

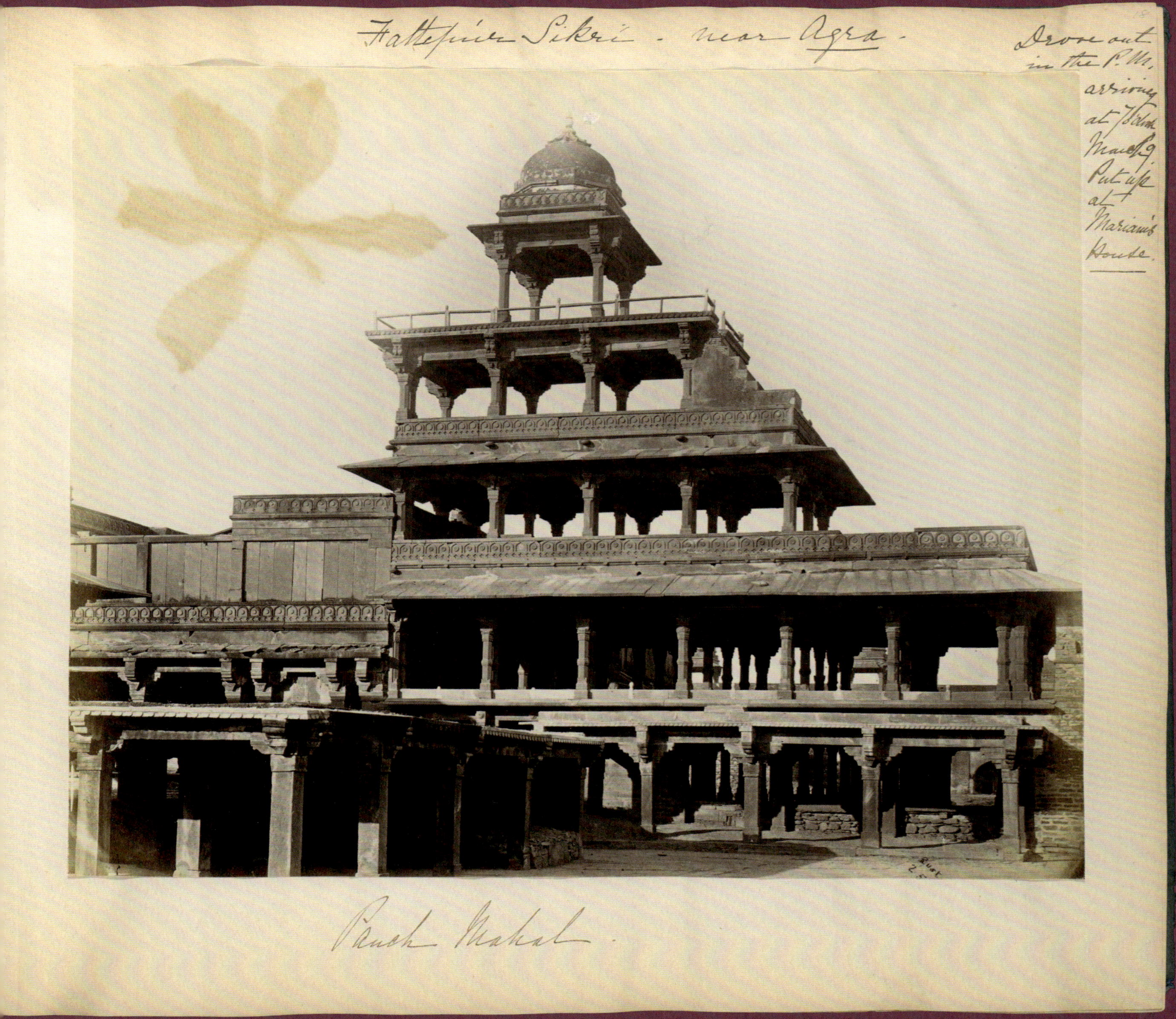
Fattepur Sikri - near Agra.
Drove out in the P.M. arriving at
March 9. Put up at Mariam's House.
Panch Mahal.

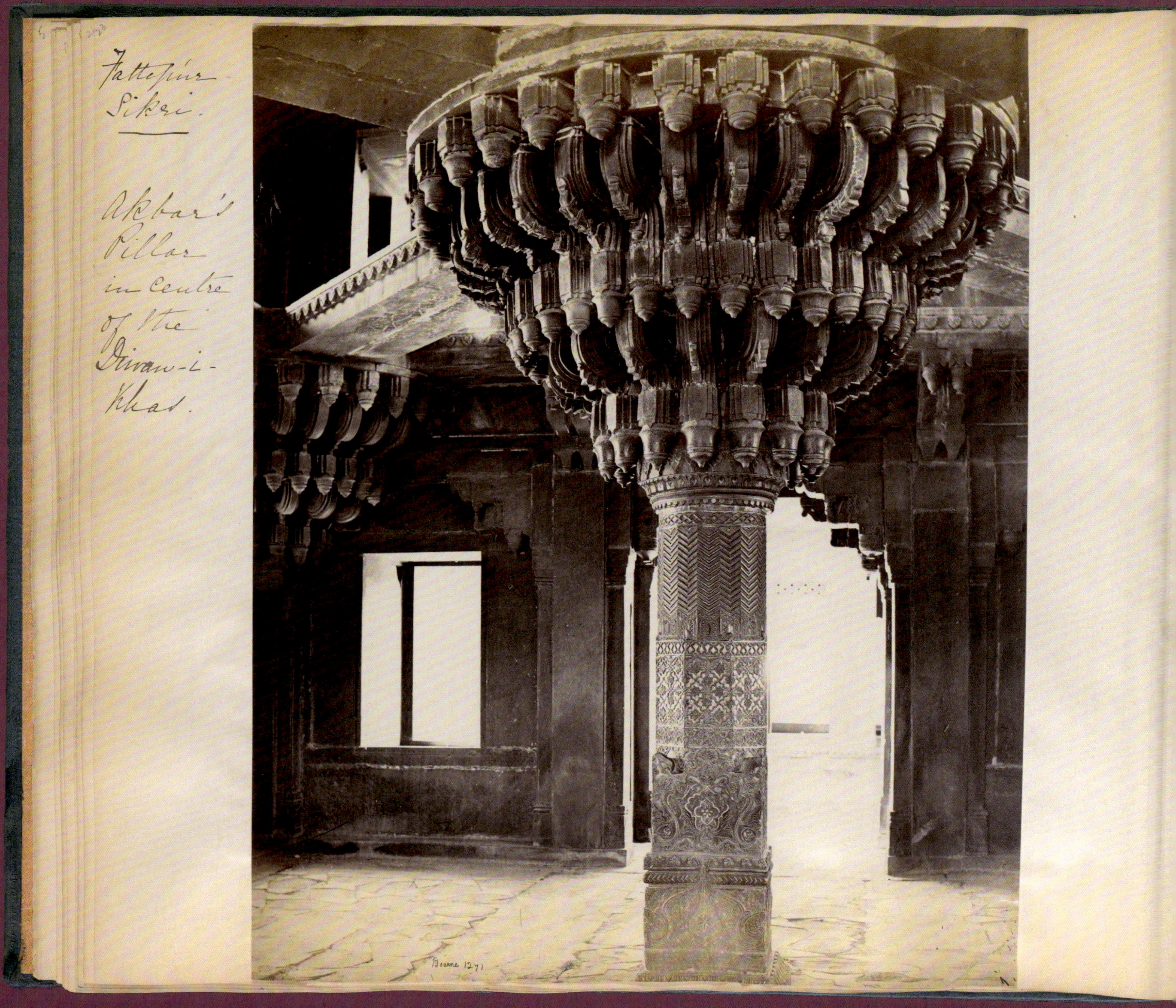
Fattehpur Sikri.
Akbar's Pillar in centre of the Diwan-i-Khas.

Fattepur-Sikri.

"Isa,
on whom
be peace,
The world
is a bridge,
pass over
it,
but build
no house
on it.
The world
endures
but an
hour,
spend
it in
devotion."

"His Majesty
King of Kings,
Heaven of the
Court,
Shadow of
God,
Jalal-uddin
Mohummed
Khan,
The Emperor."

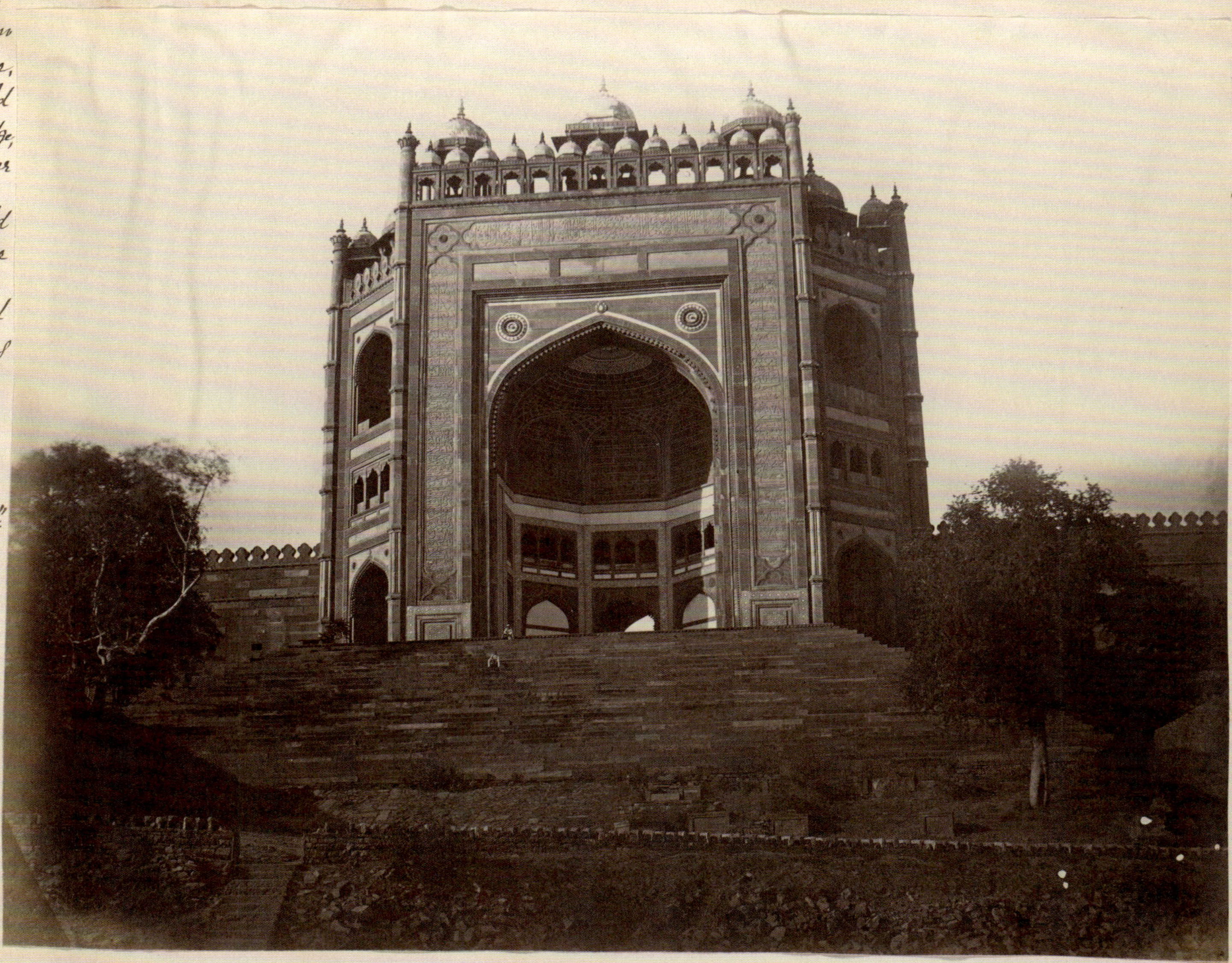

Great Gateway.

INDIA, PAKISTAN, YEMEN, AND EGYPT
Volume VI, pages 28–29, March 1884

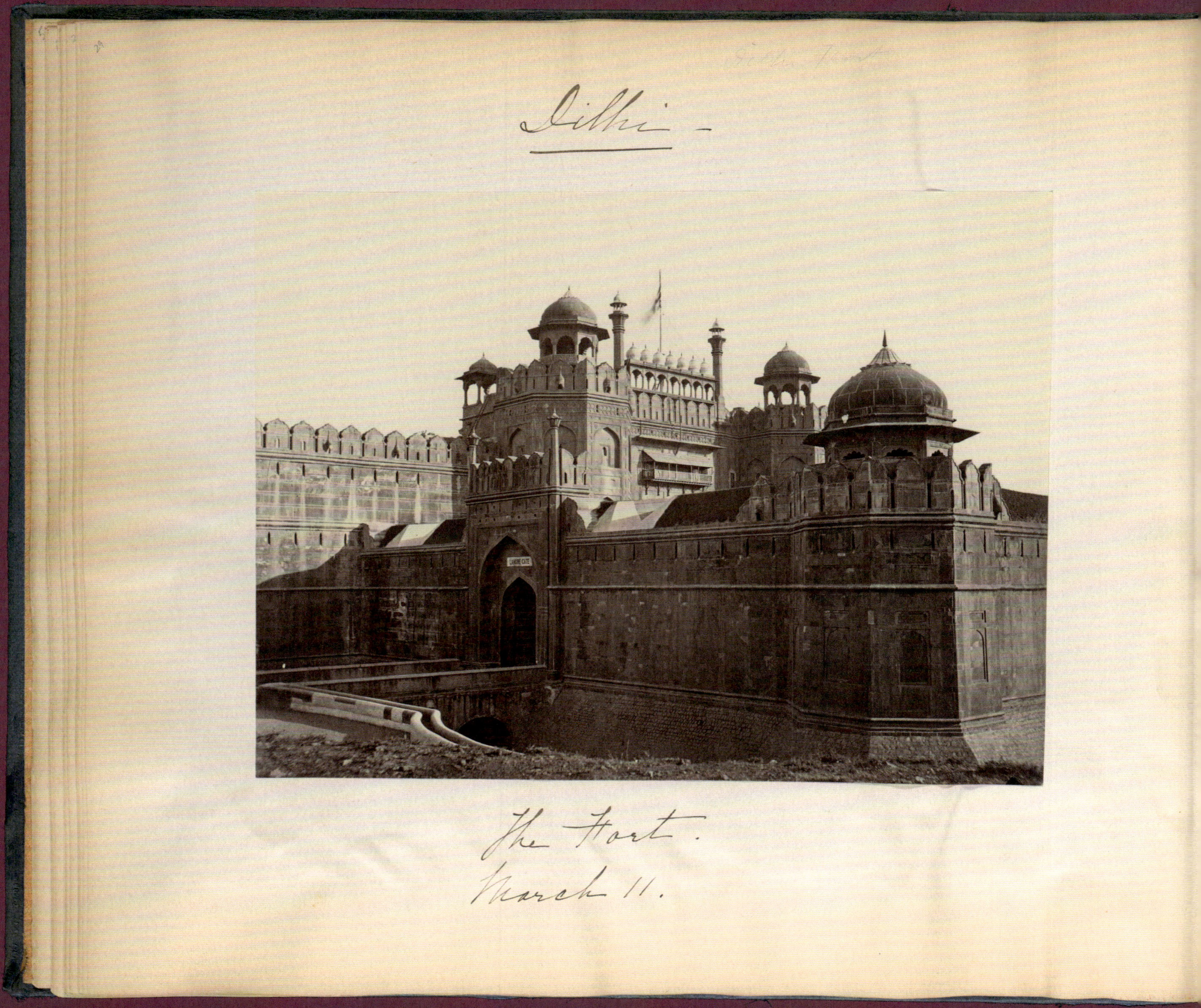

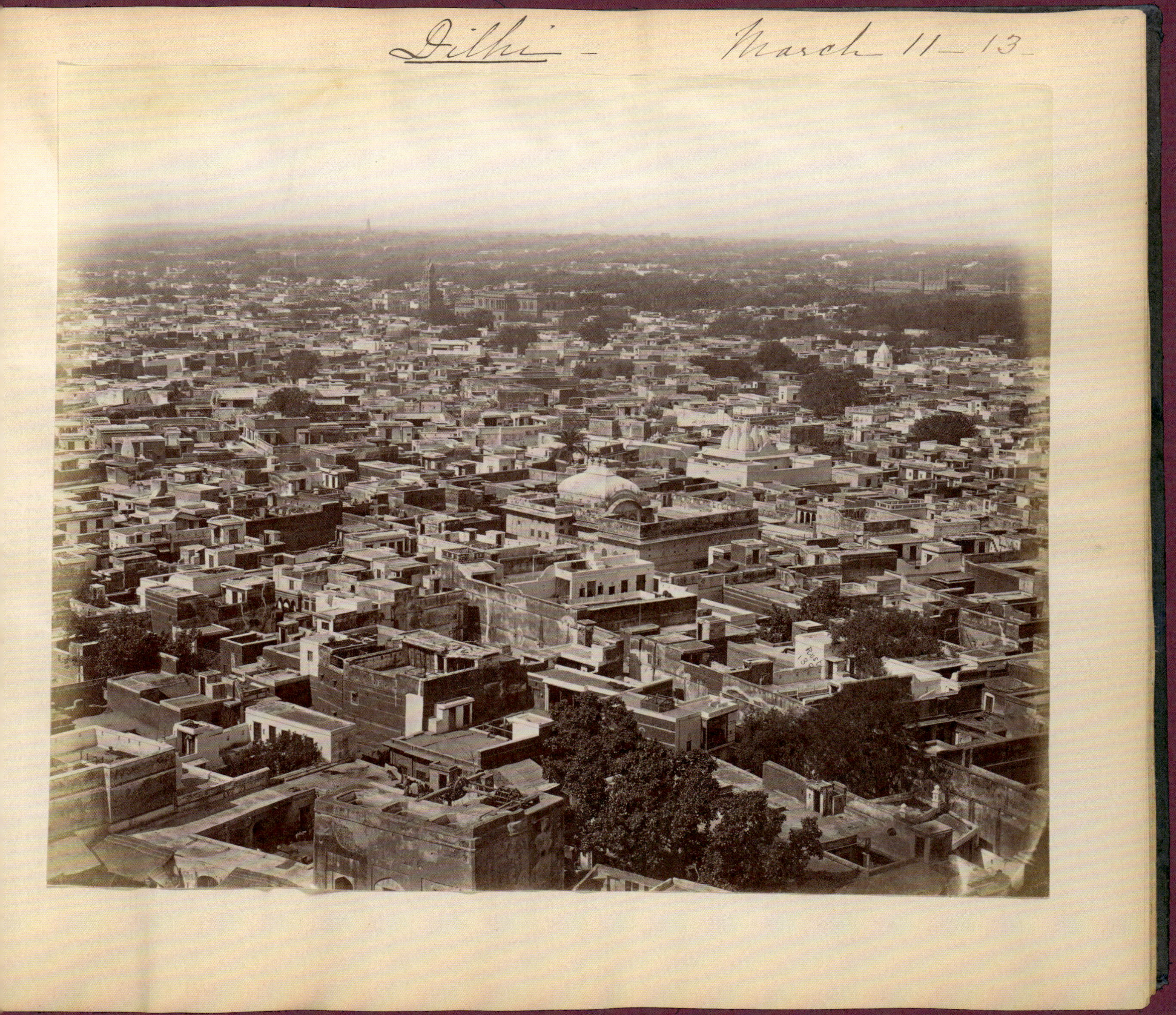
Dilhi – March 11–13.

INDIA, PAKISTAN, YEMEN, AND EGYPT
Volume VI, pages 79–80, March 1884

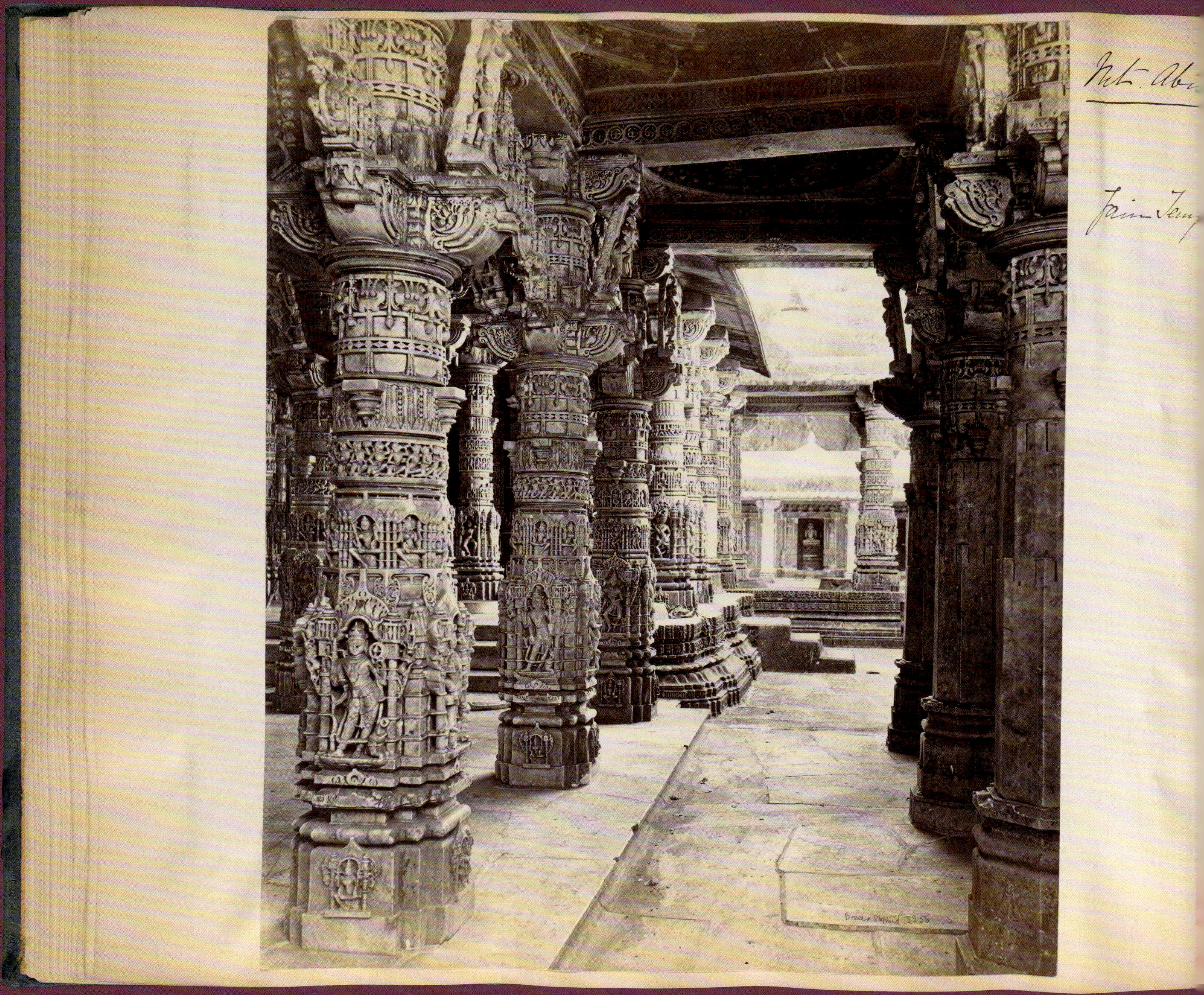

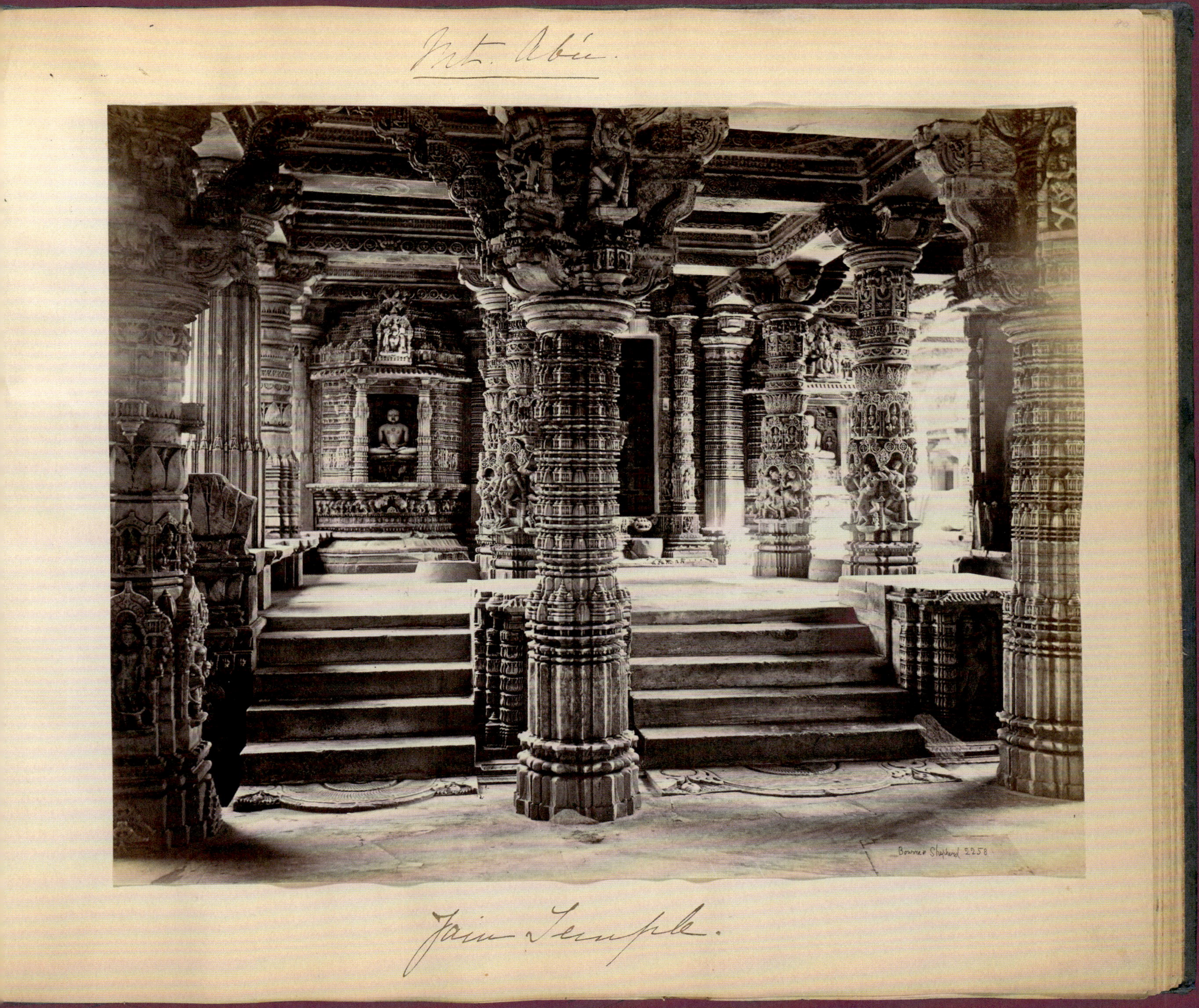
Mt. Abu.
Jain Temple.

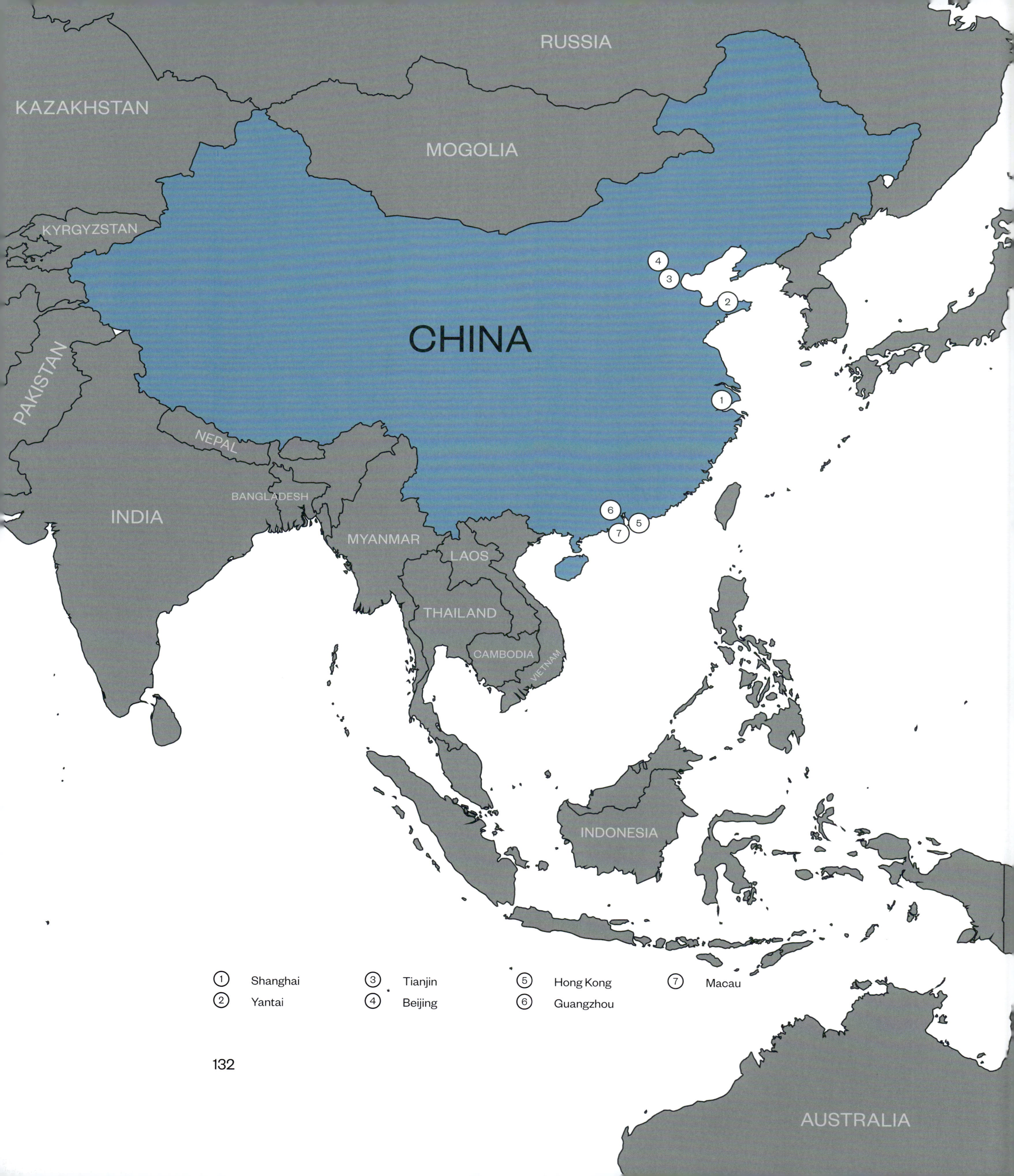
RUSSIA
KAZAKHSTAN
MOGOLIA
KYRGYZSTAN
CHINA
PAKISTAN
NEPAL
BANGLADESH
INDIA
MYANMAR
LAOS
THAILAND
CAMBODIA
VIETNAM
INDONESIA
AUSTRALIA
1 Shanghai
2 Yantai
3 Tianjin
4 Beijing
5 Hong Kong
6 Guangzhou
7 Macau

China

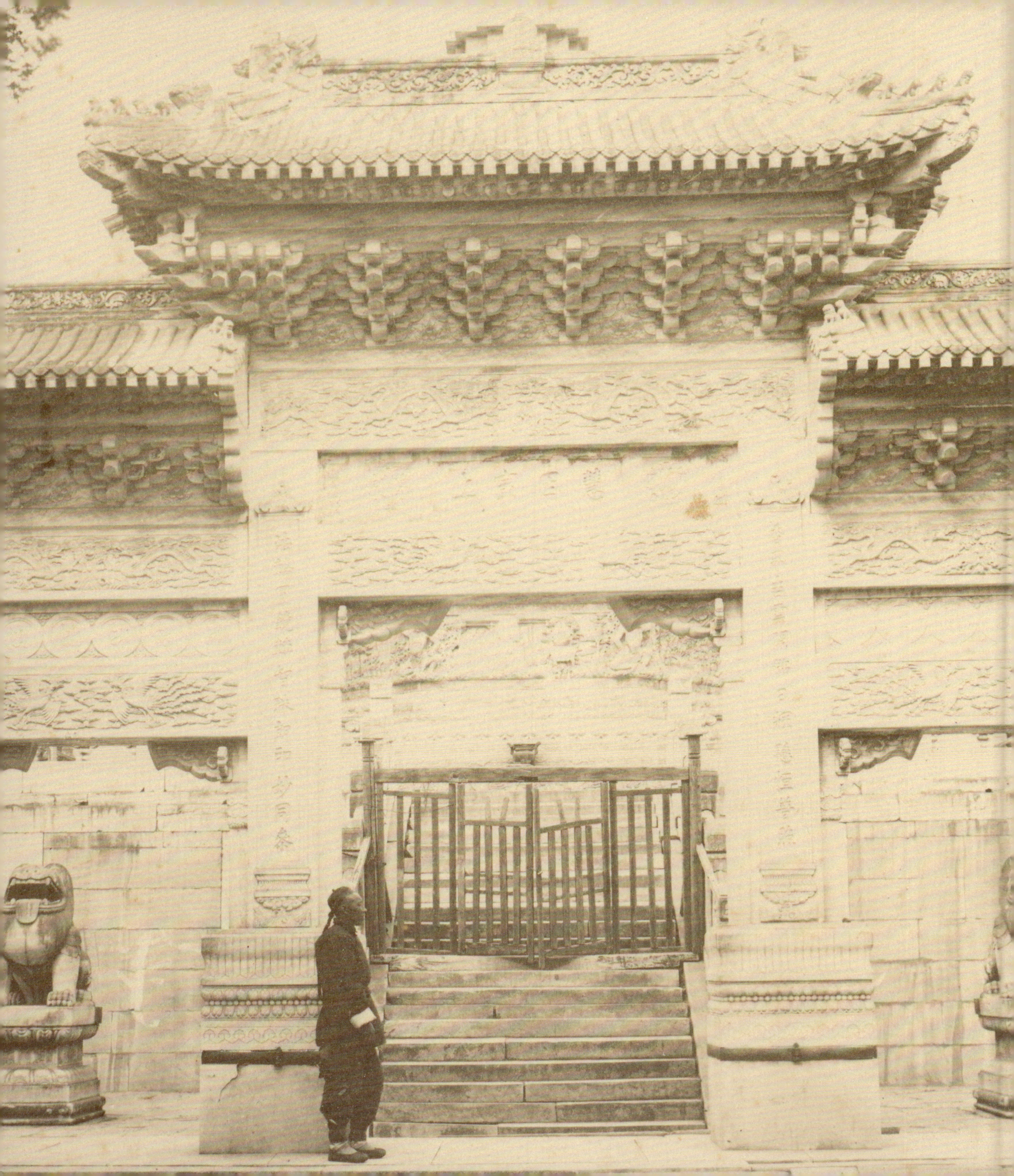

China, as Told by Isabella Stewart Gardner

Stephanie Tung

The first entry in Isabella Stewart Gardner's China album (Plate 31) is an assemblage of photographs and ephemera heralding the Gardners' arrival in the imperial capital of Beijing. A photograph of a grand, bustling city gate anchors the spread on the left, facing a long view of the city wall receding into the distance on the next page. The walls dwarf the few figures and carts in the foreground, and nothing can be seen beyond the gate. On first impression, Beijing seems to be a mysterious, well-fortified, and timeless city that rises up out of the desert. Gardner's handwritten entry—"Arrived at Tung-Chow by boat from Tien-Tsin Sept 21 at 7 a.m. and started for Peking on horses at 8:30 a.m. Got there at 2 p.m."—narrates their arduous approach and a slip of paper with the dates according to the Gregorian and lunar calendars marks the day. Great care has been taken to write the name in both Chinese and English, and a small spray of three dried yellow flowers is affixed beside the title "Peking" at the center of the page. The assemblage combines ready-made photographs from commercial studios with Gardner's travel notes and personal mementos to create a palimpsest in which sensations, observations, and imagery from different temporalities overlap and produce a vision of China unique to Isabella.

Isabella Stewart Gardner created six albums to document the 1883–84 Asia trip with her husband, Jack. After three months in Japan, they traveled to China, arriving in Shanghai by steamship, then transferring to another for their journey north to Tianjin, where they would embark on the westward trek to the capital. The placement of photographs and notes roughly follow their itinerary. The first views of Shanghai and Tianjin are in the latter third of Isabella's second Japan album. The full China album begins with Beijing, where they spent most of their trip, and includes brief stops in Hong Kong, Guangzhou, and Macau afterward.[1]

Where Jack's diary is full of numerical information—times of day, hours taken, money spent (fig. 1)—Isabella's notes and entries are more impressionistic and observational. "Ate my 1st persimmon—*delicious*," she writes, savoring a fleeting moment of experiential delight upon arriving in Shanghai.[2] In her diary entry for 7 October, Isabella writes of a beautiful day in Beijing:

Beijing, October 1883 (detail) from Isabella Stewart Gardner's *Travel Album: China, Volume III*, page 41

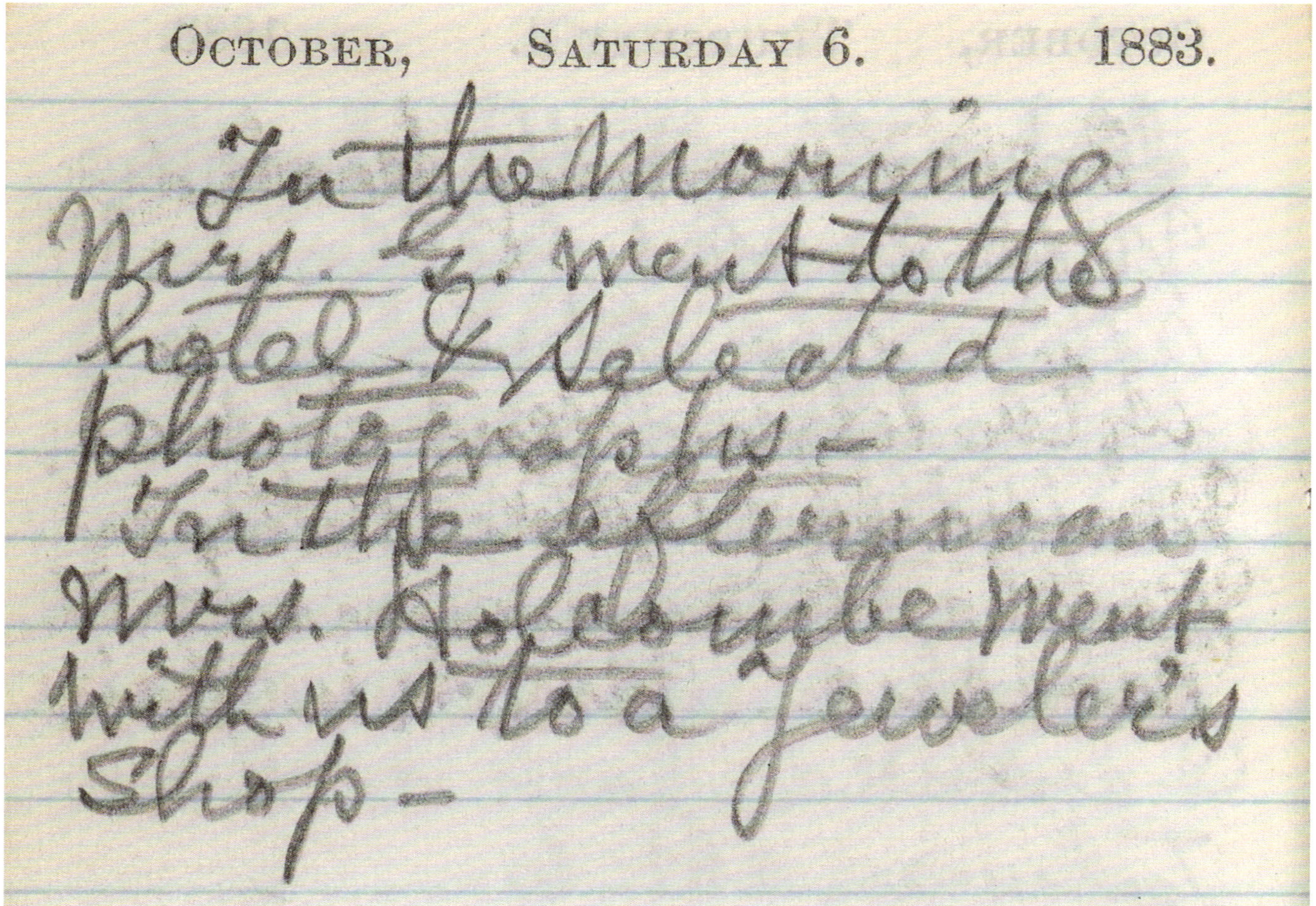
OCTOBER, SATURDAY 6. 1883.

In the morning Mrs. G. went to the hotel & selected photographs— In the afternoon Mrs. Holcombe went with us to a Jeweler's shop—

Fig. 1 John Lowell Gardner Jr. (American, 1837–1898), *Diary: Japan, China, and India, 1883-1884,* page 166

> *Wrote letters in the morning, sitting at open door, delicious weather. After tiffin Jack and I went to see the Temple of the Sun [Ritan]. Uncovered, circular platform with steps at the four sides, surrounded by a walled enclosures [sic] which is topped with green tiles. There is tiled oven, open at top where the burnt offerings are made. On the platform a very good echo. The first day in Peking without dust.*[3]

Isabella pays attention to color, air, sound, and touch, perhaps in a heightened state of awareness that comes with traveling in an unfamiliar land and culture. Her commentary pulls the reader into the moment of that particular experience, working as a prompt or mnemonic for future reminisces and storytelling. Mood and atmosphere are often emphasized over the sheer facts of their long, difficult travels. China was not readily accessible to most tourists in the late nineteenth century and proved to be the most expensive country the Gardners visited during their Asia trip. Isabella's experience was largely mediated by the guides and servants hired to accompany them. For example, during their month-long exploration of Beijing and its environs, the Gardners retained a guide, Yu-Hai; a groom, Mafou; and a maid, Amah, from Tianjin, who worked behind the scenes to interpret, cook, clean, and make sleeping and travel arrangements on the variety of boats, horses, mules, and litters that carried them. On her last day in

Beijing, Isabella notes that Yu-Hai and Amah had been tasked with packing the cases and luggage while she and Jack dined with friends. She carefully records the twenty-four Chinese dishes she sampled for luncheon and the exotic textures and sounds of an acrobatic performance—"All amusing. No scenery, only red, embroidered curtains. No end of jing-jang in the way of music."[4] China was a feast of the senses to remember, filtered through the lens of the Gardners' privilege and expectations.

In Beijing, the Gardners focused primarily on well-known tourist sites such as the Temple of Confucius, Observatory (Plate 32), Great Wall, and Summer Palace, guided by their hosts at the American legation and a guidebook that Jack had purchased in the city.[5] The rituals and visual splendor of temple and palace architecture fascinated Isabella. As Alan Chong notes, one of the inspirations for their trip was a budding interest in Asian religions, particularly Buddhism, perhaps as an alternative or counterpoint to Isabella's own Protestant upbringing.[6] The arches, pagodas, and courts that Isabella encountered were tangible representations of Buddhist beliefs, rituals, and traditions, which became integral to her perceptions of China. Writing of the Lama Temple, or Huangsi, in northern Beijing (Plate 33), she observes:

> *The latter of strange and very interesting architecture—court after court—all the Lamas in yellow cloaks and such strange yellow hats. Services going on, one of them in open court, consisting of odd chanting in very low voices, with an accompaniment of clapping together two pieces of wood. In one of their temples, a very large gilded Buddha, standing. In another beautiful Thibetan [sic] carpets and cloisonné altar decorations.*[7]

Like the great walls leading to the entrance of the city, the grand courts that hosted Buddhist, Daoist, and imperial rituals and ceremonies seemed to be timeless. In her album, Isabella includes a series of photographs that depict the arch and pathway to the Lama Temple's marble stupa, built to commemorate the Sixth Panchen Lama's visit to the Qianlong emperor in Beijing a century before. A lone Chinese figure poses in the grand, empty architectural space as a stand-in for local color or scale (Plate 33). Leaves, flowers, and a handwritten note—"A Monument to the Memory of the Teshoo Lama"—adorn the pages leading to an image of the imposing stupa structure, which highlights the relief of the Buddhist deities, the details of the intricate carving, and an open gateway (Plate 34). The photographs are picturesque, presenting a solemn, enduring view of religious practice in China with little inkling of interruption.

Isabella produced a "curated" view of China in her albums that was not entirely grounded in the present. The First and Second Opium Wars had profoundly changed the way China engaged with the world in the nineteenth century. The British and French military forced the Qing government to alter the terms of trade, legalizing opium and granting unfettered access to Chinese markets. Foreign diplomats, merchants, and missionaries soon arrived to establish settlements in select cities, including Beijing, rapidly transforming the look and feel of these places. Nearly all of Isabella's

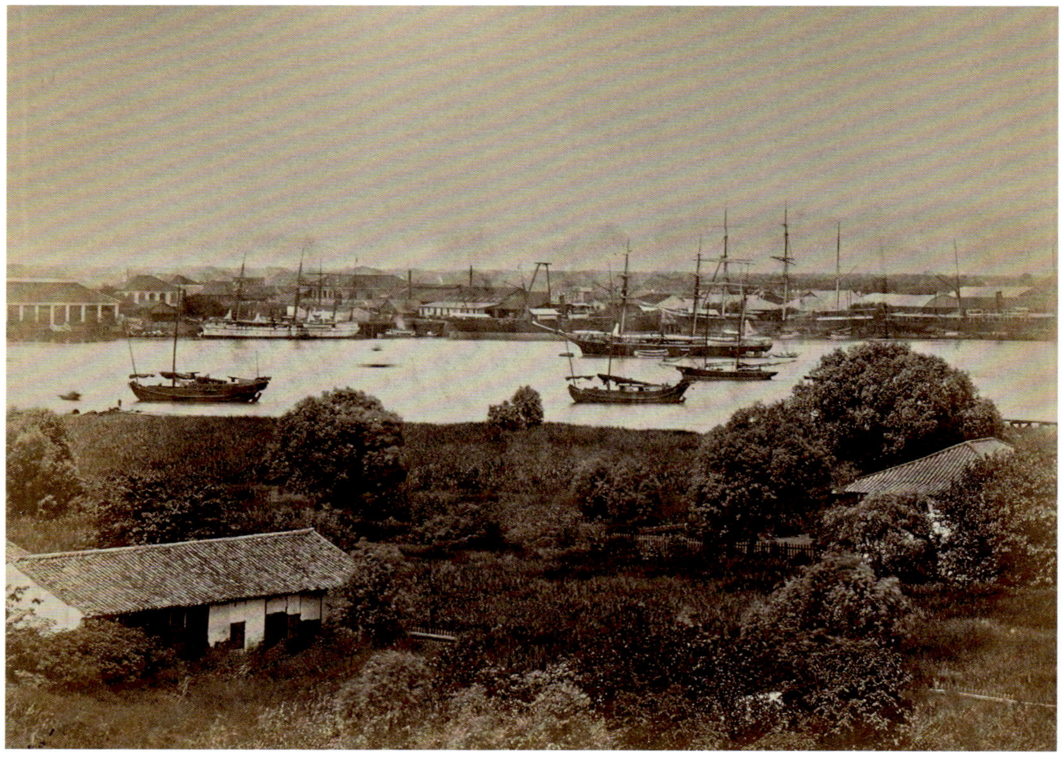

Fig. 2 Kung Tai studio (active 1870s–90s), Panorama of the Shanghai Bund (detail), 1882. Thirteen-plate panorama; albumen prints, 19 × 346.7 cm. Peabody Essex Museum, Gift of Mrs. Beverley R. Robinson, 1950, PH93

Beijing photographs have the signature "T. Child 1876" engraved in the negative (Plates 31 and 32), indicating they were from Thomas Child (1841–1898), a British gas engineer working with the Imperial Maritime Customs Service and a foreign expatriate living in Beijing.[8] The majority of Child's views were made in the 1870s as a substantial investment into his side hustle as a photographer. Sold as souvenirs, Child's photographs present a China that is tamed and ordered for Victorian consumers either living temporarily in the city or, like the Gardners, passing through. The main focus is usually a monument or architectural element; Chinese people are almost entirely absent in Child's photographs. Surprisingly, however, Isabella selected none of Child's famous views of the ruined European-style palaces of the Yuanmingyuan, a Jesuit-designed imperial complex that had been destroyed as a punitive measure by British and French forces in the war. Though she and Jack had visited the site briefly, the ruins and their status as symbols of Western influence and then destruction perhaps did not conform to the idea of China that she wished to portray in her album.

Isabella encountered a similar problem in Shanghai, which she described as "too European for local color." Shanghai was a boomtown, which Isabella perceptively compared to post–Gold Rush San Francisco.[9] A treaty port opened after the First Opium War, Shanghai developed into a bustling metropolis with specific concession areas for British, French, and American settlements. By the time of Isabella's visit in 1883, the Bund area on the banks of the Huangpu River was developing so quickly that photographers like the Kung Tai studio made panoramas every few years to capture the newly erected buildings and changes in cityscape (fig. 2).[10] These multipanel

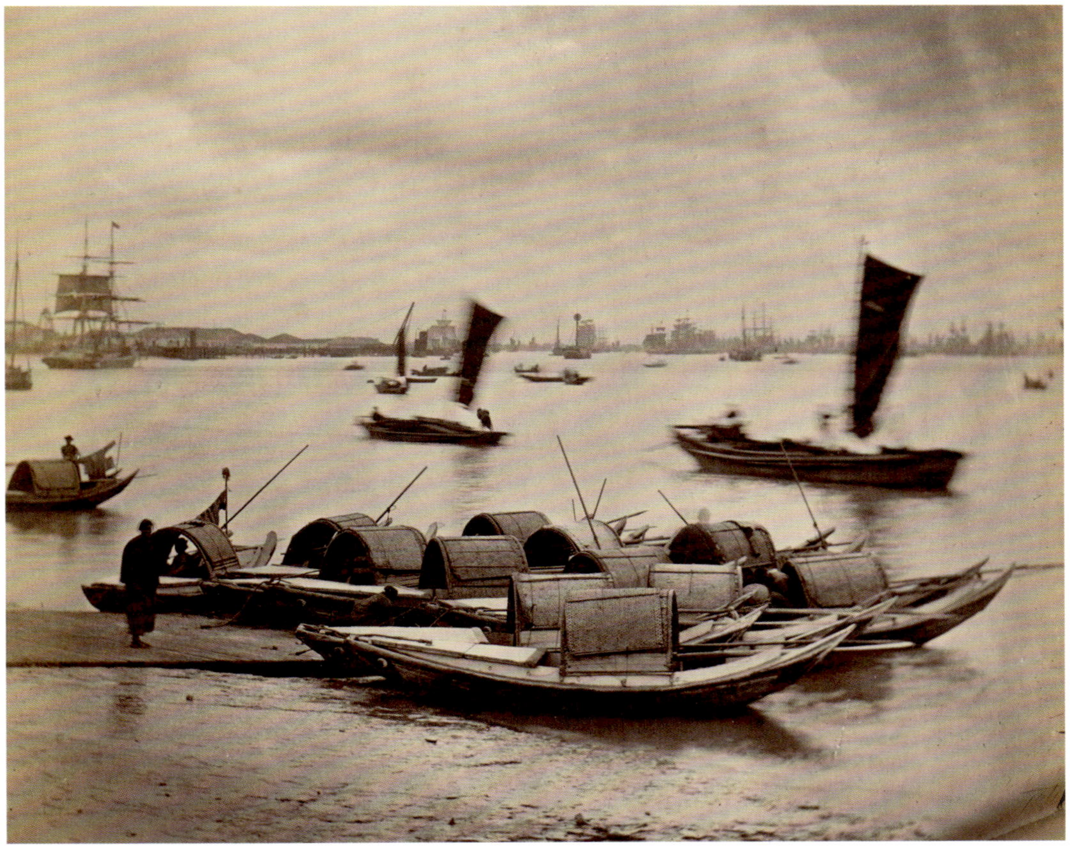

Fig. 3 Shanghai, October 1883, from Isabella Stewart Gardner's *Travel Album: China, Volume III*, page 45

panoramas were popular souvenirs for the merchants affiliated with the firms that dotted the riverbank, who bought them to convey the exciting, cosmopolitan feel of the city to family members back home. Yet it seems that Isabella was not that interested in the urban landscape; she introduces Shanghai with a single river view of a group of sampans pulled ashore, underselling the scale of trade and business then happening in the city (fig. 3). Instead, Gardner incorporated photographs of Chinese musicians, prisoners, laborers, and farmers—figures from the portfolio of "life and character studies" attributed to the studio of William Saunders (1832–1892)—into her album (fig. 4). These images were deeply influenced by eighteenth-century paintings and prints of Chinese daily life and occupations and circulated without much change through the early 1900s (fig. 5). Exotic modes of transportation and punishment were popular genres of travel art and photography across cultures, a means of understanding and reinforcing difference through the familiar.

In Hong Kong, Guangzhou, and Macau, Isabella manifested time and history in a different way. Throughout their travels, the Gardners associated with many people and sites affiliated with the earlier nineteenth-century American trade with China. Jack Gardner was the grandson of the Salem ship captain Joseph Peabody (1757–1844), who partnered with the Gardner family to finance shipping endeavors and made a fortune in the spice and opium trade.[11] Isabella's father had started his career in the New York offices of Russell & Company, the largest American trading firm in China, which specialized in tea, silk, and also opium.[12] By the 1880s, most of these firms

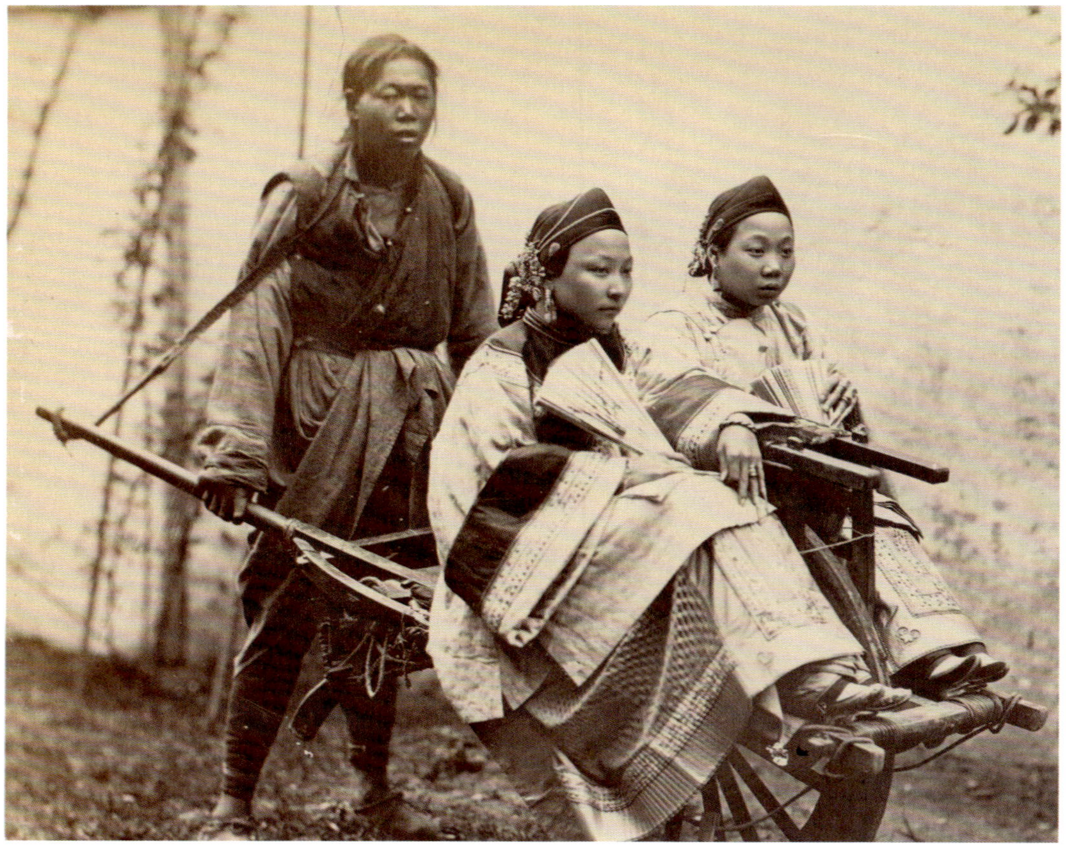

Fig. 4 Photograph of Chinese women by William Saunders (English, 1832–1892) from Isabella Stewart Gardner's *Travel Album: China, Volume III*, page 48

had transitioned into the shipping, steamship, and rail businesses, some areas of investment that Jack Gardner still maintained. While it is unclear to what extent the Gardners were aware of the scale and scope of goods associated with their families' businesses, the financial capital that made it possible for the Gardners to travel so widely was inextricably linked to the early American China trade. China was not just a place to visit for pleasure or edification, but also the source of power and wealth for many in their social circle.

Jack and Isabella most likely did not have close family ties to merchants still working in China, but the photographs and entries in their journals demonstrate the small, elite circles they moved in and the power of family connections. They stayed with the merchant E. G. Low (d. before 1880) and his wife in Shanghai, and then at the American legation on the invitation of Secretary Charles Holcombe (1842–1912) in Beijing.[13] They socialized with the principal partners of Russell & Company and the "Princely house of Jardine Matheson" in Shanghai.[14] In Hong Kong, they stayed at the Rose Hill estate, a grand house built by the Russell & Company partner Warren Delano Jr. (1809–1898) that became an informal American social hub in the city (Plate 35). Delano's daughters, Mrs. Annie Delano Hitch (1849–1926) of Shanghai and Mrs. Dora Delano Forbes (1847–1940) of Hong Kong, became Isabella's shopping companions.[15] Isabella notes that after dining with Mrs. Hitch in Shanghai, she "saw some very pretty opium boxes," and most likely purchased her own examples as souvenirs of her time in China.[16] The intergenerational connections and shared networks of Boston-based

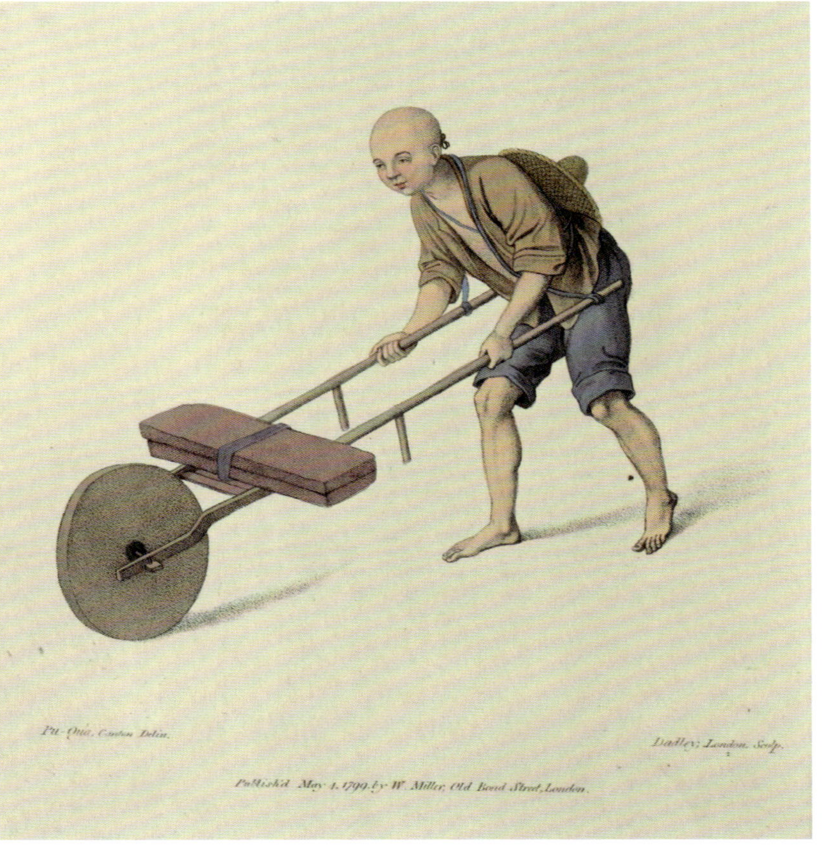

Fig. 5 "A Labourer" from George Henry Mason's
The Costume of China (detail), 1800

China trade families filled the Gardners' social calendar with tiffins, dinners, and excursions in the south. In her letters and diaries, Isabella wrote extensively about these meetings, aligning herself imaginatively, socially, and economically with other members of this elite network.

The Gardners' short trip to Guangzhou was particularly tinged by feelings of nostalgia and current events. Guangzhou was the earliest port open to foreign trade, and many American merchants' first point of contact with China in the late eighteenth and early nineteenth centuries. The Gardners visited at a time of rising tensions between the foreign merchants and the local population. Riots had broken out the summer before in response to the killings of two Chinese men at the hands of Westerners. Accompanied by the American counsel and under the protection of the Russell & Company staff, the Gardners traveled to notable sites such as the Honam Temple, Flowery Pagoda, and Howqua's Garden, the home of the Wu family, important trading partners for the Americans and at one point the wealthiest merchants in the world. "A most wonderful city. Such crowds, such movement, such color," writes Isabella in her diary. In the album, she pasted in a mix of highly specific photographs—a portrait of their guide Mak Tat-Hing alongside his calling card, a view of Shamian Island, where they stayed at the Russell & Company residences with their hosts Mr. and Mrs. Cunningham, and more recent photographs of the tents of troops and "the ruins of a house burned by the mob during the riot" (Plate 36).[17] Alongside these photographs reflecting current events are older photographs of temple architecture and city views from an unidentified photography studio. The photograph of the Flowery Pagoda with the cropped signature of the long-gone Dutton and Michaels seems to have been made from a negative or copy print from twenty years before, when the tower had not yet been renovated (Plate 37). Old and new images provide a jarring temporal juxtaposition in Isabella's photographs of Guangzhou.

The time of China in Isabella Stewart Gardner's albums varies. Contemporary China, as Isabella experienced it, was a dynamic entity comprising fleeting moments of color, sound, air, and life in the city. In contrast, she often presents a vision of a timeless China through the photographs of religious and imperial sites in Beijing or the Shanghai "life and character" studies that circulated as prints and photographs over the course of the nineteenth century. In Isabella's notes and photographs of Guangzhou, on the other hand, old and new coexist as the realities of violent colonial conflict intermingled with romantic intergenerational nostalgia. One could say that as foreign visitors, the Gardners saw China through the lens of Victorian imperialism, which colored perceptions of China as a perpetually historicized, never modern other. However, Isabella's methods of layering time in the albums demonstrate the ways in which our understanding of any place and culture is an assemblage of personal experiences, general knowledge, and chance encounters. War, trade, and travel made China accessible and visible to Isabella, especially as someone with familial ties to the China trade, but tensions between the past and present, imagined and real, permeated her perceptions. In these albums, photographs, travel notes, and personal mementos combine to form a selective, dreamy memory of a place Isabella once encountered.

1. Hawley 2009, 10; Chong 2009b, 180–233.
2. Chong 2009b, 182.
3. Ibid., 206.
4. Ibid., 209.
5. Ibid., 180.
6. Chong 2009a, 18.
7. Chong 2009b, 193.
8. Gardner noted in her diary on 6 October 1883, the day after visiting several landmarks in Beijing, that she "went with Mrs. Holcombe to hotel to order photographs." On 22 October, she "pasted photos, etc. and drove out to St. John's College before tiffin." She bought more photos in Shanghai on 24 October and in Hong Kong on 3 November. Ibid., 206, 212–13, 230.
9. Ibid., 182.
10. See, for example, the Peabody Essex Museum's albums from 1882, 1883, and 1884.
11. Downs 1968.
12. Carter 1925, 8–9. For more information about the Russell & Company's operations, see Chong and Murai 2009, 180; and Stelle 1941. The Russell & Company business records are held at Harvard's Baker Library.
13. Low was a principal in Fearon, Low, and Company, importers of tea and silk and an offshoot of Russell & Company.
14. Chong 2009b, 211.
15. Their sister, Sara Ann Delano Roosevelt, was the mother of Franklin Delano Roosevelt.
16. Chong 2009b, 182.
17. Isabella Stewart Gardner, *Travel Album: China, Volume III*, 1883, 58, Isabella Stewart Gardner Museum, Boston.

Volume III, pages 3–4, September 1883

Arrived at Tung-Chow by boat from Tien-tsin Sept. 21 at 7 a.m. and started for Peking on horses at 8.30 a.m. Got there at 2 P.M.

FRIDAY
SEPTEMBER
21
未 8th MOON 癸
21
五拜禮
月九西
日一廿
大月八華
日一廿

Peking

北京

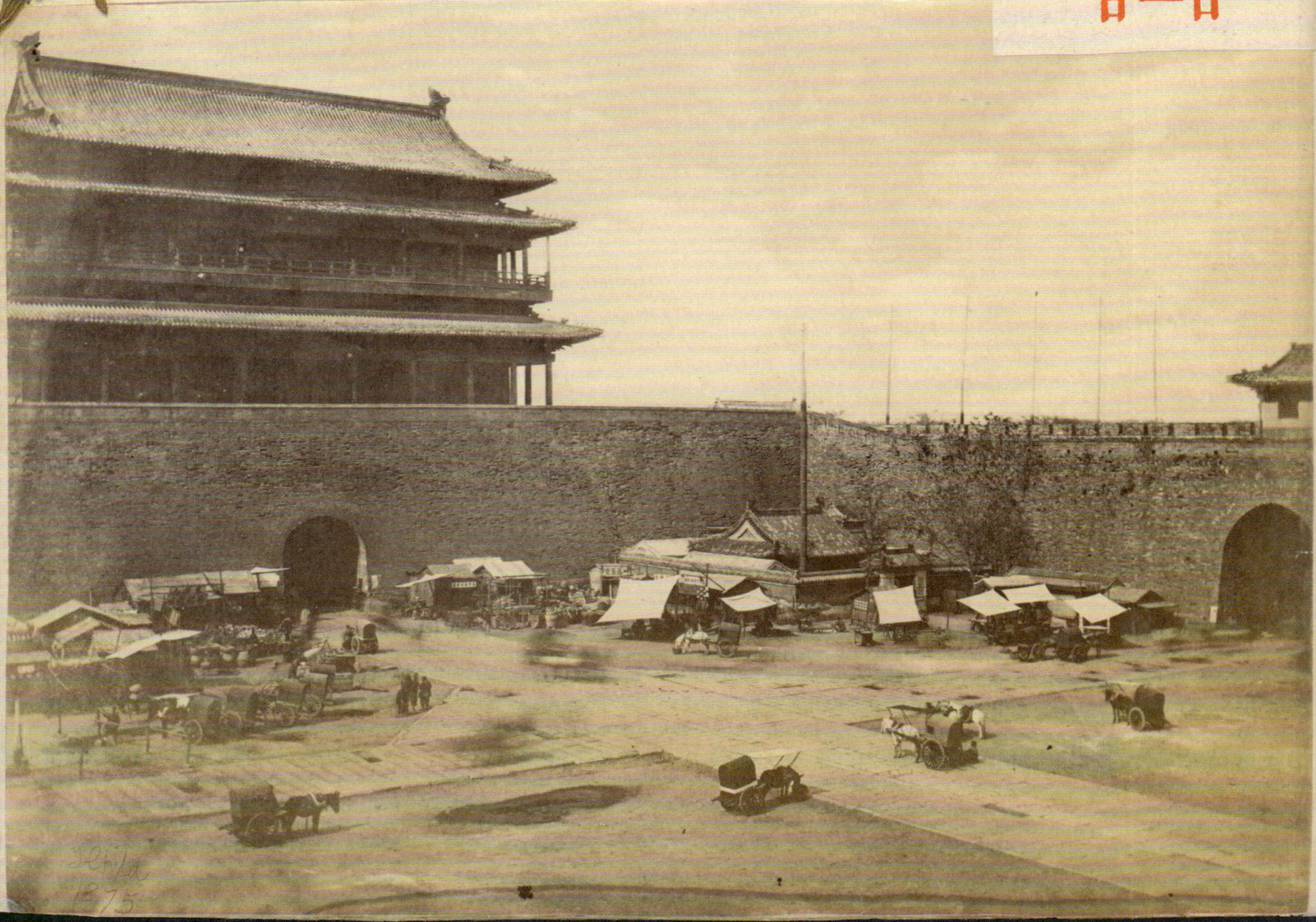

Peking — Sept. 21 — Oct. 10 —

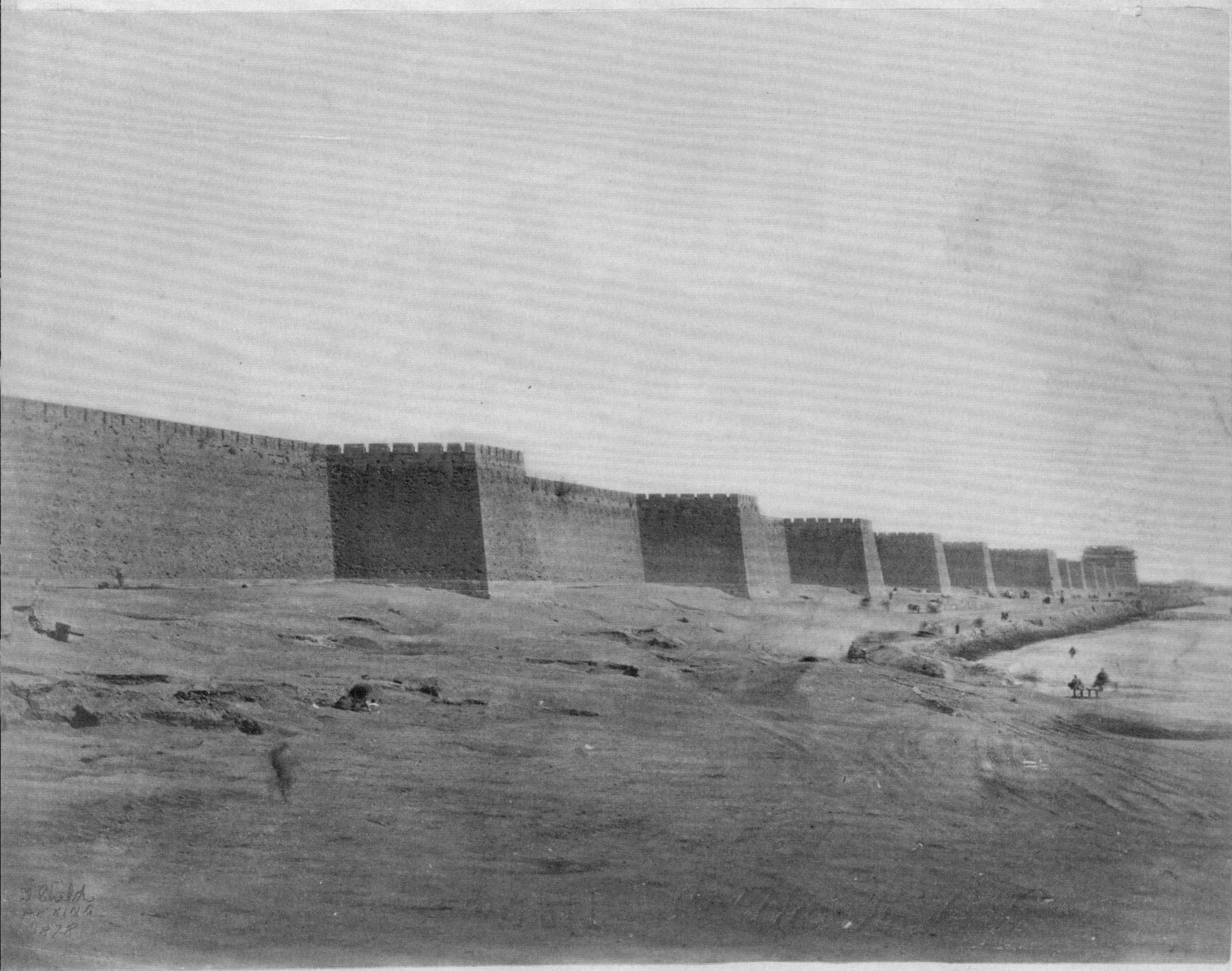

City Walls.

, pages 5–6, September 1883

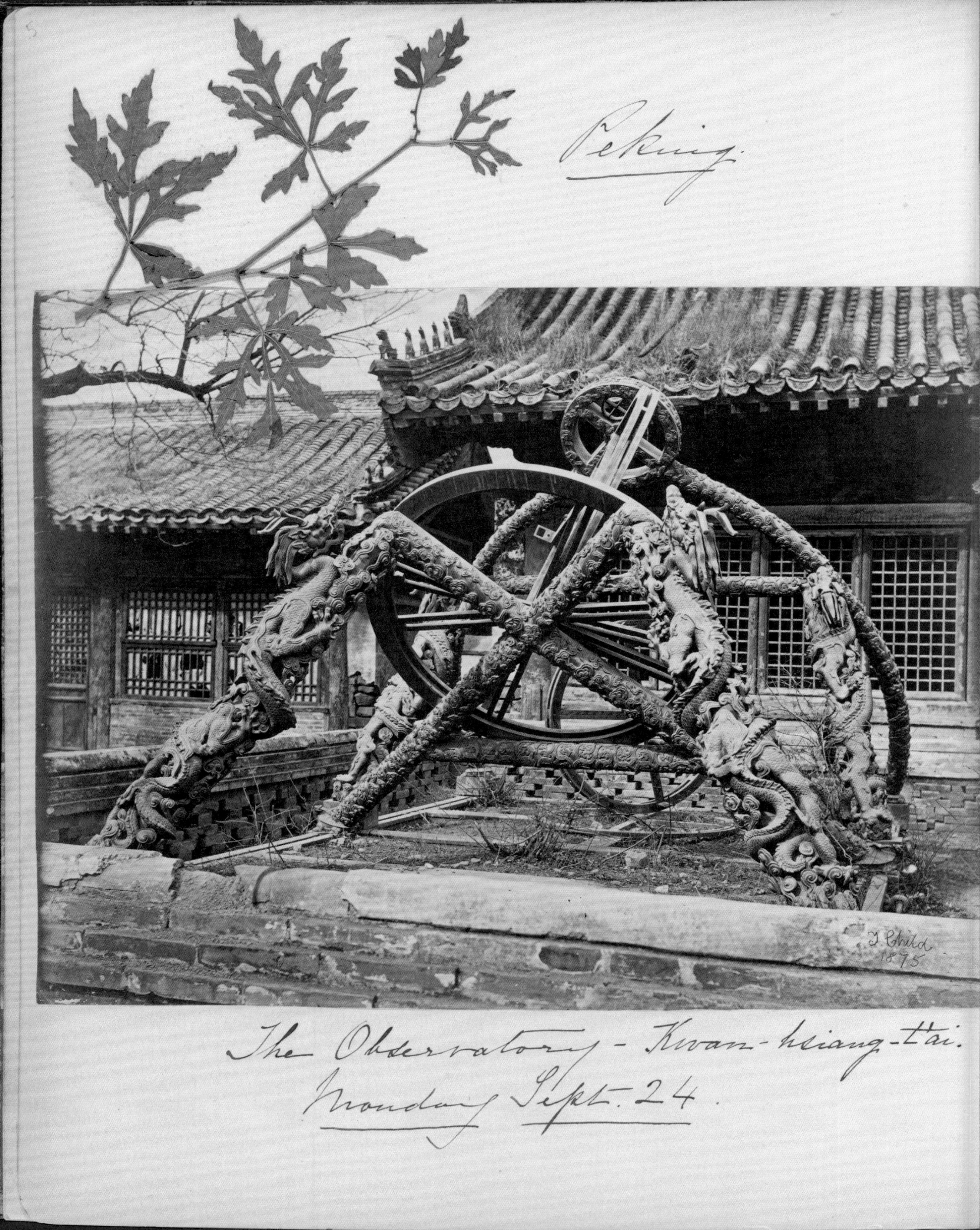

Peking

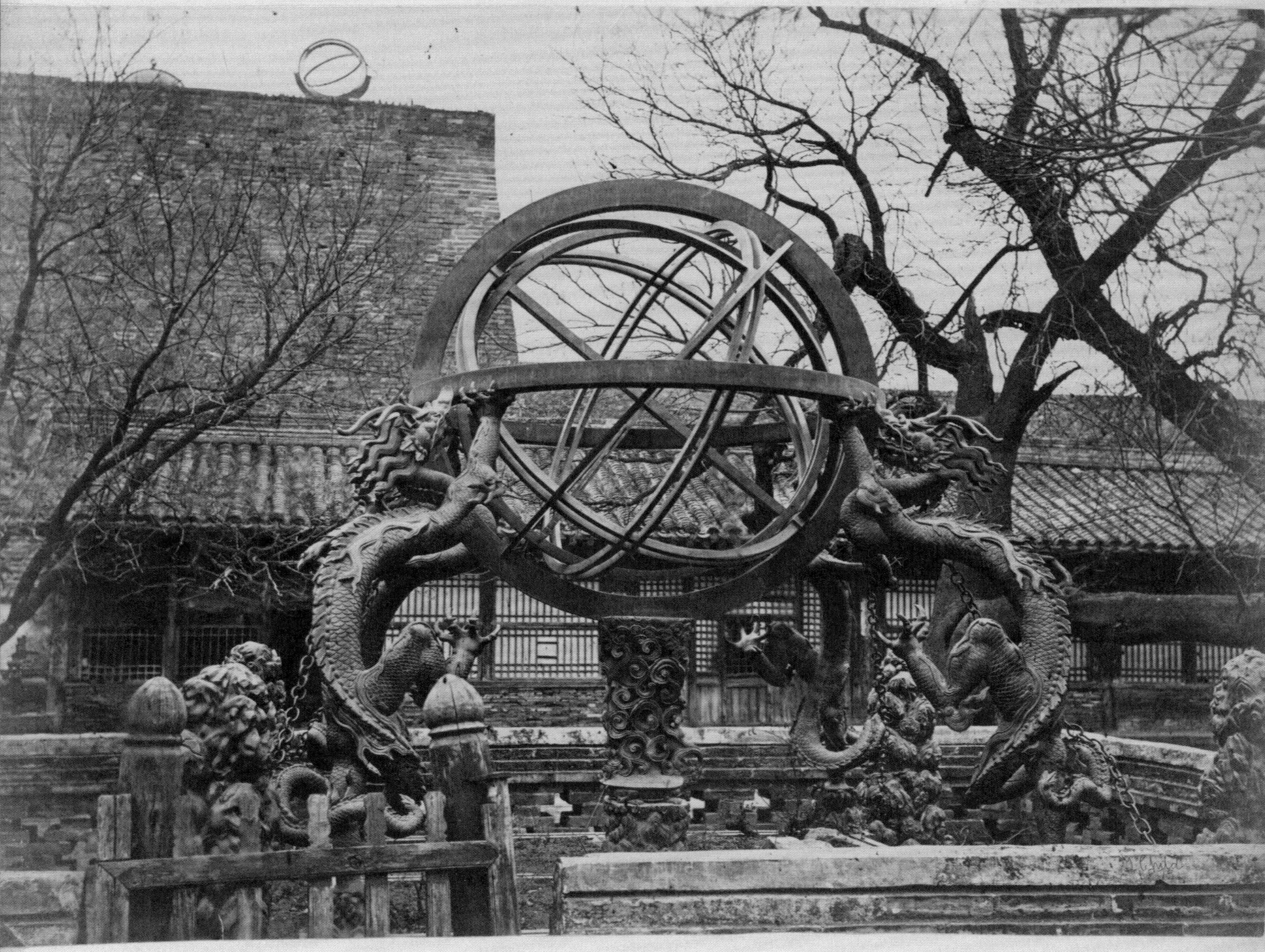

The Observatory

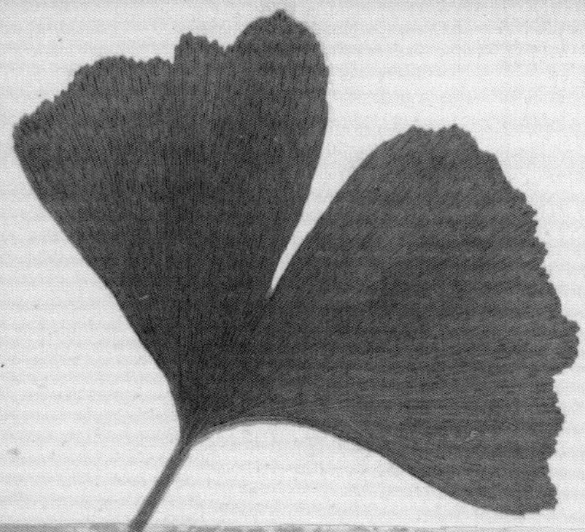

Yellow leaves
from "Temple
of the Sun".
Sunday Oct 7.

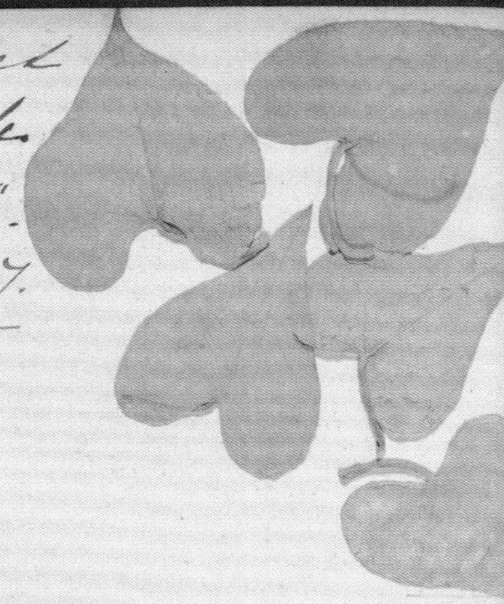

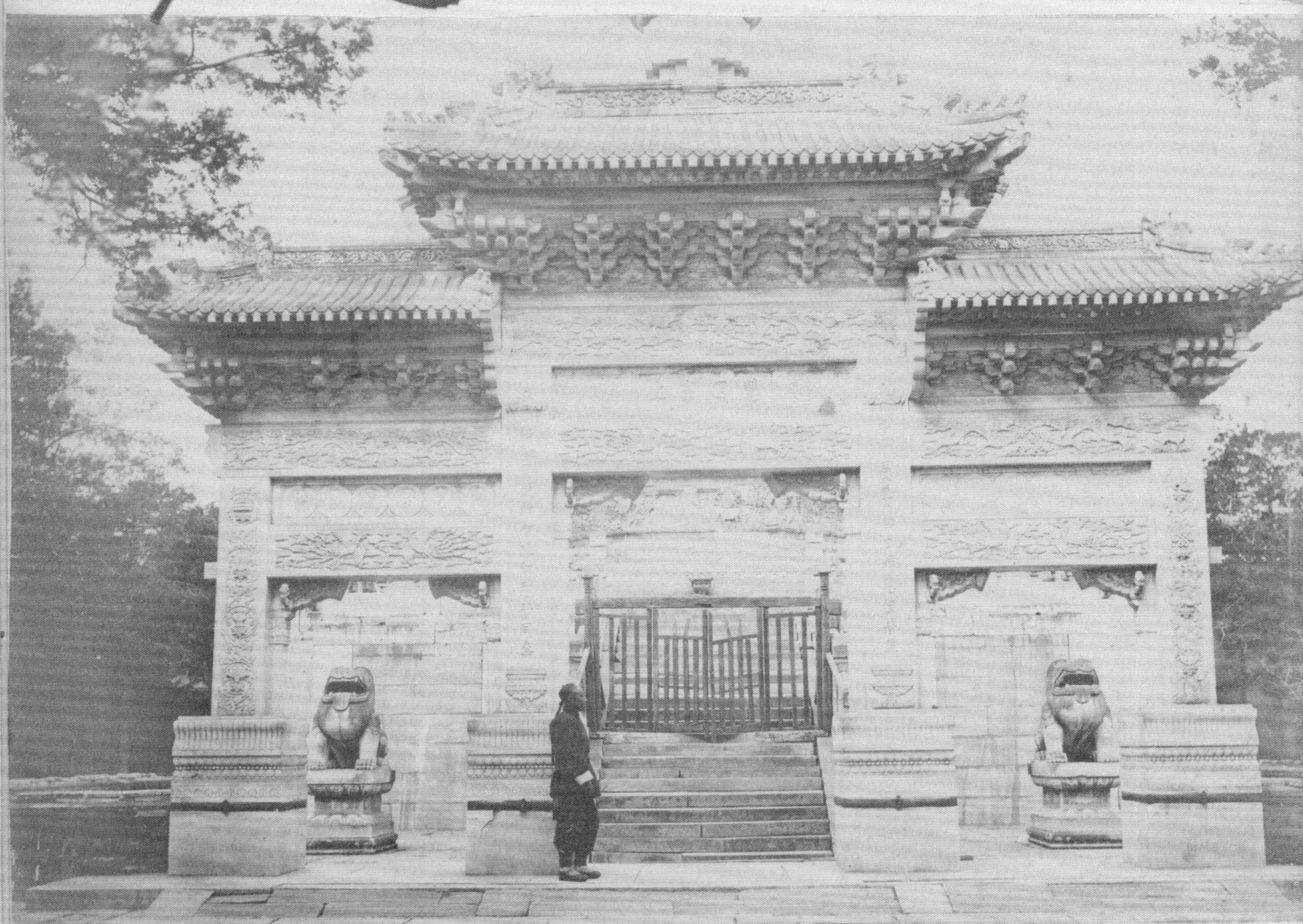

The Lama Temple "Hwang-ssu" –
A monument to the memory of the
Teshoo Lama – Oct. 5.

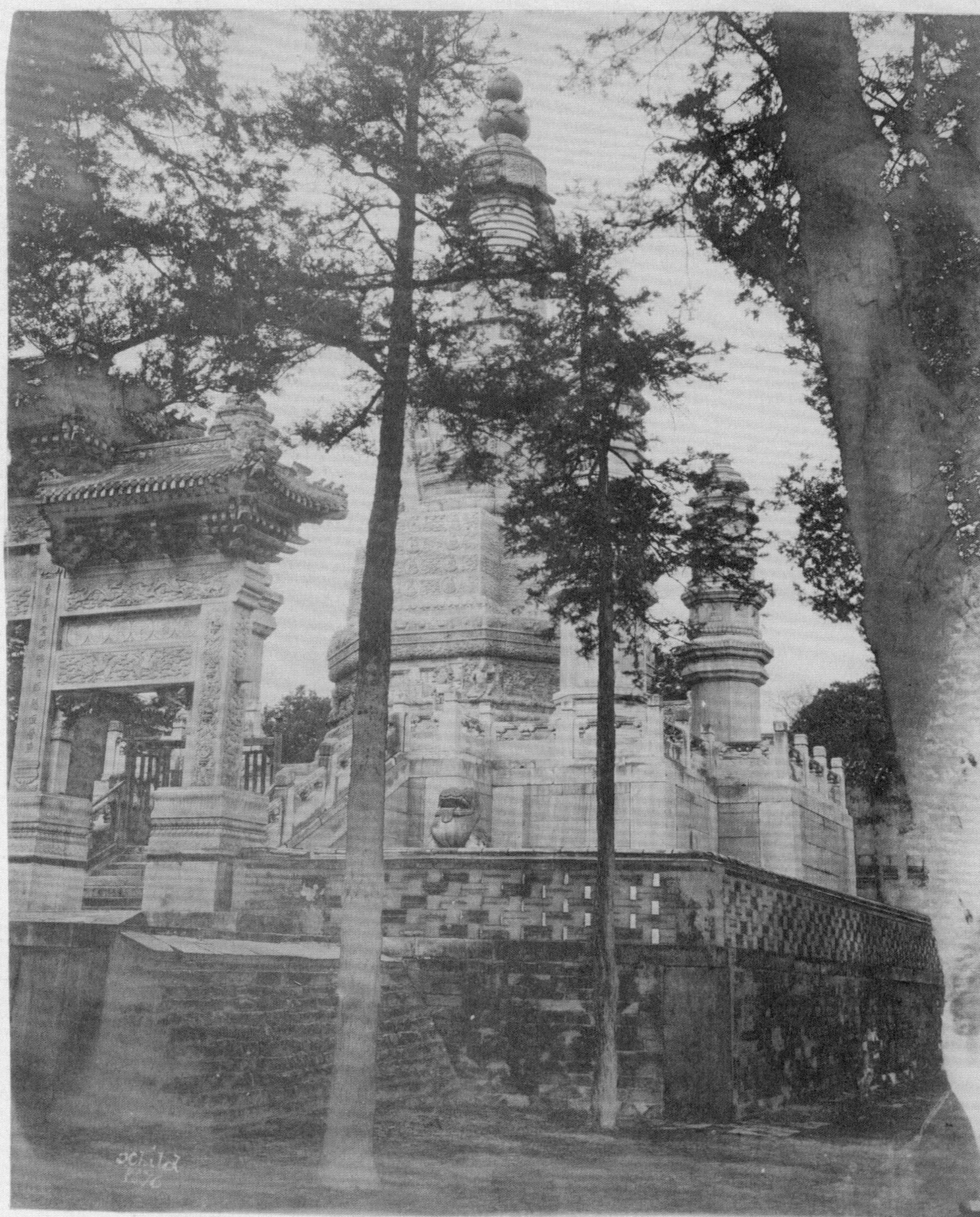

'Hwang-ssŭ - Oct. 5.

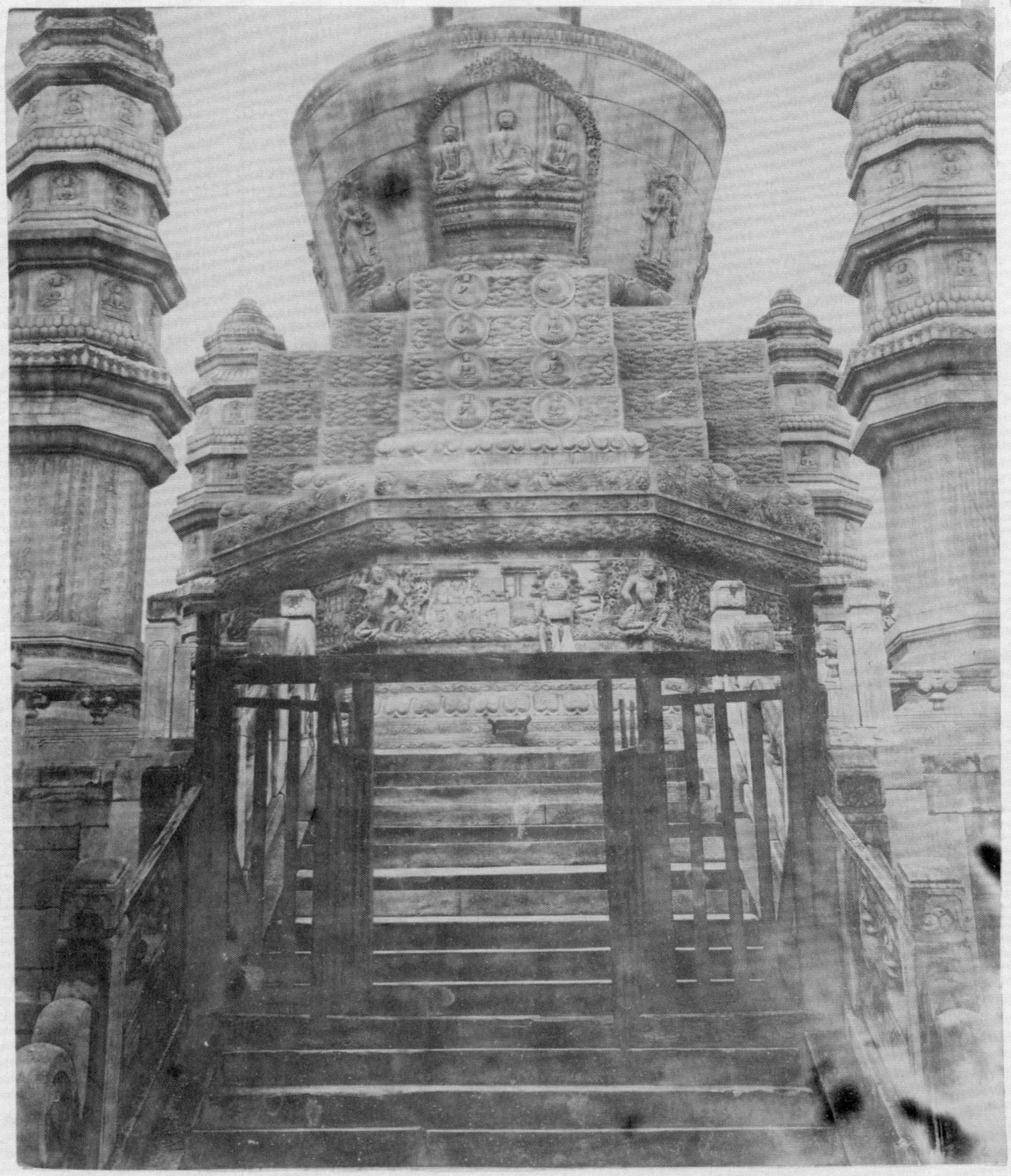

Hwang-ssu – Oct. 5.

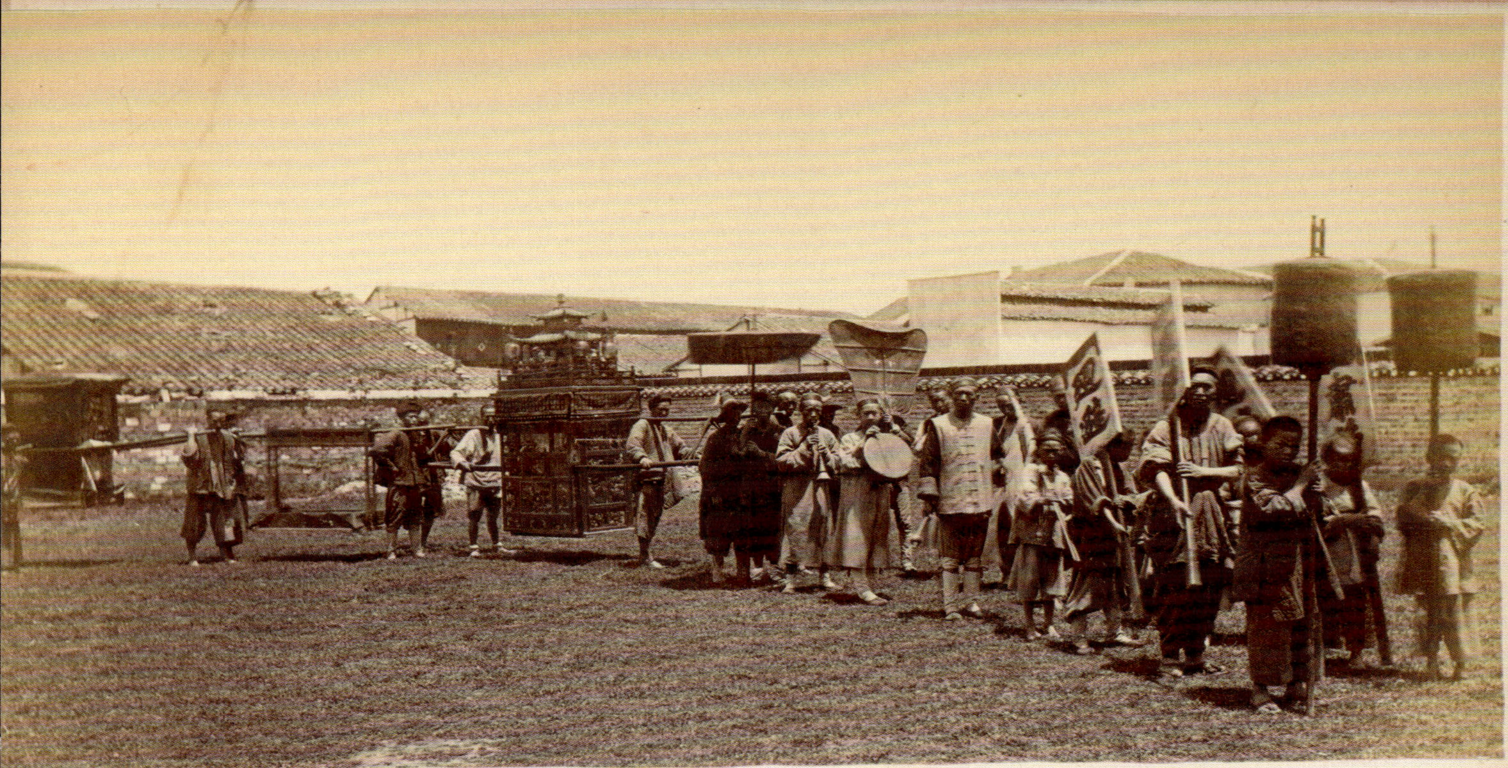

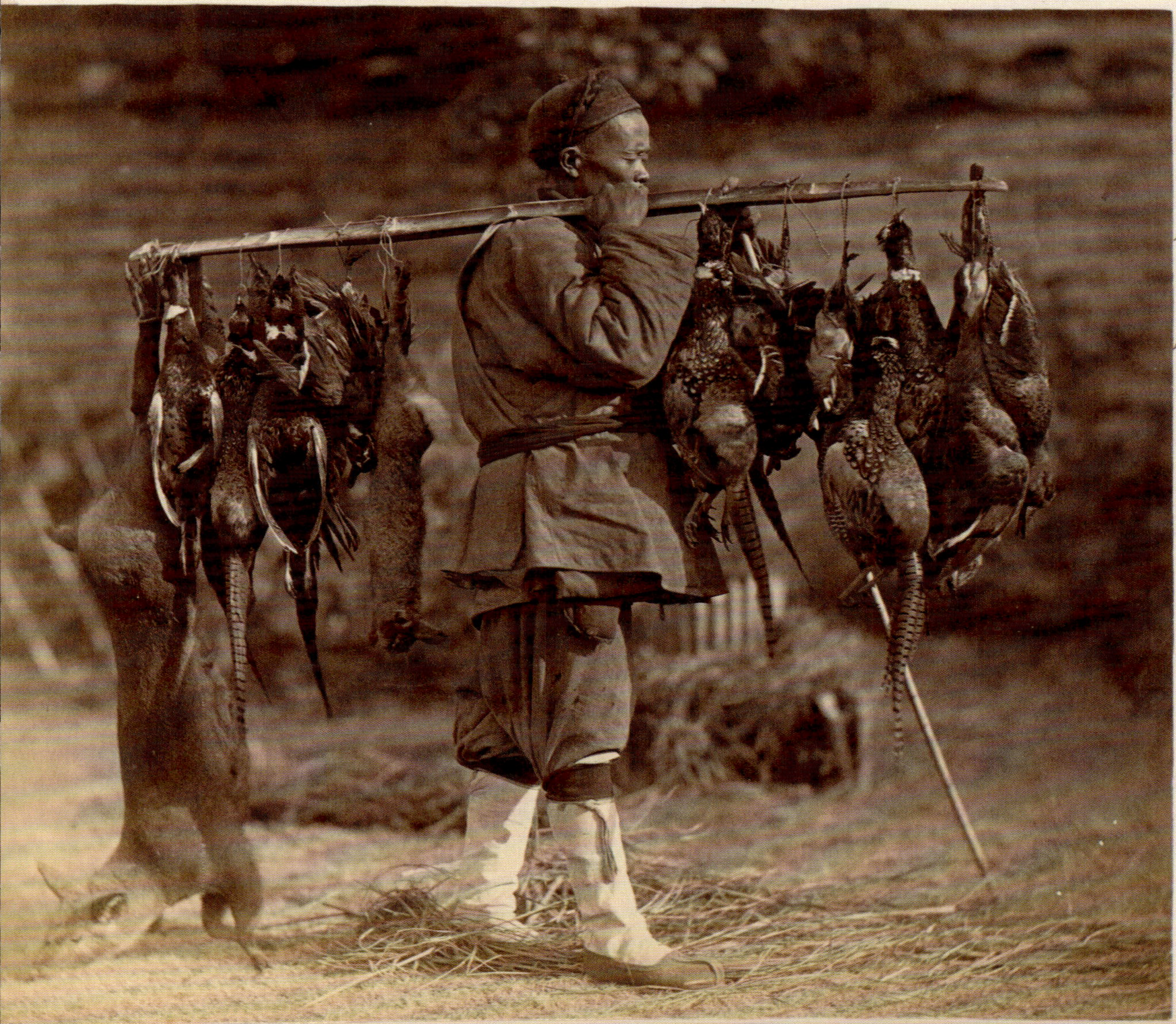

Left Peking
Wednesday
Oct. 10.
Took boats
that eve'g.
at Tung-Chow.
Got to
Tien-tsin
Oct. 12 abt
5.30 P.M.

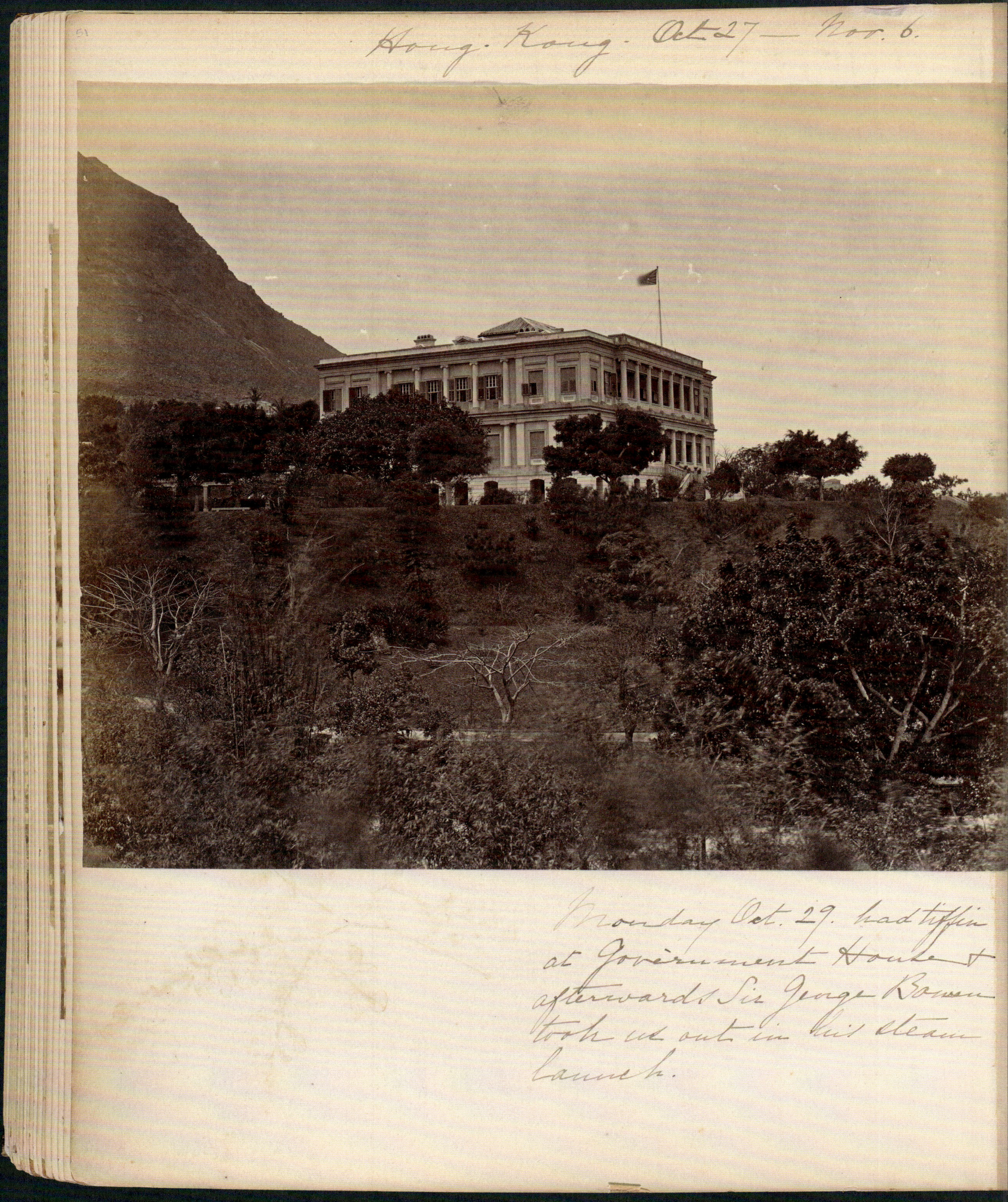

Hong-Kong- Oct 27 — Nov. 6.

Monday Oct. 29. had tiffin at Government House & afterwards Sir George Bowen took us out in his steam launch.

Government House. Hong Kong.

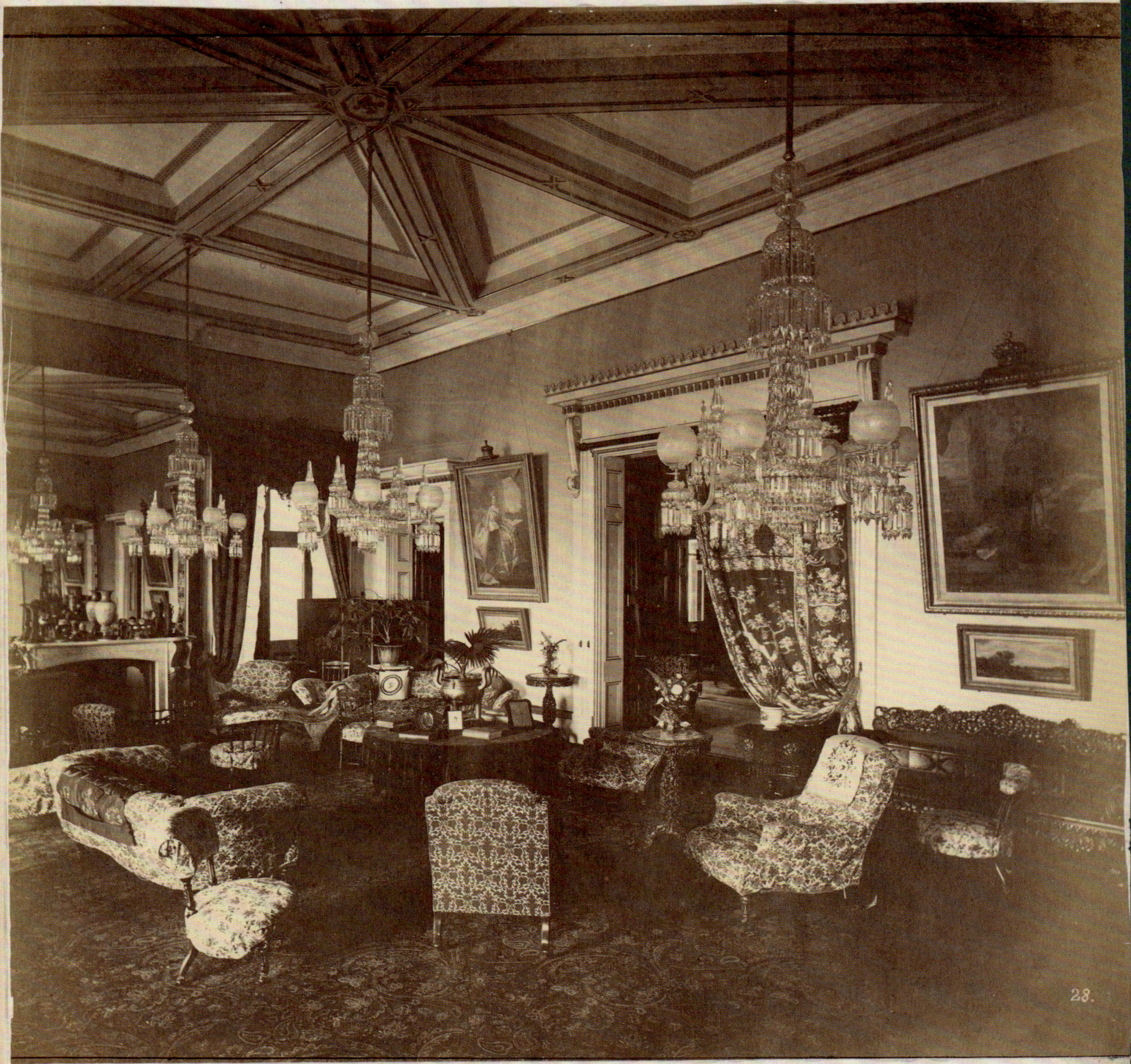

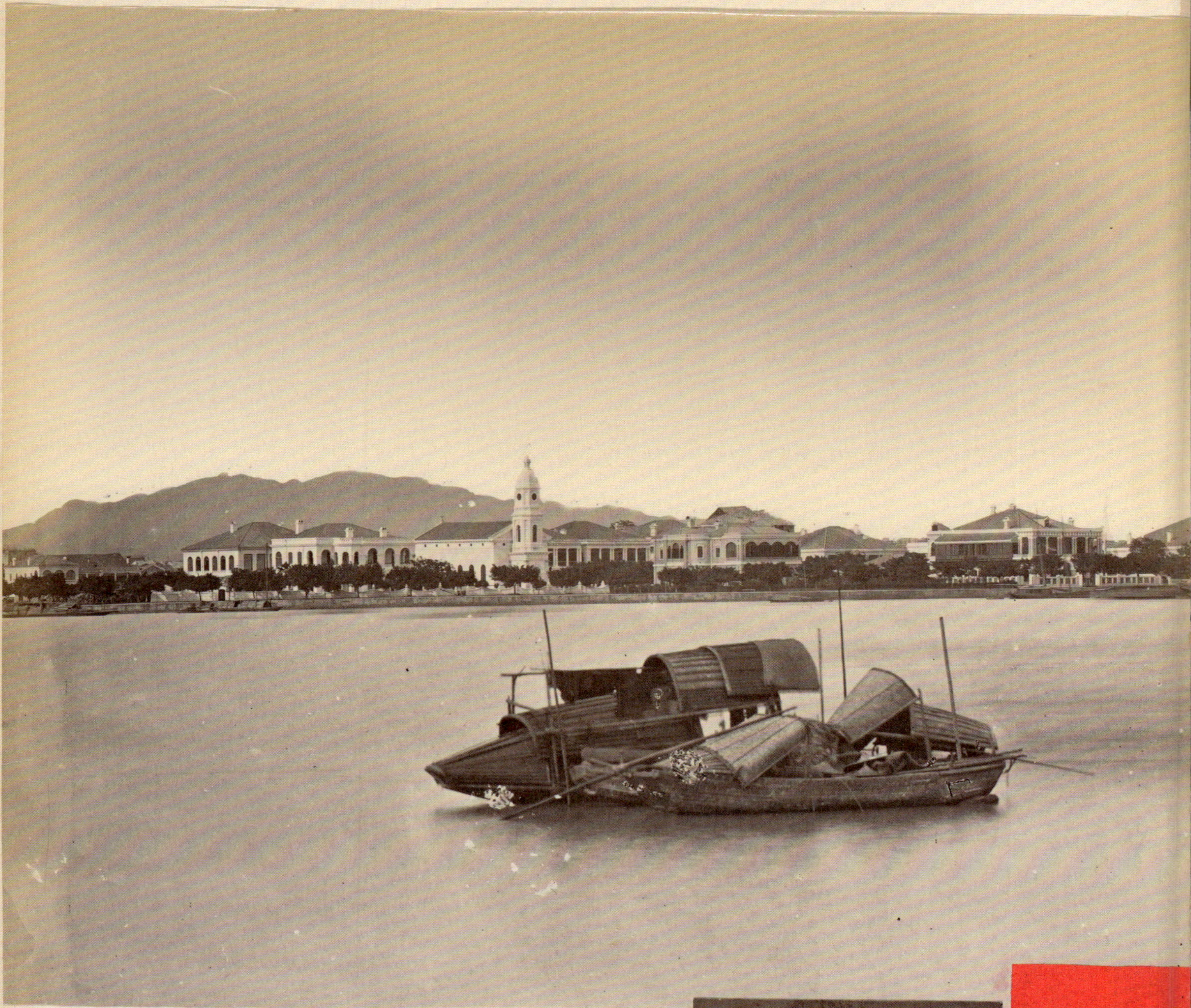

Shameen – (Canton) Oct. 30 – Nov. 2.

Went to Mr. Cunningham's (Russell & Co).

Mak Tat Hing –

Our guide & former servant of Archdeacon Gray –

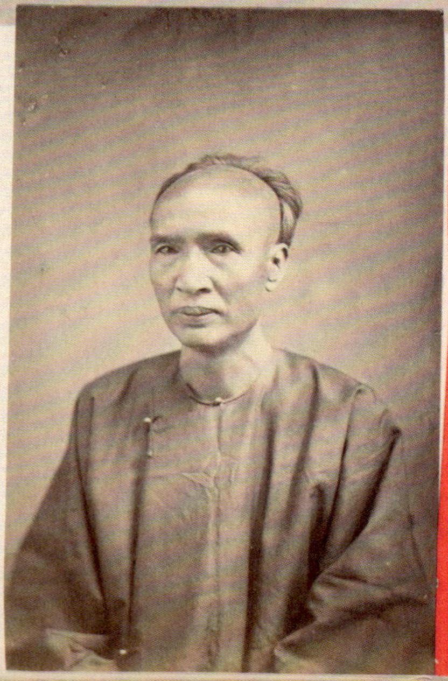

麥達卿

Shameen–

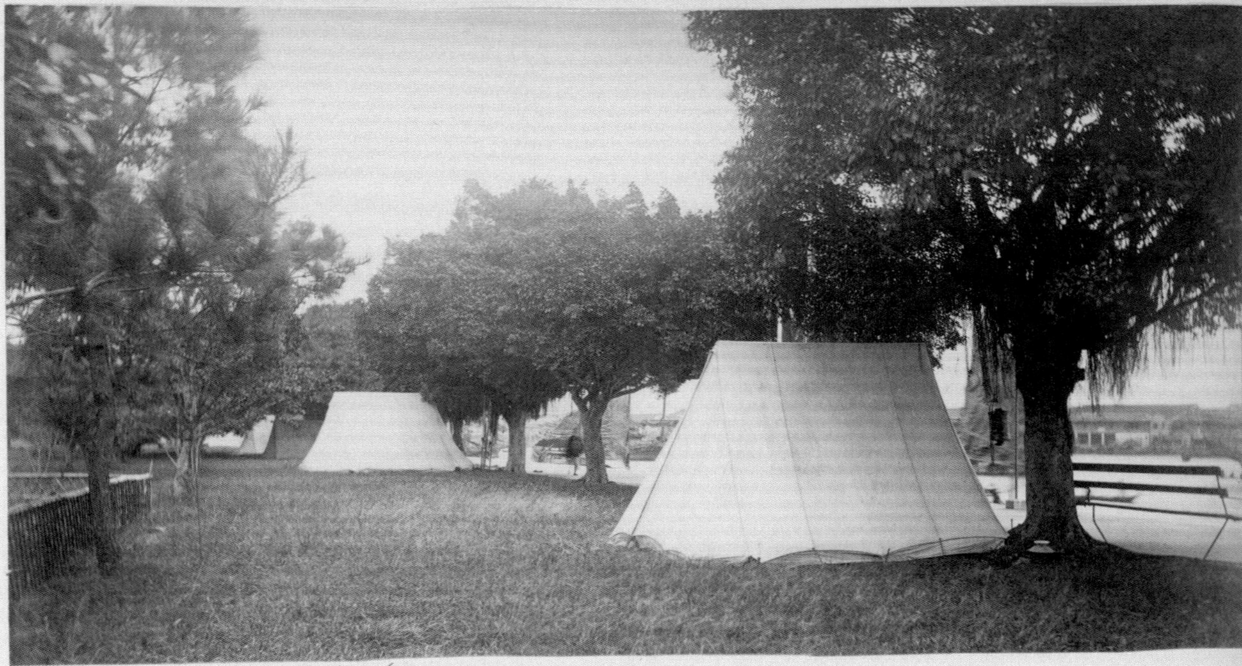

The ruins of house burnt by the mob during the riot–

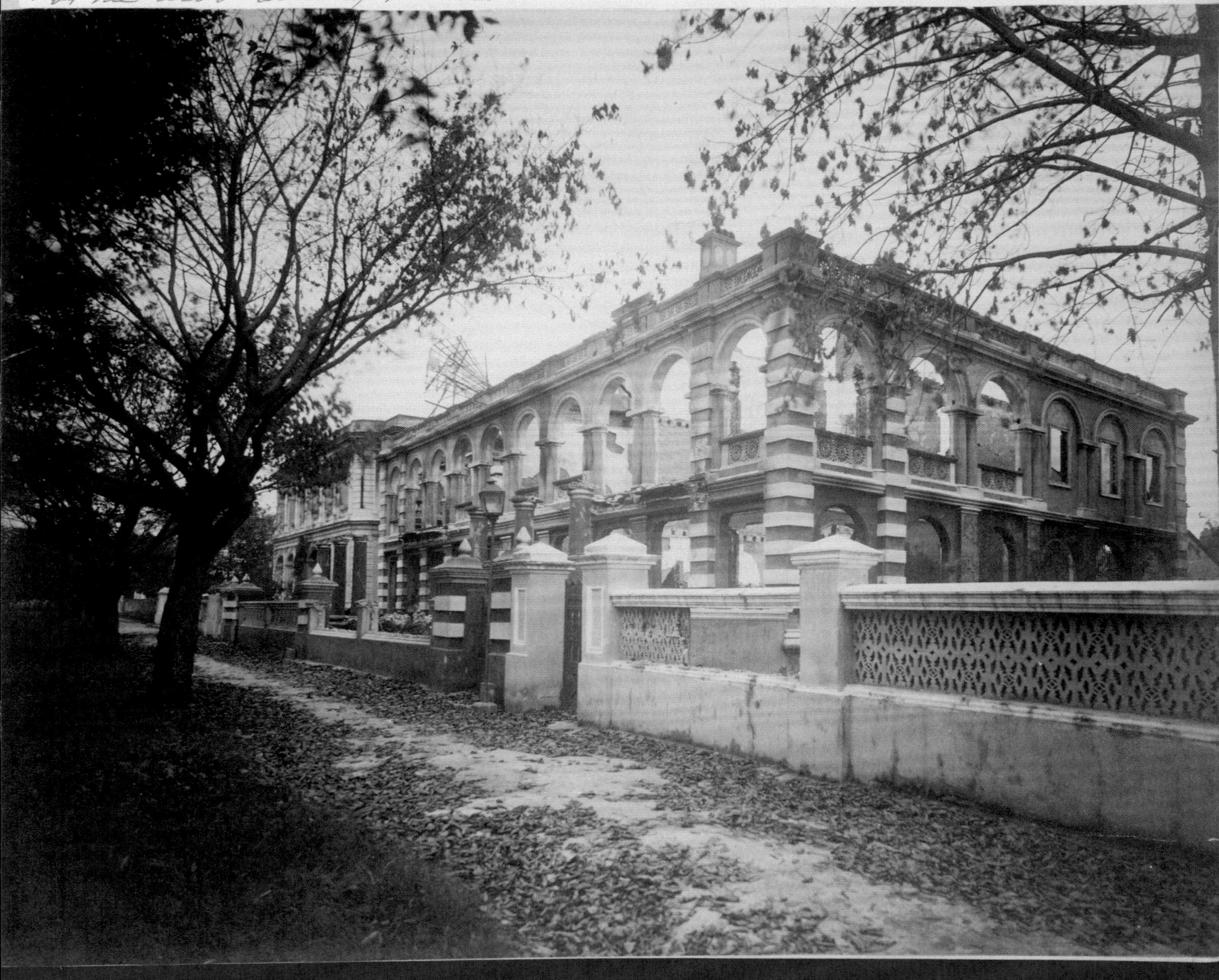

CHINA
Volume III, pages 61–62, October 1883

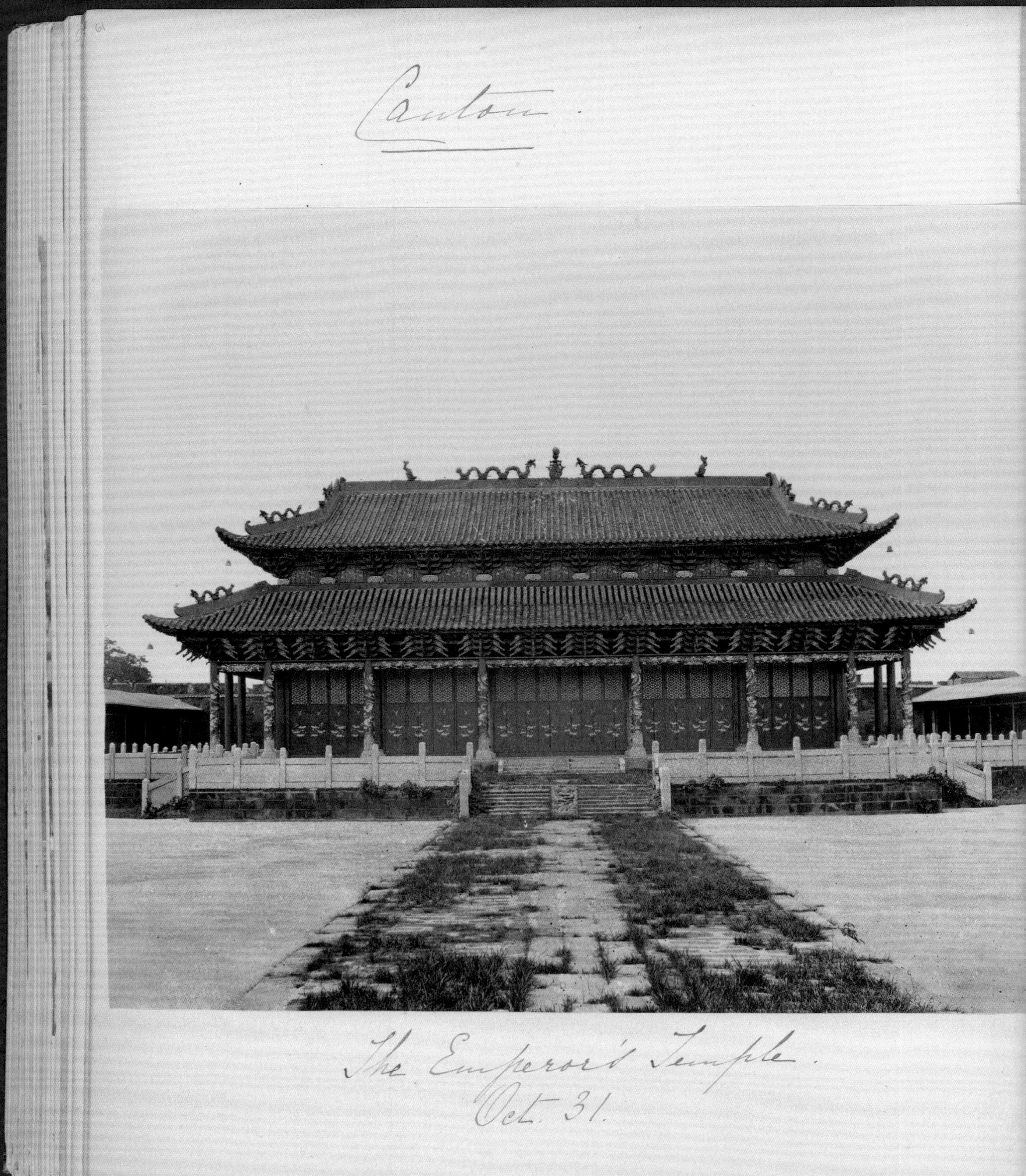

Canton.

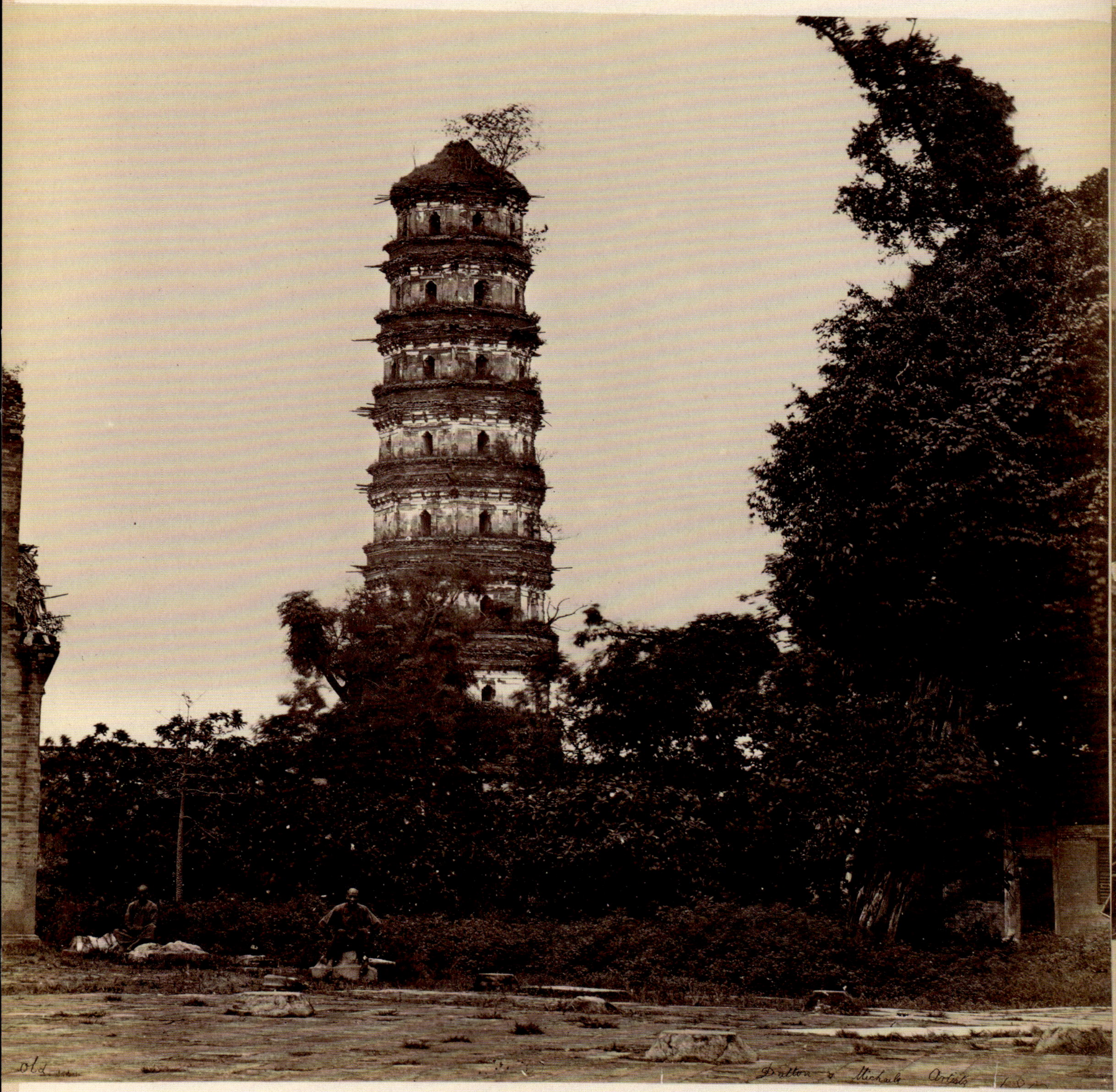

Flowery Pagoda. Nov. 1.

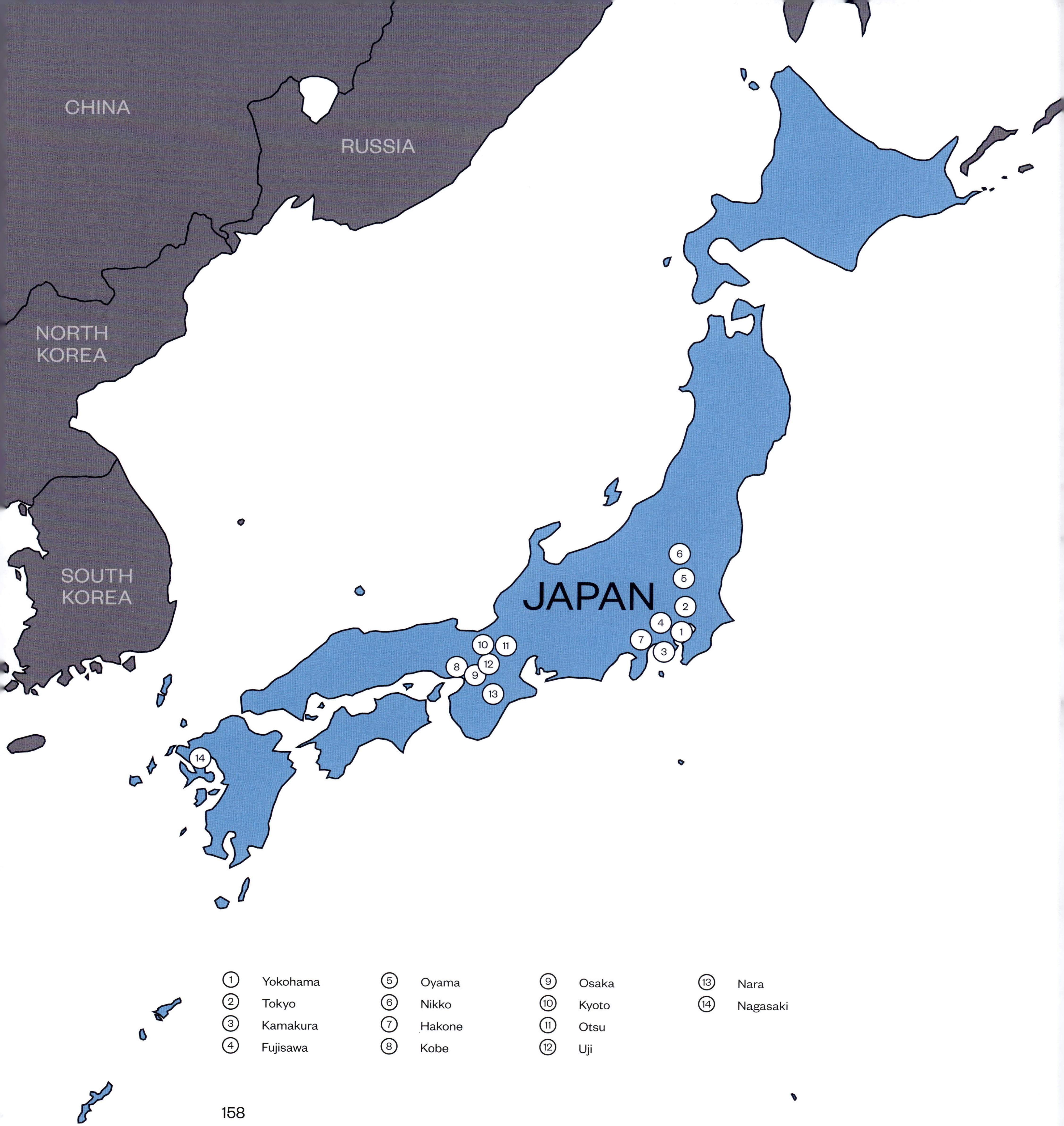
CHINA
RUSSIA
NORTH KOREA
SOUTH KOREA
JAPAN
1 Yokohama
2 Tokyo
3 Kamakura
4 Fujisawa
5 Oyama
6 Nikko
7 Hakone
8 Kobe
9 Osaka
10 Kyoto
11 Otsu
12 Uji
13 Nara
14 Nagasaki

Japan

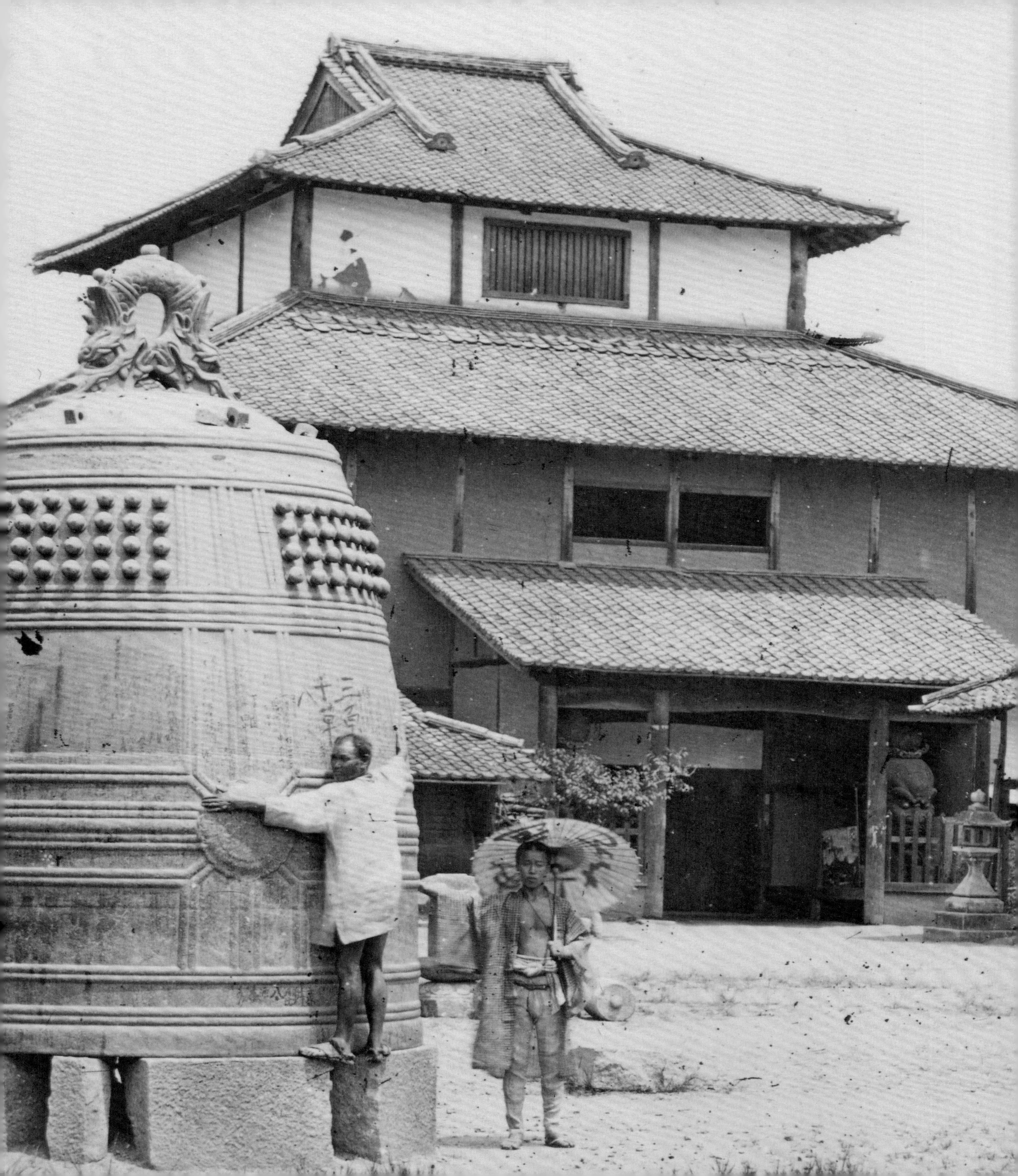

Deceptive Intimacy: Isabella Stewart Gardner's 1883 Japan Travel Albums

David Odo

I first encountered Isabella Stewart Gardner's travel albums of photographs of Japan when they were displayed in the museum's then small, temporary exhibition space in the spring of 2009.[1] Not being an Isabella aficionado at the time, I was mainly unaware of her extensive travels in Asia in the early 1880s, and I was surprised and delighted to see fine examples of the albums on display. The few images of Japan that were on view certainly left me wanting to see more, and I was grateful for *Journeys East: Isabella Stewart Gardner and Asia*, the excellent book published in conjunction with the exhibition, edited by Alan Chong and Noriko Murai. The book not only made additional images accessible but also established a scholarly basis for a much richer understanding of Gardner's deep interest in art, which so demonstrably extended beyond European and American works, for which she is well known, into Japanese and other Asian traditions.[2]

This latest examination of Gardner's travel albums of Japan, part of the museum's *Fellow Wanderer: Isabella Stewart Gardner's Travel Albums* exhibition, certainly appeals to me as a scholar of nineteenth-century photographs of Japan. But beyond the subject matter, I have always been intrigued by the idea that such albums are forms of self-fashioning, especially in the early decades of commercial photography before most individual travelers could easily take their own photographs. Travel albums can tell us more about their creators than what they purport to display. Not exactly visual diaries, albums can nevertheless be read simultaneously as highly documentary and personally revealing objects. They are not artists' notebooks but can still be highly creative in their expression. In a contemporary museum context, we often exhibit them as works of art in their own right. What to make of them? And more specifically, what to make of Gardner's travel albums of Japan?

In her travel albums, Gardner achieved a personal documentary style that is deceptively intimate, seemingly allowing viewers to travel along as she made her way but without revealing anything of her inner world. All these years later I do not recall

Kyoto, August 1883 (detail) from Isabella Stewart Gardner's *Travel Album: China and Japan, Volume II*, page 14

exactly which photographs I saw when I eagerly peered into the display case to view pages from the albums during the 2009 exhibition, but I distinctly remember a woman next to me exclaiming to her companions that she felt as if she was in the presence of Isabella herself: "It's like being there with her!" Sadly, I lacked the foresight to interview the excited fan, but her delight was palpable as she looked closely at the photographs, vicariously experiencing at least something of Gardner's Japanese sojourn. Clearly, she felt that the album was an intimate object.

Initially, I had a similar reaction. Gardner organized and pasted photographs onto pages in the albums, which she then lightly sprinkled with notes in her own hand. The combination of text and image seemed to open a window into her voyage. On reflection, however, I have a different interpretation. Now, it seems to me that although we learn details such as where she stayed, with whom she lunched and when, Gardner's notes communicate little if anything of her thoughts and emotions. Instead, her notes seem to have arisen out of an impulse to document and record facts about her time in Japan, but little else:

> *Left Nikkô at 8 a.m. Thursday July 26; arrived at Yokohama abt. 5 P.M. Friday July 27. On the road everyone busy with silk worms.* (Plate 40)

This impulse appears to be mirrored in her choice of photographs, which similarly seem assembled for the purpose of leading viewers through a set of destinations and activities, but feel somehow generic, straight off the shelf, as indeed they were. The personal seems to be largely missing from the photographs. As Chong has observed:

> *It was almost expected that a traveler of the period would drop into a studio to have a portrait taken, perhaps in local garb, and Isabella did exactly this on earlier and later trips. The Asian albums contain very few images of friends and perhaps only one of Jack Gardner, which leaves her albums strangely impersonal compared to those of other travelers. Although Isabella was always very controlling of her image, her total absence in the visual record of the year-long Asian journey is a mystery.*[3]

Also missing from the album are clear indications of her eventual interest in Japanese art. Years later, after she had achieved renown as one of the most important collectors (of European art) of her time, she branched out into Asian art. Now that we know this, the albums invite us to consider what her interest in photographs might reveal about her emerging art-collecting practice. When I examined the photographs in Gardner's albums, however, I was surprised to see, as Chong has noted, that "there are very few images of the type of Asian art that she later purchased."[4] Instead of a visual preview of her future art-collecting practice, what might we expect from Gardner's albums? I would argue that the albums present viewers with other possibilities for discovery, if only we agree to recalibrate our expectations of them. The Japan albums afford viewers an opportunity to explore nineteenth-century Japanese photographic practice and its intersections with emerging global tourism, with its attendant colonialist attitudes firmly attached.[5] Importantly, it is also a chance to understand how Gardner actively participated in the global practice of collecting photographs and creating travel albums, rather than understanding her exclusively as an elite collector and art connoisseur.

TOURISM IN JAPAN

Isabella Stewart Gardner and her husband, Jack, departed San Francisco on 31 May 1883, onboard a steamer bound for Yokohama, Japan, for a tour of Asia including China and India, that lasted more than a year.[6] They were to spend three months in Japan, where they purchased many souvenirs but with two exceptions did not intentionally collect works of "fine art" (figs. 1 and 2).[7] The sojourn undoubtedly had a major impact on her thinking and aesthetic sensibility, however; she had a Japanese-style garden landscaped at the Gardner family estate in Brookline, Green Hill, when they returned to Boston (fig. 3).[8]

The Gardners were among a select group of wealthy global travelers who were able to take advantage of relaxed political restrictions and improved travel infrastructure early in the modern history of tourism to Japan. Starting in the mid-1850s, foreign businessmen, military officers, educators, and diplomats began living and working in Japan to an unprecedented extent, thanks to changes forced on the country by the United States and Western powers, but under restrictive conditions imposed by the Tokugawa shogunate. The first commercially run steamships—operated by the Pacific Mail Steamship Company—began sailing from San Francisco, arriving in Yokohama in January 1867. In 1869, the Suez Canal, which connects the Mediterranean Sea to the Red Sea, opened to shipping, greatly reducing distance and travel time between the North Atlantic and the Northern Indian Ocean. The transcontinental railroad in the United States was completed, creating, for the first time, a continuous railway line from the Atlantic to the Pacific coast. Japan became a stop on Thomas Cook's round-the-world tour in 1872, which further increased access to the formerly nearly inaccessible country. Alongside these and other infrastructural changes, the waning of the Tokugawa shogunate's political power and the rise of the new, more internationally minded Meiji government (established by the imperialist Meiji Restoration of 1868) created conditions of comparatively easy access to the country for foreign tourists for the first time. This was especially true after 1880, when certain domestic travel restrictions were lifted for foreigners.

It is important to note here that although late-nineteenth-century foreign tourists might have thought they were "discovering" Japan, they were only the most recent visitors in what is perhaps the world's longest tradition of mass (albeit domestic) tourism.[9] Pre-seventeenth-century travel in Japan seems to have existed primarily in the form of religious pilgrimages and occupational necessity. In the more modern sense of the term, tourism was widespread by the early Tokugawa period (1603–1868).[10]

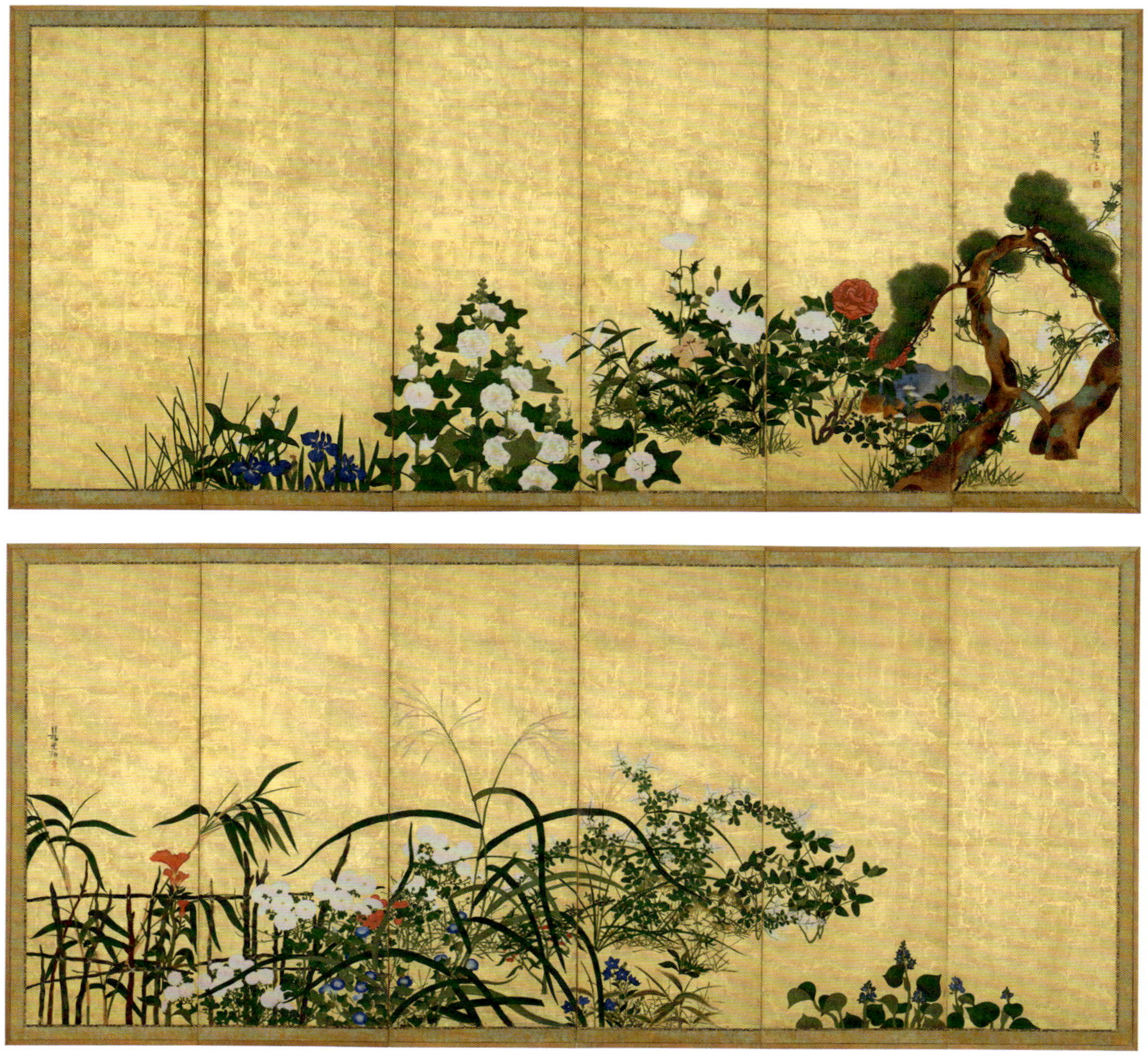

Figs. 1 and 2 Japanese, Edo period, *Spring* and *Autumn*, early 19th century. A pair of six-panel screens; ink, colors, and gold on paper, each 156 × 361 cm. Isabella Stewart Gardner Museum, Boston

Factors such as the flowering of urban culture, prohibitions against foreign travel for Japanese and travel to Japan for foreigners, and the rise of a market economy contributed to the establishment of domestic tourism on a large scale in Japan. Sights and travel itineraries, including of "famous places" (*meisho*) both sacred and secular, such as temples and shops (which in turn sold "famous products," *meibutsu*), were standardized through illustrated guidebooks and stand-alone woodblock prints. These texts and images (unintentionally) created a blueprint of sorts for later, foreign tourism.

Over the course of several decades, government leaders engineered radical changes to nearly every aspect of Japanese life, transforming what had previously been a mostly isolated feudal country into a powerful modern nation. By the time the Gardners visited Japan in the early 1880s, the tourist trail, enabled by the changes mentioned above, was firmly established. As Allen Hockley has described, the term "globetrotters" was in use in Japan to describe tourists by at least the early 1870s as the country became a favorite destination for Western travelers:

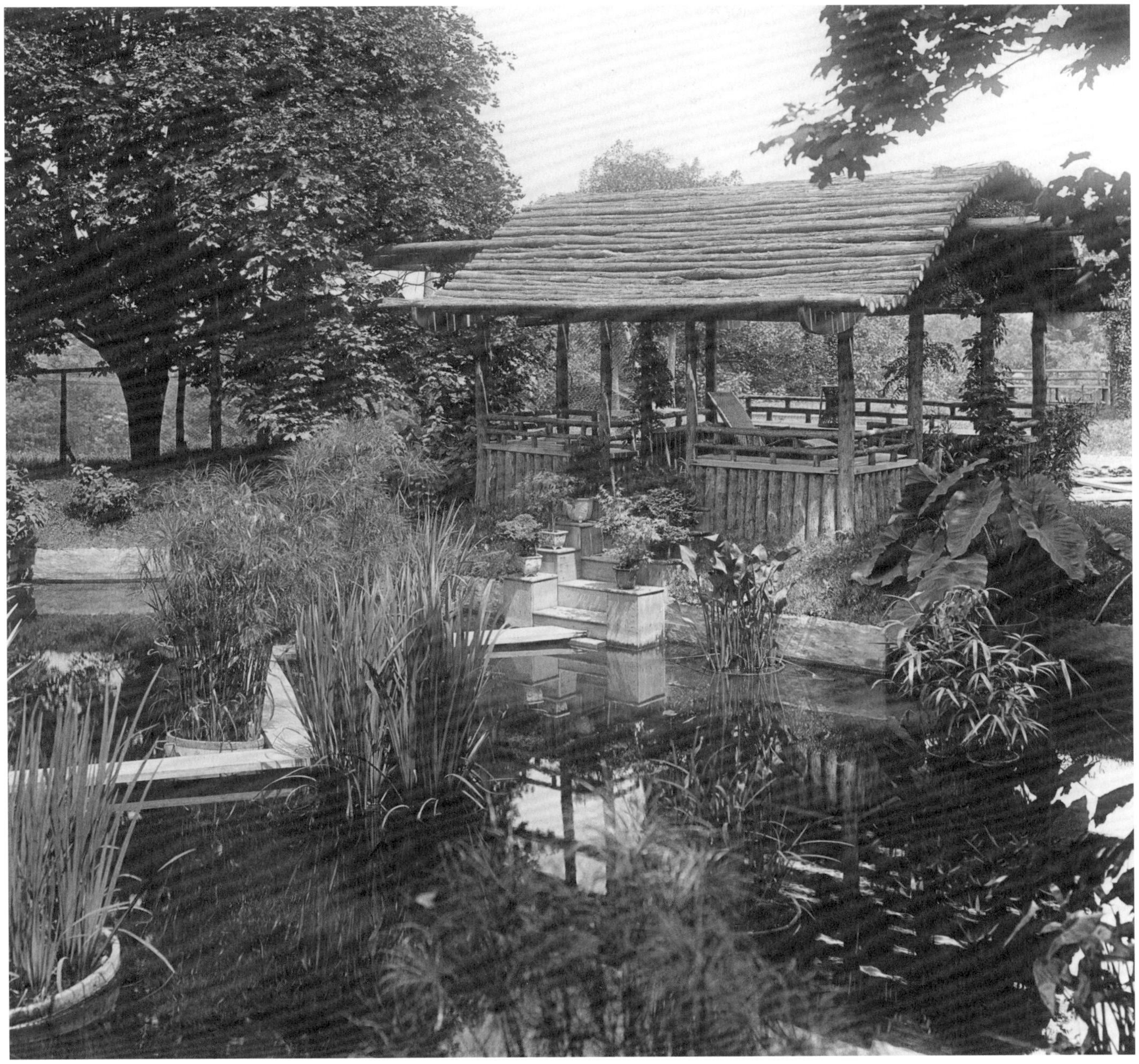

Fig. 3 Thomas E. Marr (Canadian–American, 1849–1910), Japanese Garden at Green Hill, about 1900

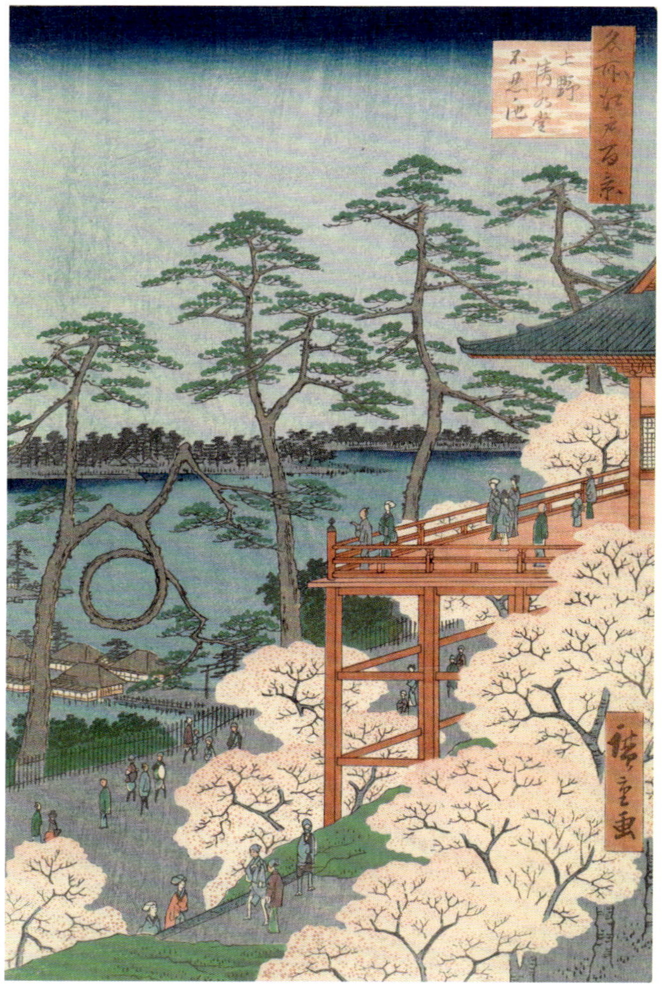

Fig. 4 Utagawa Hiroshige I (Japanese, 1797–1858, artist) and Uoya Eikichi (Japanese, active 19th century, publisher), *Kiyomizu Hall and the Shinobazu Pond at Ueno (Ueno Kiyomizudô no ike)*, from the series *One Hundred Famous Views of Edo (Meisho Edo hyakkei)*, 1856. Woodblock print (nishiki–e); ink and color on paper, 37.3 × 25.7 cm. Museum of Fine Arts, Boston. William S. and John T. Spaulding Collection (21.10407)

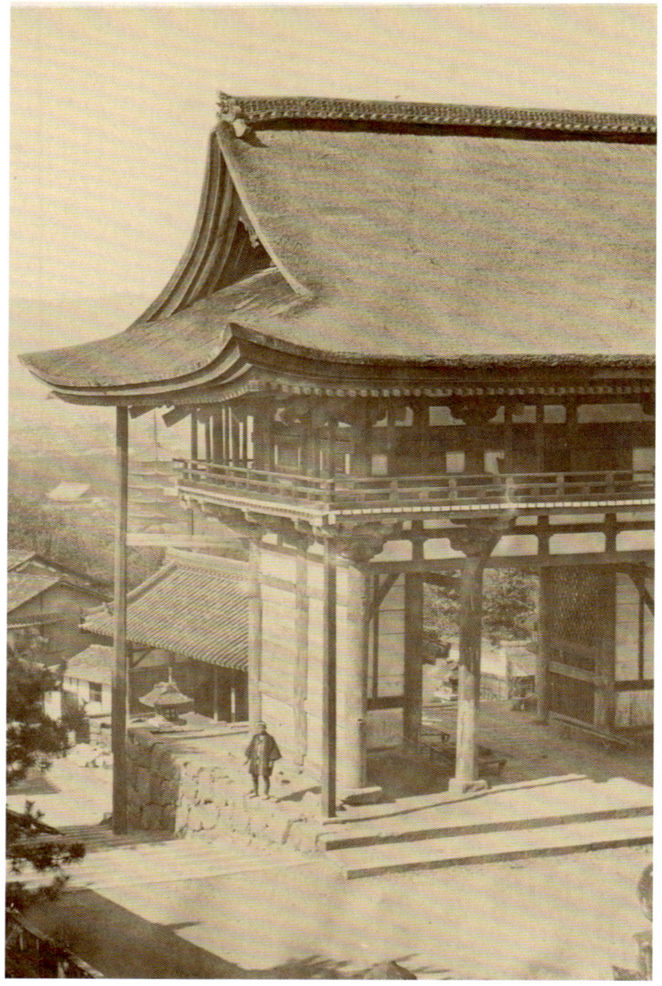

Fig. 5 Kyoto, August 1883 from Isabella Stewart Gardner's *Travel Album: China and Japan, Volume II*, page 22

> *Globetrotters, as they were called, arrived in the treaty ports in ever increasing numbers, stayed in new hotels built especially for them, visited scenic spots and famous places that they had read about in guidebooks, newspapers, or accounts written by other travelers, then moved on to other ports of call. Dozens of commercial photography studios boasting large inventories catered specifically to globetrotter tastes and sensibilities. Photographs bound in luxurious lacquer-covered albums became the most popular souvenirs globetrotters collected to memorialize their visits to Japan.*[11]

Nothing could be more central to a foreign visitor's stay in Japan than the purchase of photographs. Photographs not only memorialized a tourist's visit; they also created, as one late-nineteenth-century guidebook author put it, an "index of what is worth seeing in Japan."[12] In true globetrotter form, Gardner returned to Boston with nearly two hundred individual photographs from commercial photography studios and other vendors purchased during her visit to Japan (Plate 39). Although she does not seem to have collected any of the lacquer-covered albums popular among her contemporary travelers, her photo-souvenir purchases place her squarely within tourist conventions of the time. Gardner's decision to create—during the course of her travels—her own illustrated albums, pasting in her own selection of loose photographs together with other ephemera she collected, rather than purchasing the more common lacquer-covered photograph albums, strongly suggests her desire to construct her own narrative.[13]

PHOTOGRAPHY IN MEIJI JAPAN

By the time the Gardners arrived in Japan in 1883, photography was a firmly established industry, catering to the demands of a seemingly ever-growing tourist market (Plate 41). The commercial market for photographs was pioneered by Western photographer entrepreneurs in the 1860s and 1870s, including Felice Beato (1832–1909), Raimund von Stillfried (1839–1911), and Adolfo Farsari (1841–1898), among others. Beato established one of the first commercial studios in Japan when he arrived in 1863, riding his wave of success as a war photographer (in the Crimea, for example) and commercial photographer in Asia (including India and China).[14] He established his studio in Yokohama, which would become the first port of entry for throngs of Western tourists sailing to Japan from San Francisco just a few years later. "He learned that viable markets were to be found wherever Westerners confronted Eastern civilizations and cultures … [and] acquired a sense of what Westerners found attractive about Eastern culture."[15] He and his students and rivals created such a successful and dynamic industry that Japanese souvenir photography generally came to be known as *Yokohama shashin* (Yokohama photographs).

Although Beato was indeed a pioneer in his understanding of how to create a winning combination of subject matter and aesthetic presentation that appealed to Western tourists, he was far from unique in his ability to sell vast amounts of product. Following Beato's lead, various entrepreneurial photographers established studios and curio shops throughout the treaty ports and other destinations accessible to Westerners. These businesses hawked both album-bound and loose, individual photographs, both uncolored and hand-tinted. By the 1840s, very soon after the introduction of the daguerreotype in 1839, European photographers first introduced the practice of coloring photographs.[16] Beato may or may not have been the first photographer in Japan to bring colored photographs to market, but he is undeniably largely responsible for popularizing them. The practice soared to new heights in terms of aesthetic and technical achievement, as well as commercial success in the Japanese market. Beato's style, which took full advantage of the delicate qualities of natural, water-soluble pigments, was at its best ethereal and subtle, contrasting with typical heavier-looking European treatments made using oil paints. Colorists were probably recruited from the woodblock print industry, which was then in decline as photography

Fig. 6 Tokyo, June 1883 from Isabella Stewart Gardner's *Travel Album: Japan, Volume I*, page 29

Tokio - Tuesday June 26. Shopped all da
& dined with W. S. B.

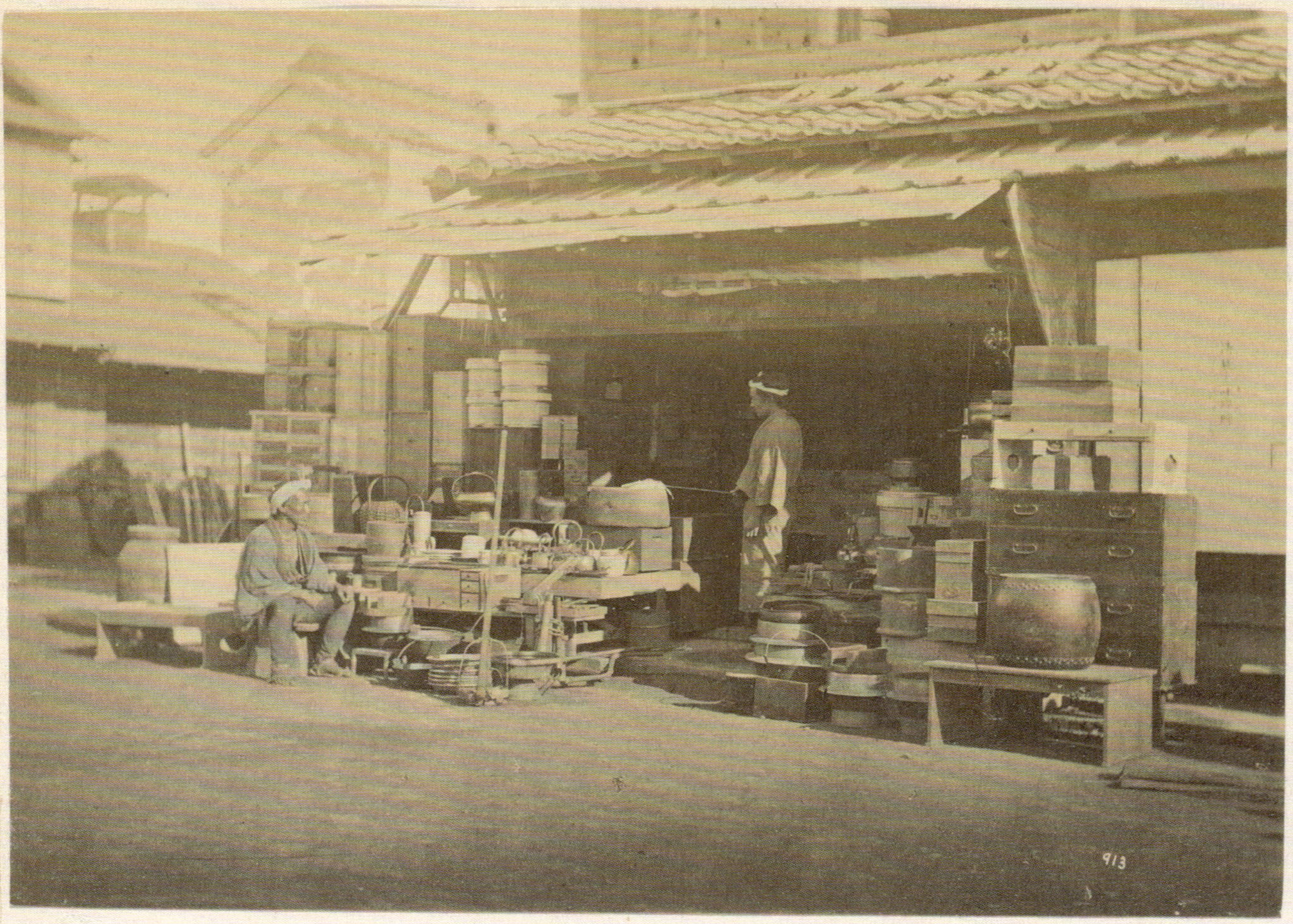

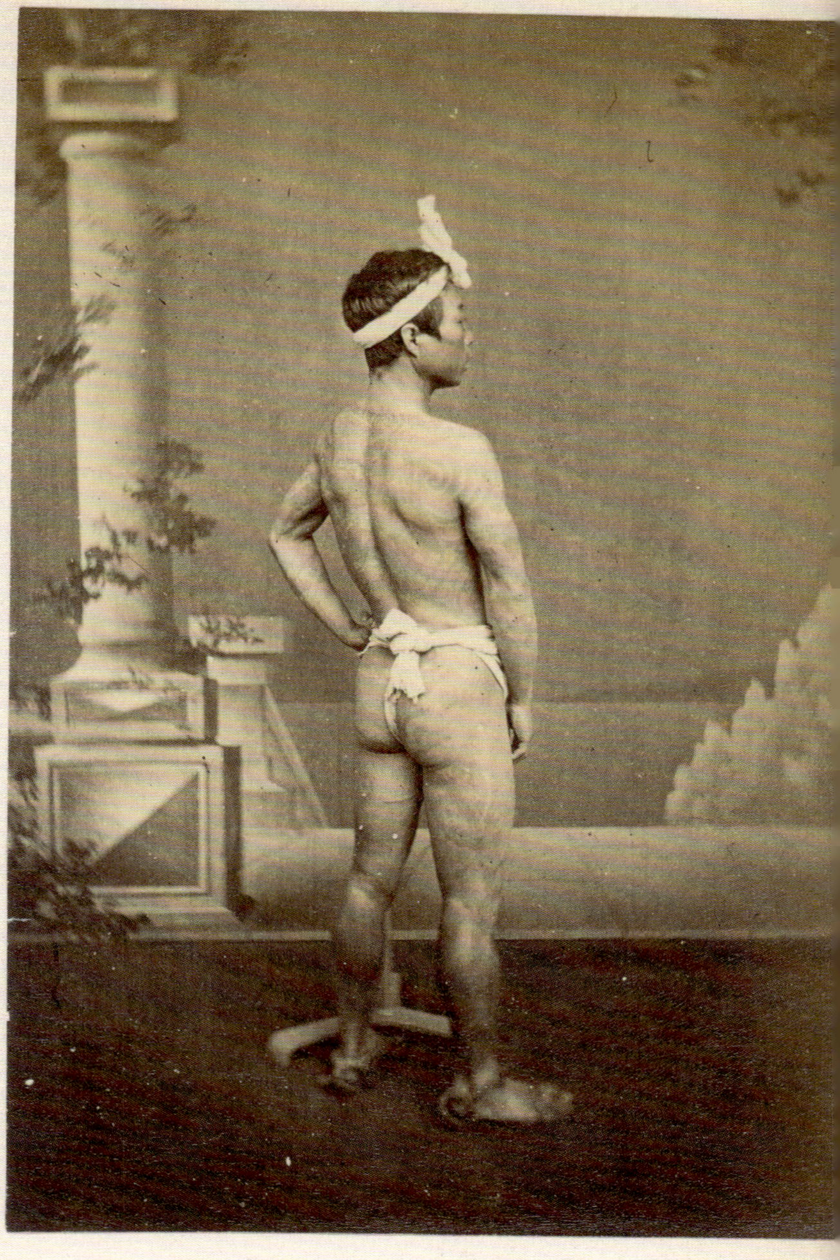

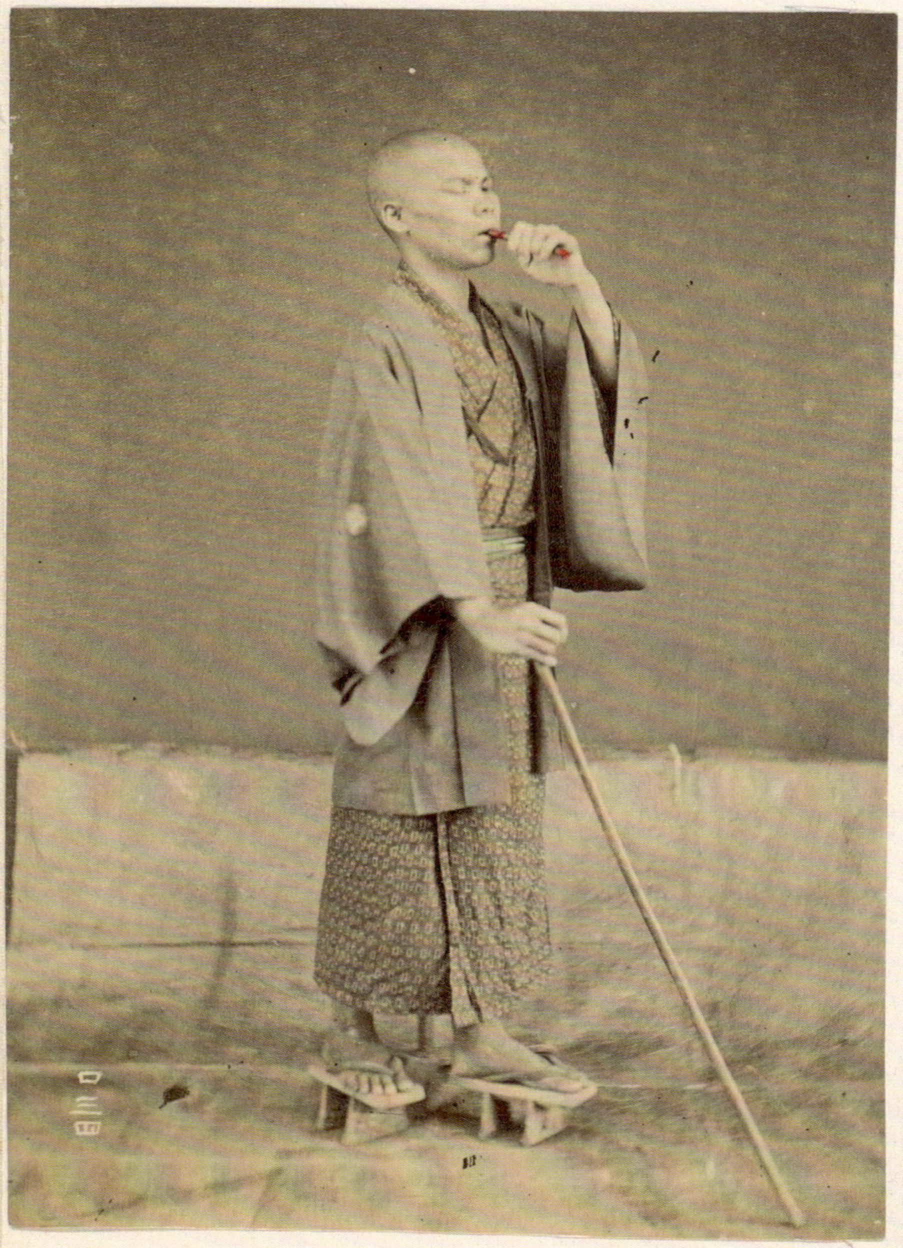

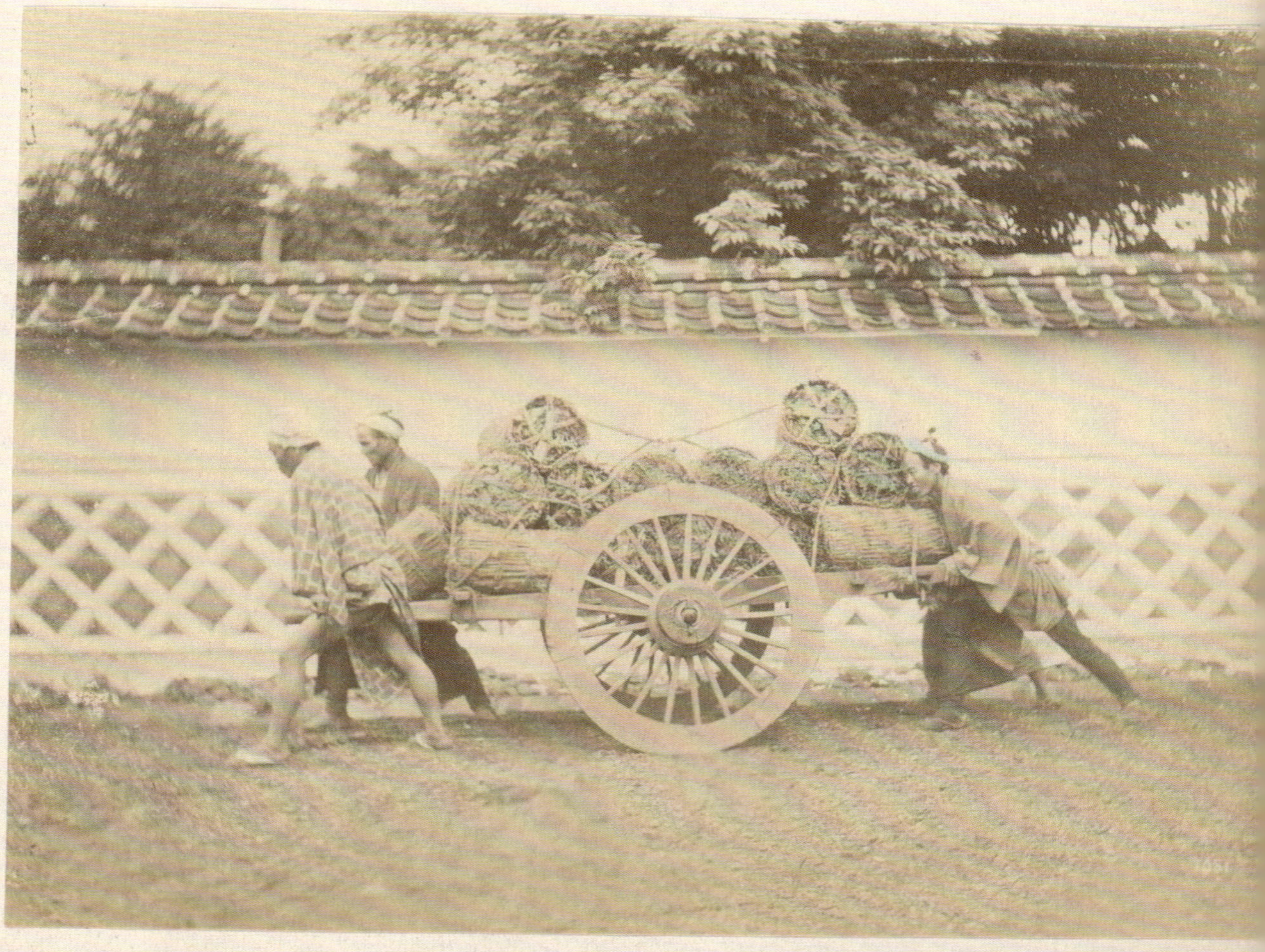

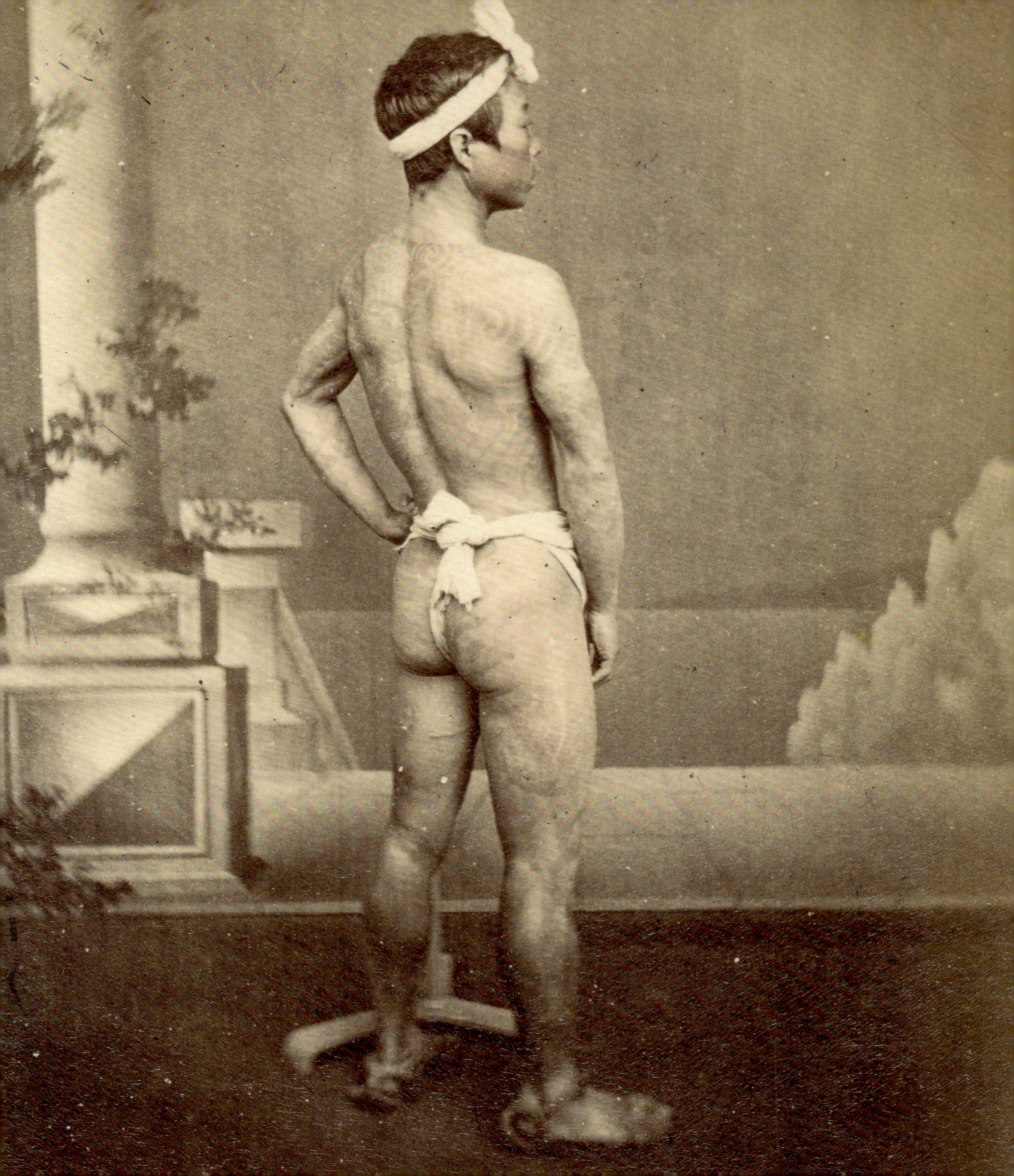

surged in popularity. This meant that there was a ready supply of highly skilled artisans who could transfer skills they had learned in the woodblock print trade to create an aesthetic on the surface of photographs that consumers highly desired.

The themes of *Yokohama shashin* reflected a complicated blend of Western fantasy and a desire for the "exotic" that appropriated longstanding Japanese visual tropes from the *meisho* tradition. This process turned them into stereotyped images for Western consumption, some of which are still powerful in today's visual field. Subjects such as temples and shrines, cherry blossoms, Mt. Fuji, tattooed men, geisha, and samurai dominated tourist albums in Gardner's time, and are likely just as familiar to today's readers as symbols of Japan and Japanese culture as they were in the late nineteenth century. These same subjects can be found in the long history of Japanese visual culture, especially in woodblock prints, which had started to penetrate Western aesthetic fields as well. Japonisme, or the craze for "things Japanese," particularly art and design, can be seen in the artistic and other cultural output among Western modern artists in fields as diverse as painting, literature, and theater. Many Impressionist and Post-Impressionist painters, for example, collected Japanese prints and were significantly impacted by their experience of Japanese art in both compositional and stylistic ways.[17] The photographs Westerners purchased in Japan, and indeed the lacquerware, ceramics, textiles, and other souvenirs they took home with them, have their reference points in the historical *meisho* tradition in Japanese tourism.[18]

When Western photographers reproduced images of traditionally famous Japanese sites, the change in context and media from woodblock print to photograph amplified the stereotyping effect of the tourist gaze rather than merely transliterating print to photograph. The truth-value of the photograph, a medium that convincingly replicated reality for many viewers, seemed to provide visual evidence to Western viewers that what they were seeing represented facts of Japanese life. Many tourist photographs directly appropriated perspective, composition, color, and other design elements from popular woodblock prints. Images of famous temples and shrines, actor portraits, landscape views, and other established scenes made their way into tourist albums via mass-produced photographs (figs. 4 and 5). Of course, in some ways, these were accurate, in the sense that Mt. Fuji or a famous temple captured by Beato's camera was indeed a true element of the Japanese cultural and environmental landscape. However, photographs of Japanese people and cultural activities were created for Western consumption rather than with a documentary ethos, as contemporary readers would likely understand it today. For example, tourist photographs of geisha (highly trained, highly skilled entertainers) were typically (but not always) in fact images of sex workers, who were often the only women who could be convinced to model for the camera (Plate 42). Even more alarming, the "geisha" image came to stand for Japanese (and perhaps Asian) women in general, and arguably continues to negatively impact Asian American women and Asian women in the diaspora even today. Japanese men did not escape the photographic tourist gaze, either, with depictions of samurai in full armor proving very popular, even well after the warrior class had been disenfranchised

Fig. 7 Tattooed Man, June 1883 from Isabella Stewart Gardner's *Travel Album: Japan, Volume I*, page 29

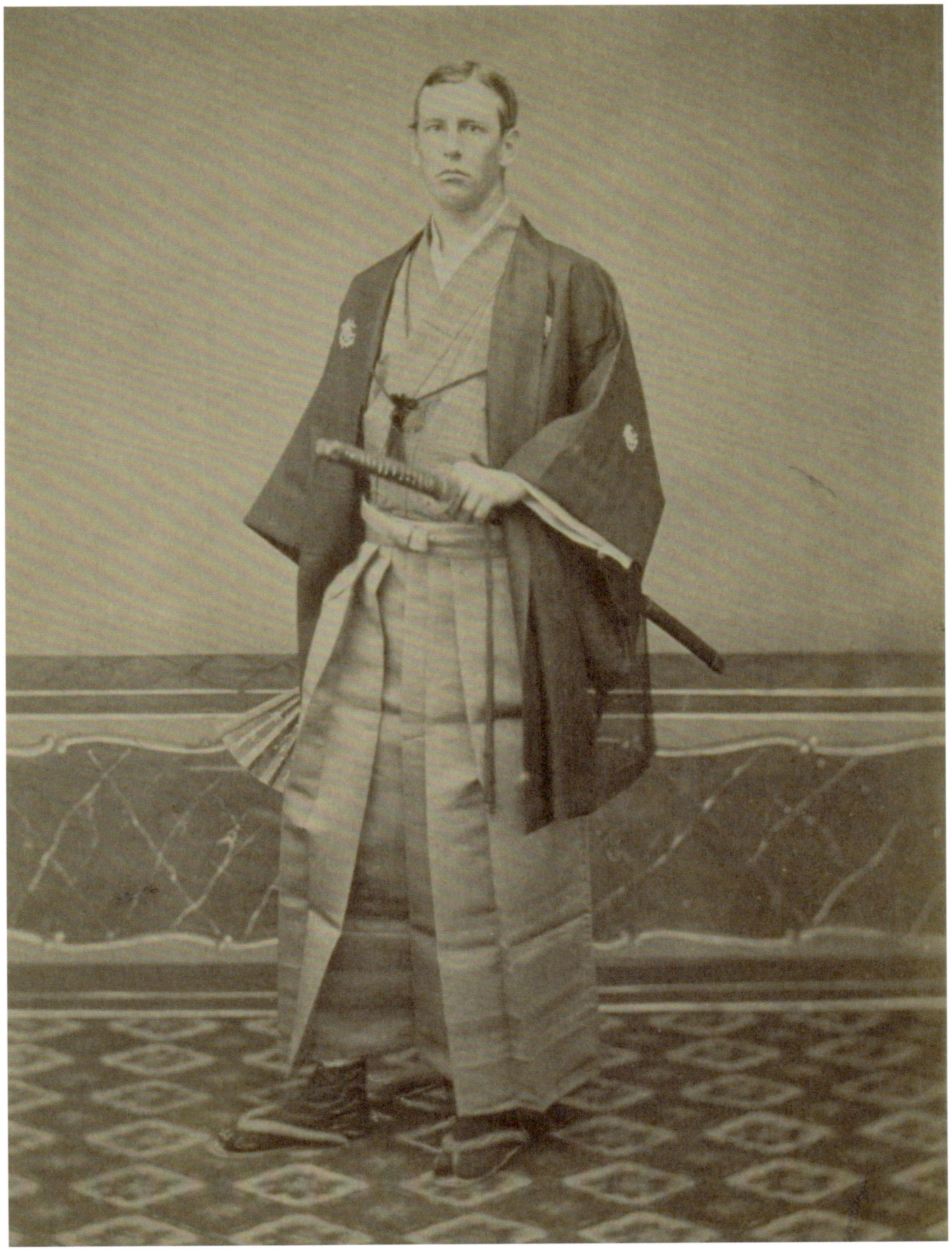

Fig. 8 Raimund von Stillfried (Austrian, 1839–1911), Charles Longfellow in formal Japanese attire, 1872. Albumen print, 14 x 9.5 cm. National Parks Service, Longfellow House, Cambridge, Massachusetts

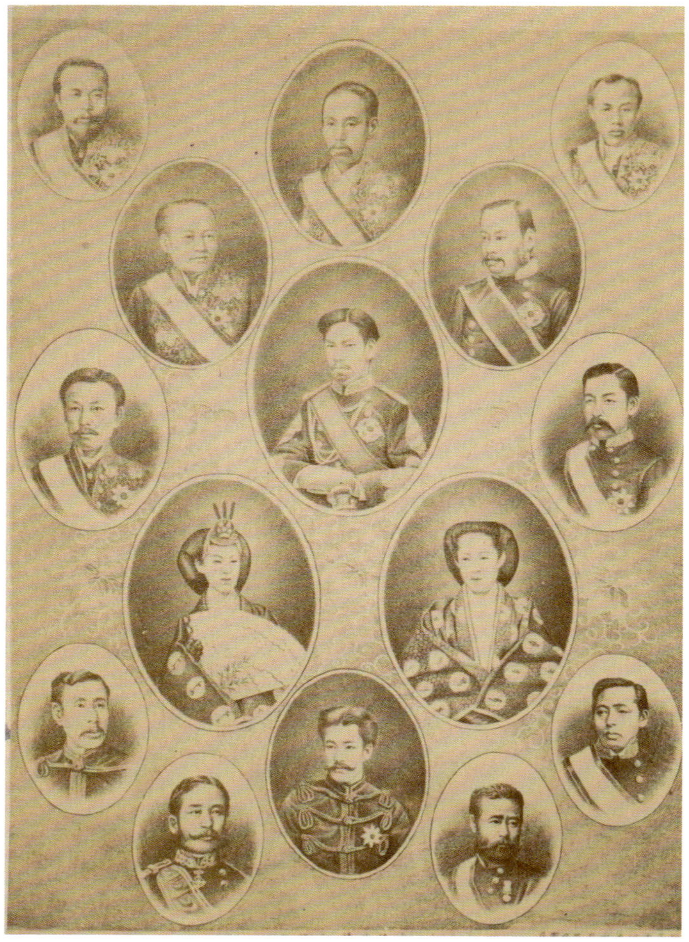

Fig. 9 Emperor Meiji and the Imperial Family, June 1883 from Isabella Stewart Gardner's *Travel Album: Japan, Volume I*, page 4

in the modern political order and thus impossible for a Western tourist to encounter in real life. Tattooed laborers were another popular subject, even after the Japanese government outlawed the practice for Japanese citizens in 1872. A close examination of different photographic prints of the same image of tattooed subjects can reveal that colorists in fact created many of the "tattoos" ostensibly pictured in souvenir photographs (figs. 6 and 7).[19]

The industry thrived thanks to the lucrative marketplace catering to foreigners that had been developing in government-designated treaty ports. The ports, starting with Shimoda and Hakodate in 1854, and Kanagawa and Nagasaki in 1859, were opened by the Tokugawa government initially to the United States and later to other Western nations. However, successful photography studios established by Western photographers in the 1860s and later were ultimately acquired or eclipsed by their Japanese protégés (and in turn, *their* protégés) and a few Chinese photographers, when photographers such as Kusakabe Kimbei (1841–1934), Tamamura Kozaburo (1856–1923?), and Shimo'oka Renjo (1823–1914) began to dominate the market. Indeed, by the end of the nineteenth century, there were essentially no foreigner-run studios remaining in Japan. Nonetheless, the changeover from Western to Japanese producers had virtually no effect on the nature of the images offered for sale. This was owing in part to the durability of photographs: the same tourist images could be reproduced and sold again and again over several decades. For example, when a photographer closed his studio, he could easily sell his stock of images to another photographer, who in turn sold the prints to tourists. And because the photographs conformed to stereotypical Western notions of "traditional" Japanese culture, they remained saleable even as the gap between representation and reality grew increasingly wide. Japanese photographers, for their part, were no less willing to peddle these photographs than their predecessors. Pioneering Japanese photographer Shimo'oka Renjo might have regretted late in his life that his production and sales of tourist photography contributed to a stereotyped image of Japan,[20] but he, along with numerous other Japanese photographers, willingly supplied the voracious Western demand for images of Japan.

The Gardners and other visitors to Japan in the 1880s traveled the well-trodden path to photography studios and curio shops in pursuit of photo-souvenirs (Japan Plate 1). Some travelers commissioned bespoke photographs during their visit, including portraits, sometimes in Japanese dress—an act Christine Guth has characterized as kind of cultural cross-dressing—in addition to the "ready-mades" available off the shelf.[21] Interestingly, Gardner appears to have commissioned very few photographs during this trip and none of herself or her friends in Japan, in contrast to fellow wealthy globetrotters, for whom this was very common practice.[22] Indeed, Gardner does not often insert her own image in any of her travel albums, including the albums of Asia; she appears in "peasant costume" in her first travel album to Scandinavia and the Russian Empire, and in some snapshots in her Spain album. Her personal candids went into her guestbooks and snapshot albums, which she illustrated with photographs and ephemera.[23] Bostonians well known to Gardner, such as William Sturgis Bigelow (1850–1926) (with whom the Gardners socialized during their stay in Japan) and Charles Longfellow (1844–1893), commissioned photographs in Japan, including by the famed Austrian photographer Baron Raimund von Stillfried, mentioned above (fig. 8).[24] What the Gardners and other globetrotters might not have realized was that, in addition to the robust trade in souvenir photographs, a domestic market aimed at Japanese consumers was also booming. There was an entire parallel industry—often worked by the same photographers who sold *Yokohama shashin* to tourists—operating throughout the country, catering to local customers. Like Western tourists, Japanese clients, too, flocked to photo studios. Instead of purchasing stereotyped views of Japan, however, they, like their Western counterparts at home in Europe and North America—not to mention all over the world, including in colonized places—were sitting for portraits and having cartes de visite made for friends and business associates.

Sebastian Dobson has discussed a fascinating small guide to having one's picture taken, written by Tokyo photographer Matsuzaki Shinji for the Japanese public. Published in 1886, a few years after the Gardners' trip to Japan, Matsuzaki's *Dos and Don'ts for the Photographic Customer* covers numerous aspects of the sitting process, including facial expression (a slight smile is best), hairstyles (the appearance of comb lines left in the hair should be ignored), make-up (heavy application is discouraged), clothing (with separate advice for Western and Japanese attire), intoxication (wait until you've sobered up), and the photographic appearance of skin color:

> *It is a habit among the Japanese that many of them request [the photographer] to make their face appear whiter. However, this is an exceedingly unreasonable request.... Even Westerners of the white-skinned race are not fond of an excessively white complexion.*[25] *When the picture becomes too white and the shading naturally fades, it becomes just like a woodblock print, and it goes without saying that the innate beauty as well as the restrained elegance of the sitter are lost as a result.*[26]

Perhaps the most amusing—and self-serving rule—in Matsuzaki's pamphlet is the admonition that customers should not quibble over price. He argues that prices are coming down anyway, and that haggling will only result in the photographer's working in a "desultory mood." And worst of all, "Asking to have the price of such an important photograph reduced is tantamount to demeaning one's self."[27] Truly, Matsuzaki wants his readers to believe, it is for their own good not to try to lower the photographer's fee.

The domestic market in photography comprised not only individual consumers but also, and crucially, the Japanese government, which itself was actively involved in the development of the industry, which it believed would aid the country's nation-building project and colonial ambitions. The Meiji government (1868–1912) considered photography to be "one of the seven items in the 'standard paraphernalia of civilization and enlightenment' (*bunmei kaika nanatsu dōgu*)—along with newspapers, the postal system, gaslights, steam engines, exhibitions, and dirigible balloons."[28] The government posited that taking advantage of the new medium of photography would help it modernize the country, document the nation's landscapes and cultural treasures, and even stake colonial claims. This drive to photographically record people, places, and things was so pervasive, in fact, that Kinoshita Naoyuki has characterized the photography of Meiji Japan overall as driven by a desire for the "documentary."[29]

Kinoshita's characterization draws our attention to numerous government photography projects, beginning in the early 1870s, which were conceived and carried out in the heat of this desire to "objectively" document—record and catalogue—objects of material culture, landscapes, stages of industrialization, architecture, notable people, and other things considered important. A pattern begins to emerge in this period wherein government officials increasingly commission photo-documentation projects, including those with very high budgets, and do so within the context of their attempts to become a modern, colonial nation. In 1871, for example, the government began a massive project to photo-document the newly colonized northern island of Hokkaido, then and now home to the indigenous Ainu people, just a few years after its incorporation into the Japanese state in 1869. The government established its prefectural capital at Hakodate, which was one of the first two treaty ports opened in 1854 (along with Shimoda), and began constructing schools, hospitals, shrines, and temples, and at the same time encouraged settlement from Japan. The capital was moved to Sapporo in 1871, and Tamoto Kenzō (1832–1912) was hired to produce photographs of the newly minted seat of government, documenting colonization efforts generally, including recording the building of various kinds of infrastructure, making ethnographic images of Ainu subjects, and photographing newly emerging towns and villages. The photographs illustrated written reports and were sent to Tokyo in May 1872, where they were publicly exhibited.[30]

The year 1872 was also when the government first commissioned photographic portraits of Emperor Meiji. Despite government efforts to control and limit their distribution, as well as the production of subsequent portraits, the images were reproduced in various formats and widely circulated.[31] One such image was collected by Gardner, which she pasted on an initial page of her first Japan album (fig. 9).[32] It is a small pamphlet with a photorealistic image of Emperor Meiji surrounded by members of the imperial family and notable political figures. Although photographs (including hand-tinted versions) of the imperial couple were widely available and appear in many tourist albums of the era, Gardner does not appear to have purchased these; instead, she included this card, which was produced for domestic Japanese consumption. Below the card, in her own hand, Gardner reproduced a translation of a classical Japanese poem by Kakinomoto no Hitomaru (active 689–700), as well as an excerpt of a text by the British scholar of Japan Basil Hall Chamberlain, which characterizes the Japanese people as a nation that believes in its divine origins, and thus the divinity of the emperor and by extension, all Japanese people.[33] The inclusion of this particular text and image suggests that Gardner had at least some understanding of imperial power in Japan, and even if not explicitly, the entanglement of photography within that power structure.

Also in 1872, the government contracted the prominent, Yokohama-based commercial photographer Raimund von Stillfried to continue the photo-documentation. In a period of less than two months, von Stillfried, along with three Japanese assistants who were government staff photographers, carried on with the work started the previous year by Tamoto, photographing some identical views and buildings to demonstrate the progress of the area's development over time. He also produced panoramic images, Ainu studies, and town scenes, some of which were displayed in Japan's pavilion in the 1873 International Exposition in Vienna. By 1882, when the agency tasked with "developing" Hokkaido for the Japanese state was decommissioned, several thousands of photographs of the colonization of Hokkaido had been produced. Some images were eventually recycled in aid of governmental efforts to recruit potential (Japanese) settlers to the island.[34]

Following the photography project in Hokkaido, increasing numbers of governmental photographic projects were undertaken throughout the Meiji era, including a five-month-long survey of old shrines, temples, and their sacred objects conducted in western Japan in 1872, which was the first such survey to be undertaken in the country. The government commissioned Yokoyama Matsusaburō (1838–1884) to carry out the photo-documentation and he produced hundreds of photographs,

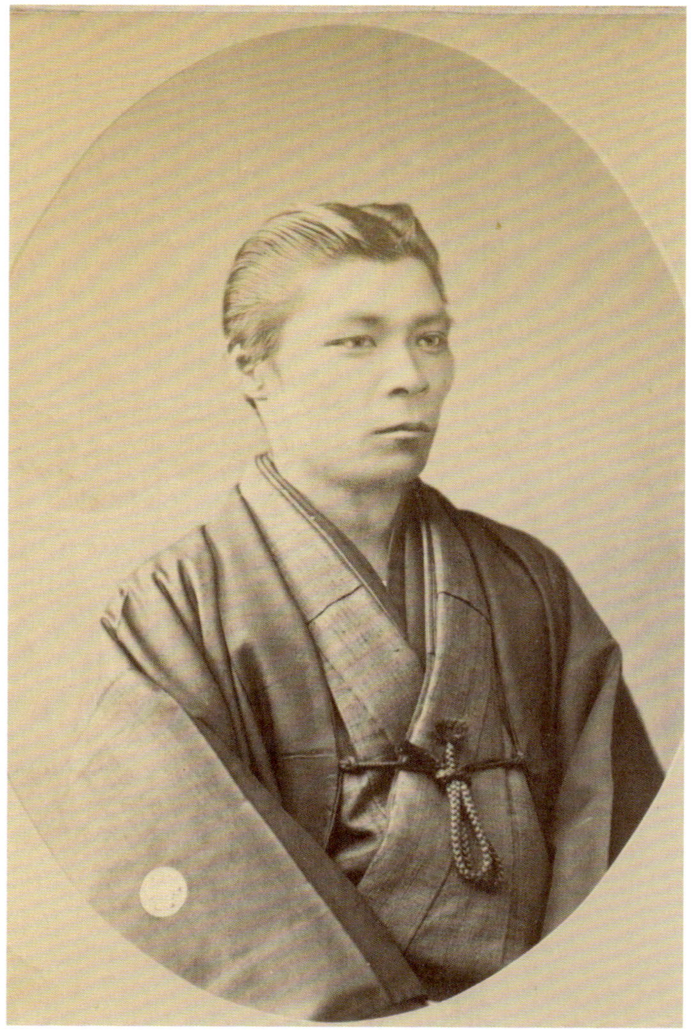

Fig. 10 "Captain Swain," perhaps the captain of the *Tokio Maru*, August 1883 from Isabella Stewart Gardner's *Travel Album: China and Japan, Volume II*, page 58

including images of many sites that would become globetrotter favorites, such as the Todaiji Temple, visited by the Gardners in 1883. Another example is a project implemented by the Department of the Treasury to produce photographs, which were converted into illustrations for the national currency. A subsequent survey of cultural treasures occurred in 1888.

ISABELLA'S "CAPTAIN SWAIN"

It is perhaps fitting, then, that Gardner closes out the second of her two Japan travel albums with pages that include both souvenir photographs and a single carte de visite (fig. 10). The carte de visite is a domestic photograph identical in format to ones she received from friends and acquaintances in Boston and on her travels in Europe.[35] The others are typical *Yokohama shashin*, identical in format to the ones with which she populated the rest of her album. The carte de visite is an image of a certain "Captain Swain," perhaps the captain of the *Tokio Maru*, on which she and Jack sailed from Hakodate via Nagasaki onward to Shanghai on 31 August 1883, presumably given to her by the gallant swain himself. It is not surprising that a woman of Gardner's position would have received this memento from a ship's captain; the more salient points are that the photograph itself is a material connection to the domestic Japanese world of photographic practice and sociality, to which the image connects Gardner, however fleetingly and tangentially. This object can be seen as a connecting fiber between the globetrotter world and a lived Japanese reality, in contrast to the Western fantasy embodied in tourist photographs.

An element of Japanese reality is materialized not merely in the physical object that is the carte-de-visite photograph of Captain Swain, but also in what is pictured in the image. At first glance it is an unremarkable portrait of a well-presented Japanese man: a handsome, young, well-dressed individual who understands how to present himself to the camera. The comb lines in the man's expertly coiffed hair, held firmly in place possibly by camelia oil—the very sort of markers/lines Matsuzaki Shinji claimed were acceptable to Westerners, even if undesirable to many Japanese women customers—are clearly visible here.[36]

Gardner was surely playing with naming—or rather nicknaming—this man "swain," which, in the parlance of the time, referred to a beau or male admirer. But Gardner's application also indicates a misnaming, a colonialist attitude of refusing to learn local/indigenous names for either people or places, instead renaming them with more familiar, European (in this case, English) names, sometimes with derision, sometimes with a condescending affection. We can't know with any certainty what Gardner's relationship was with the man pictured in the photograph, for neither her album nor her diaries reveal any intimate details that might serve as clues here. If we look closely at the carte-de-visite portrait of "Captain Swain" in the context of Gardner's collecting practice, its initially endearing appearance yields clues to the complicated location of 1880s Japan as an object of the tourist gaze. At the same time, it is a Japan that insists on itself as an active subject, building its own colonial ambitions, eventually creating its own othering gaze with no less catastrophic results for its victims than those subject to Western colonialisms. However playful the moniker might appear to be, we must simultaneously understand that it could just as easily have been a sign of Gardner's lack of respect for the man's personhood and culture, an index of a colonialist, tourist gaze.

This essay has greatly benefitted from the critical reading of Casey Riley and Diana Seave Greenwald, to whom I offer my sincerest thanks. I am also grateful to Doris Lin for her indispensable research assistance.

1. *Journeys East: Isabella Stewart Gardner and Asia*, which was on view at the Gardner Museum from February to May 2009, exhibited two albums with photographs of Japan, v.1.a.4.6 and v.1.a.4.7. The first album, v.1.a.4.6, was previously displayed from April to August 2003 in *The Making of the Museum: Isabella Stewart Gardner as Collector.* Readers may be interested to see digital versions of the Japan travel albums here: Volume I https://www.gardnermuseum.org/experience/collection/16089 and Volume II https://www.gardnermuseum.org/experience/collection/17285.
2. Chong and Murai 2009.
3. Chong 2009a, 20.
4. Ibid.
5. I have discussed elsewhere the insidious connections between nineteenth-century tourism and the "crypto-colonialism" of Japan in relation to early photography. Please see Odo 2015, especially 78–86.
6. After Japan, the Gardners visited China, Vietnam, Cambodia, Indonesia, Singapore, Malaysia, Burma (Myanmar), and India.
7. Thanks to Diana Seave Greenwald for bringing the exceptions to my attention. The Gardners purchased two screens: *Spring* (P29w7.a) and *Autumn* (P29w7.b). For more, see https://www.gardnermuseum.org/experience/collection/15435 and https://www.gardnermuseum.org/experience/collection/15436.
8. McKenna 2020.
9. Ishimori 1989, 192.
10. Ivy 1995, 32; Graburn 1983, 50.
11. Hockley 2010b.
12. Sladen 1891, quoted in Hockley 2010b.
13. Lacquer-covered albums were often sold with photographs preinserted, although customers could also customize them with their own selections.
14. Felice's brother, Antonio, was also a successful commercial photographer. Riley 2015a notes that Gardner purchased several photographs by Antonio for her Egypt diary.
15. Hockley 2010a.
16. Louis-Jacques-Mandé Daguerre introduced his daguerreotype process of photography to the world in 1839, via the French government, which shared it freely with the public, thus making it the first widely available photographic process. Although there were several contemporaneous (and competing) technologies, the daguerreotype is often credited as being the first practicable photographic process, which provided a method for permanently fixing an image on a light-sensitive silver-coated metallic plate.
17. See Emery 2020 for an illuminating discussion of Japonisme and the little-discussed intersections with the movement and racism, anti-Semitism, and misogyny.
18. Guth 2004, 36.
19. Odo 2015, 40–42.
20. Iwasaki 1988, 25.
21. See Guth 2004, 121–41 for a fascinating discussion of the many implications of foreigners dressing in Japanese clothing in late-nineteenth-century Japan.
22. Chong 2009b, 104.
23. Riley 2020a.

24. Many of these collections were donated or sold to Boston-area institutions such as the Museum of Fine Arts, Boston, the Peabody Museum of Archaeology and Ethnology, Harvard University, and the Longfellow House–Washington's Headquarters National Historic Site.
25. Dobson links Matsuzaki's admonitions about the appearance of skin color in photographs to contemporaneous debates about color and identity in Asia.
26. Matsuzaki 1886, translated by Dobson 2017, 213.
27. Ibid., 215.
28. Dower 1980, 9.
29. Kinoshita 1999, 5.
30. Ibid., 59.
31. Alice Tseng has persuasively argued that despite the primacy of photographs of Emperor Meiji, it was in fact the visual representation of his successor, Emperor Taisho (1879–1926; r. 1912–26), that established the aesthetics of the modern Japanese monarchy. See Tseng 2020 for a full discussion of this fascinating issue. Luke Gartlan has found that Raimund von Stillfried illicitly produced the first photographs of Meiji, which prompted the government to allow the emperor to sit for his official portrait by Uchida Kuichi (1844–1875). Gartlan 2016, 83–99.
32. It follows the first portrait in the album, also printed on a pamphlet. The portrait is a drawing of Prime Minister Iwakura, who died during the Gardners' sojourn in July 1883.
33. Chamberlain 1880, 88.
34. Dower 1980, 9.
35. My thinking about this carte de visite has benefitted greatly from a discussion with Makeda Best, especially with regard to the implications of misnaming subjects.
36. It is intriguing to consider the possibility that the sitter read Matsuzaki's guide and followed his advice.

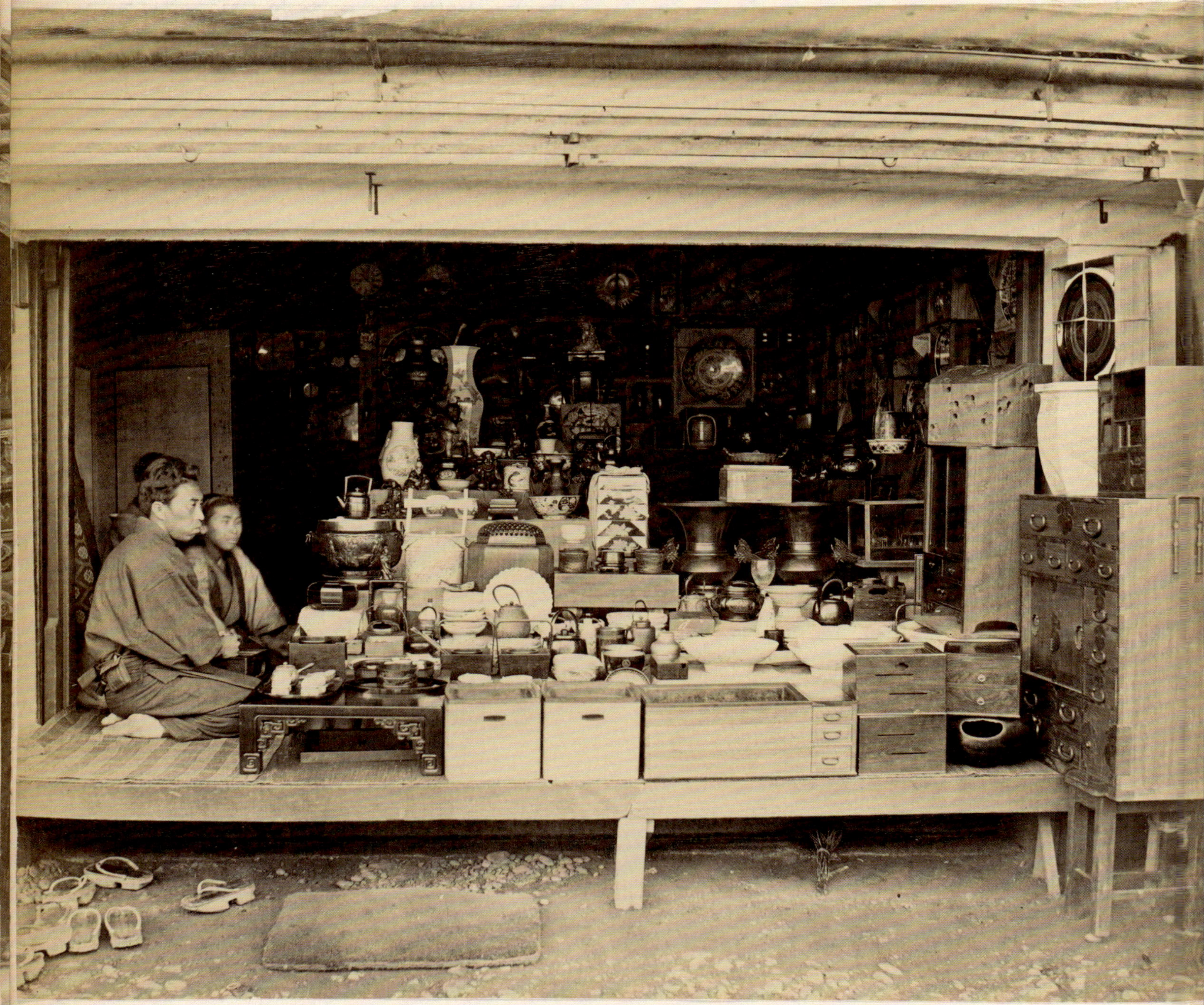

Tokio. Tuesday June 26.

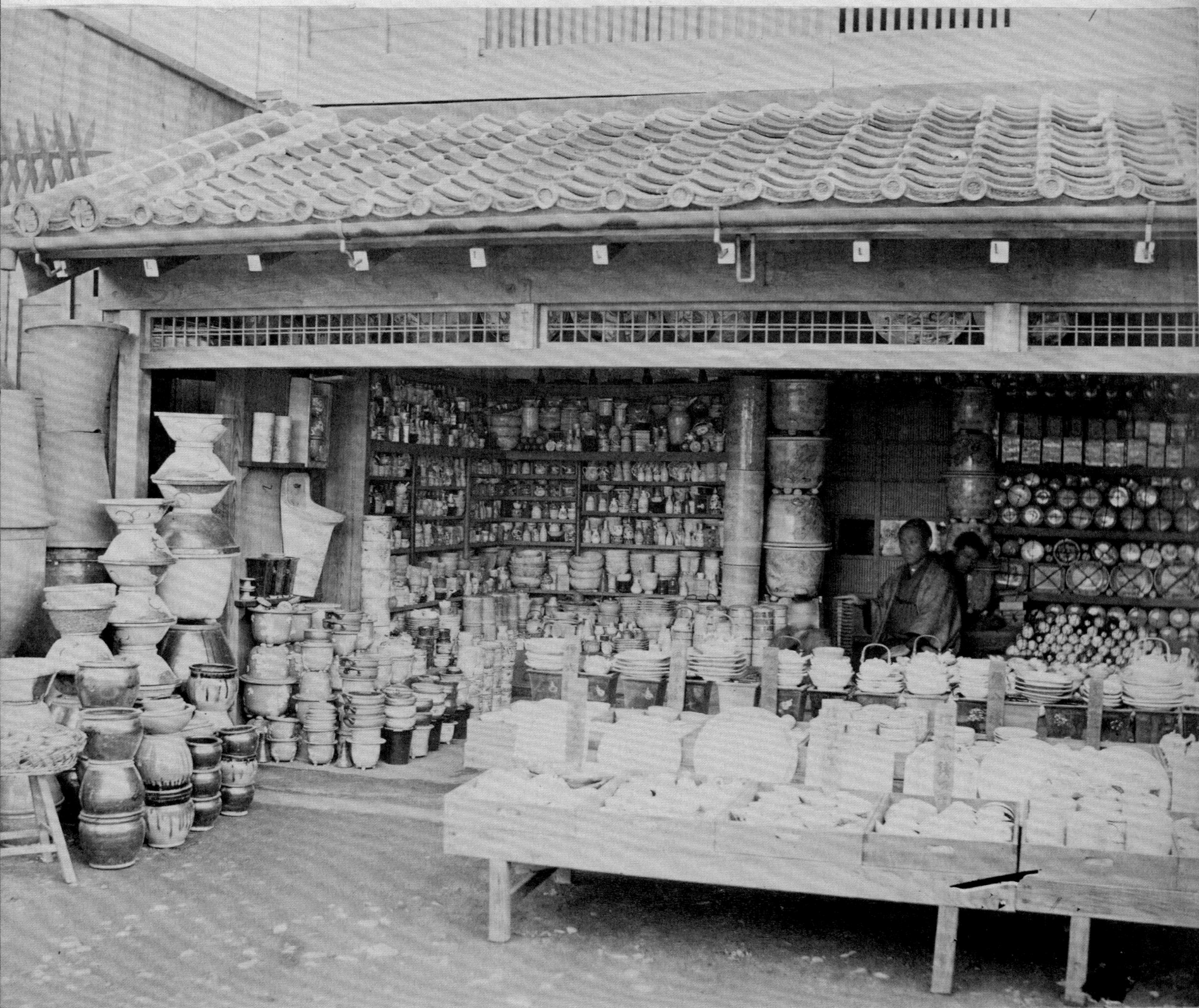

Tokio June 26.

Yokohama.

Friday June 29.

Took a drive through the Country beyond race course to Buddhist Temple, then to Fishing village.

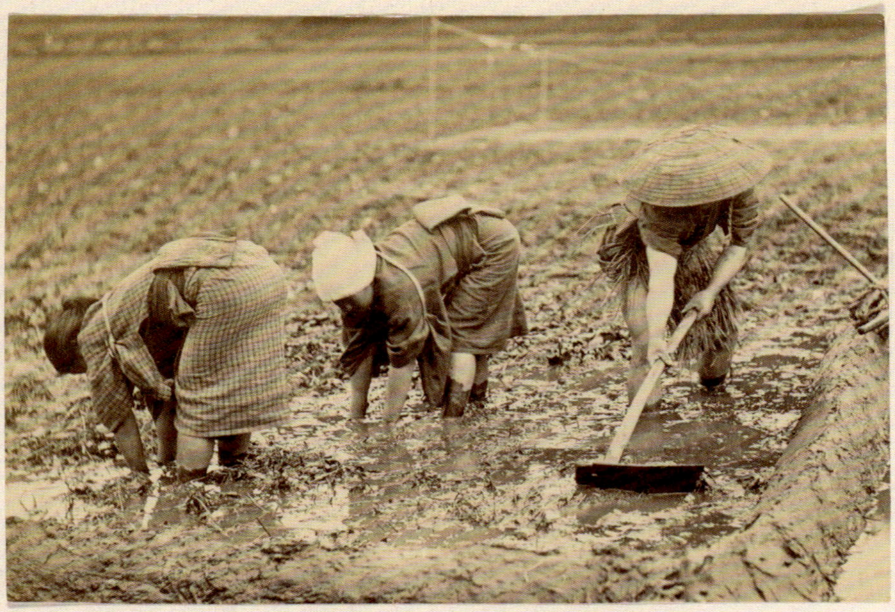

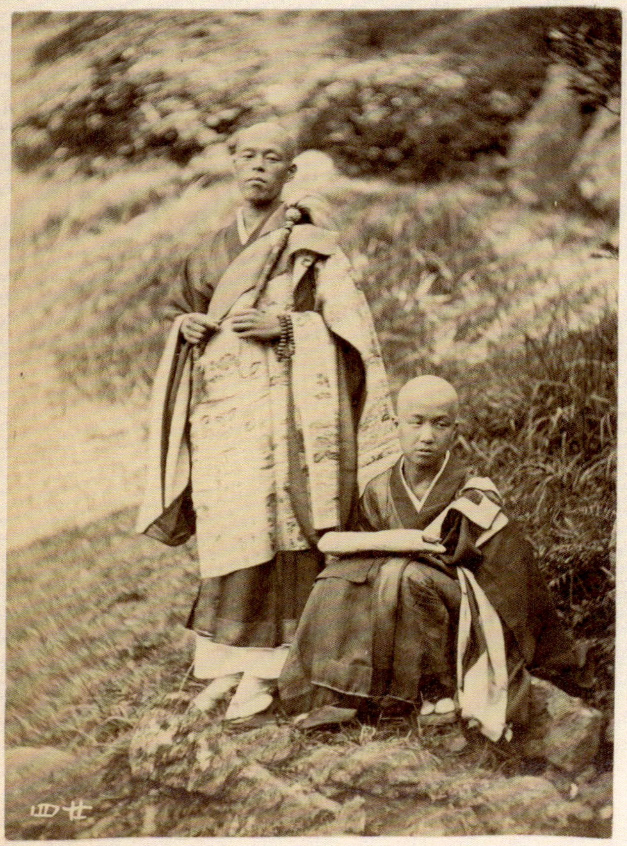

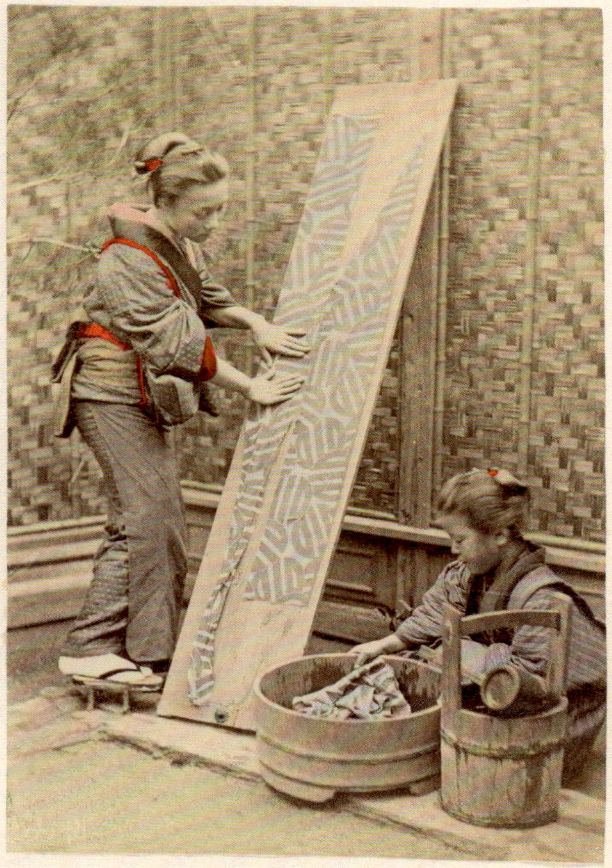

Saw religious procession in Tokio Thursday June 28

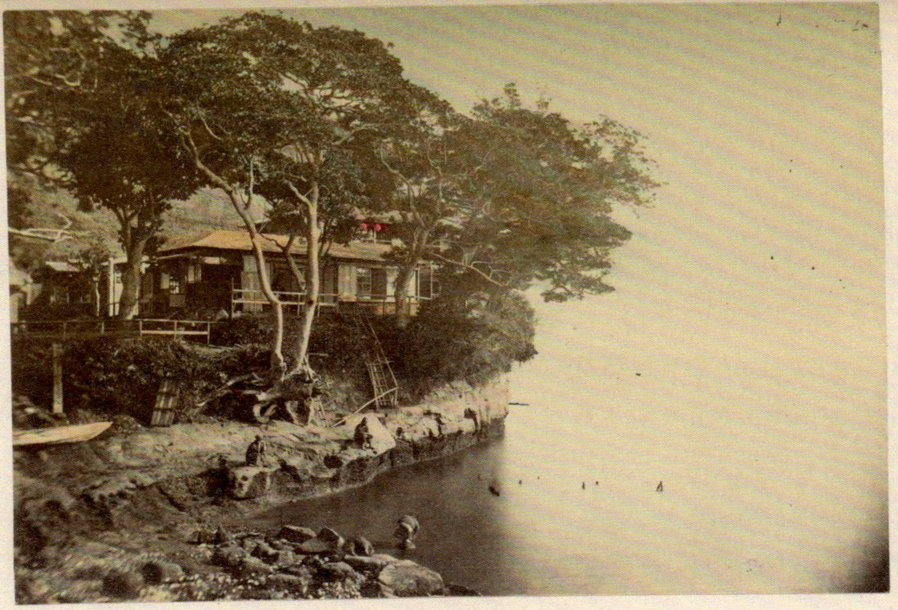

Yokohama
Friday June 29.

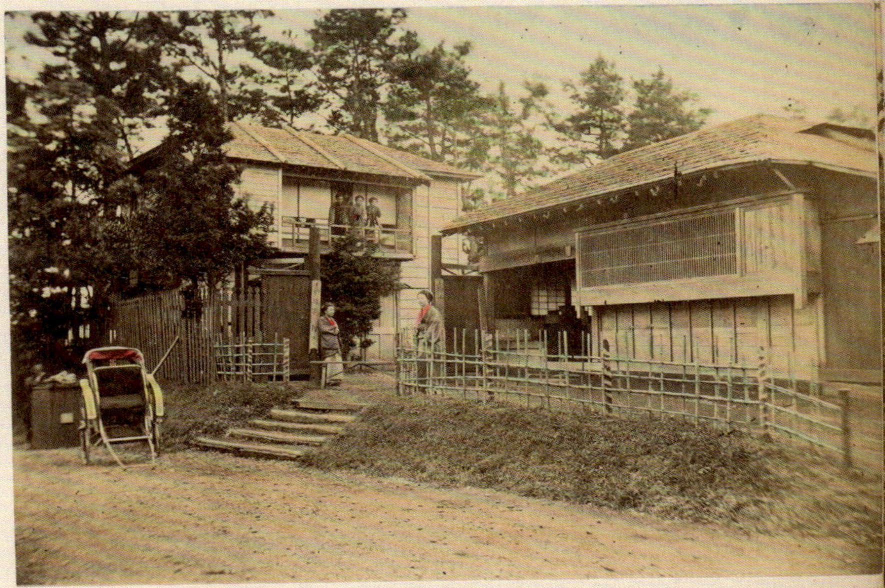

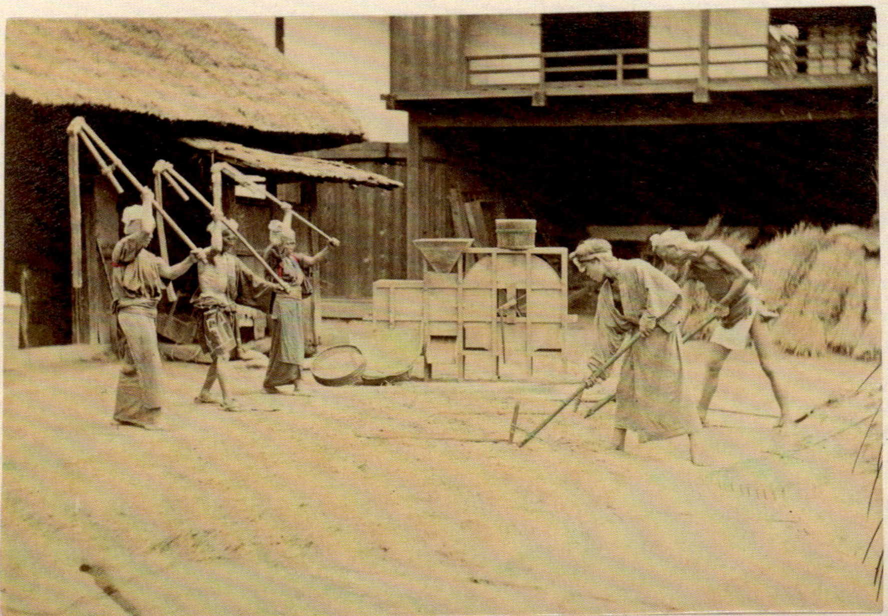

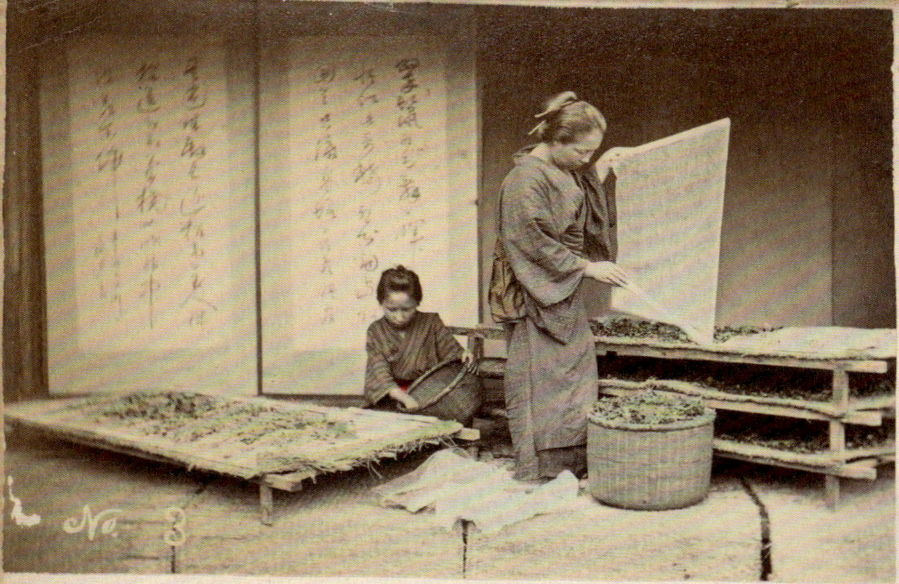

Left Nikkō at 8 a.m. Thursday July 26; arrived at Yokohama abt. 5 P.M. Friday July 27–
On the road everyone busy with silk worms.

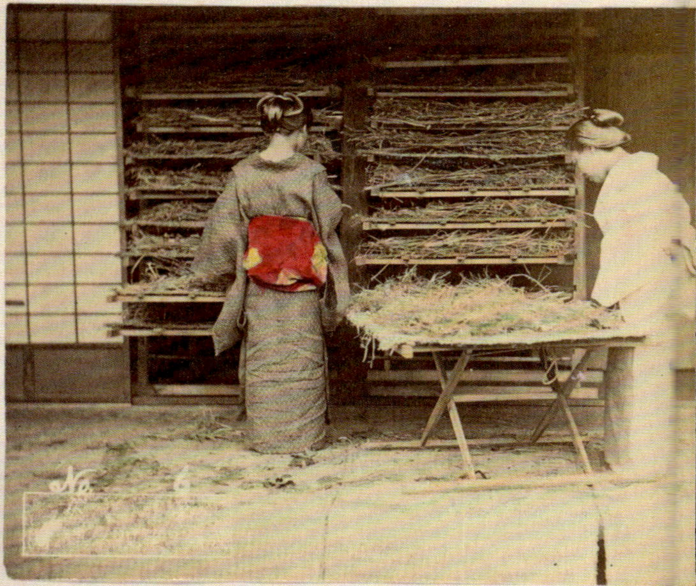

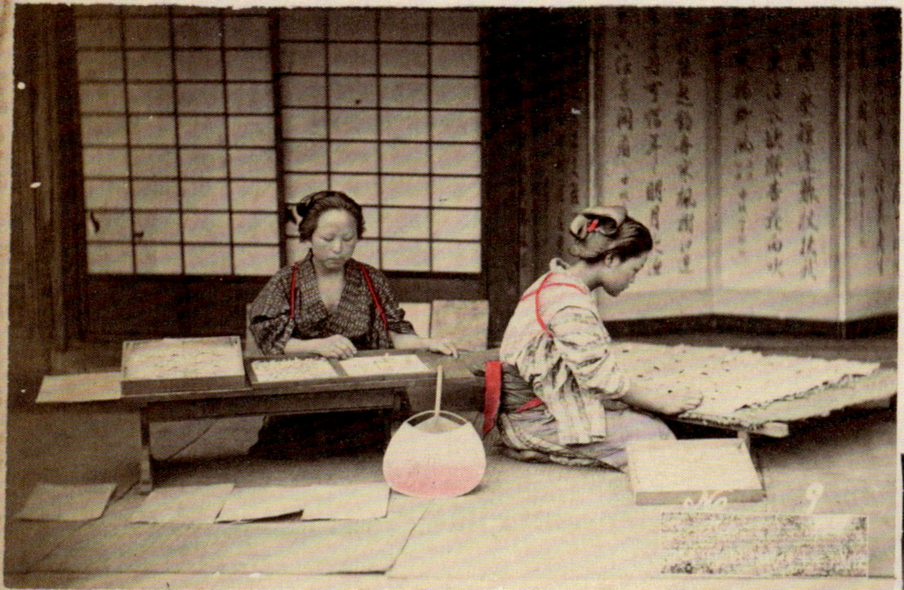

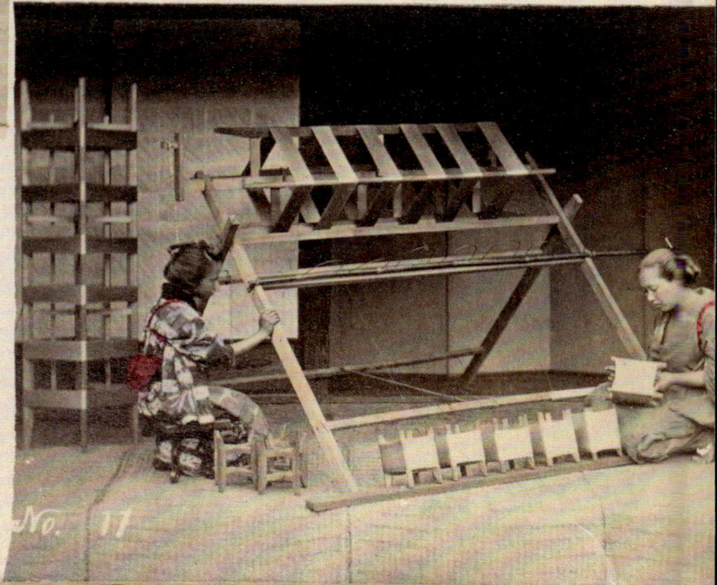

Left Yokohama Thursday August 2. for Miyanoshita in carriage with 4 horses. Mrs. Ackley + Mr. Mignard. Started abt. 3 P.M. Got to Sammai Bashi abt. 11.30 P.M., where we took Cangos & went up by torchlight to Miyanoshita where we arrived at 2 a.m. – Hotel Fuji-ya. –

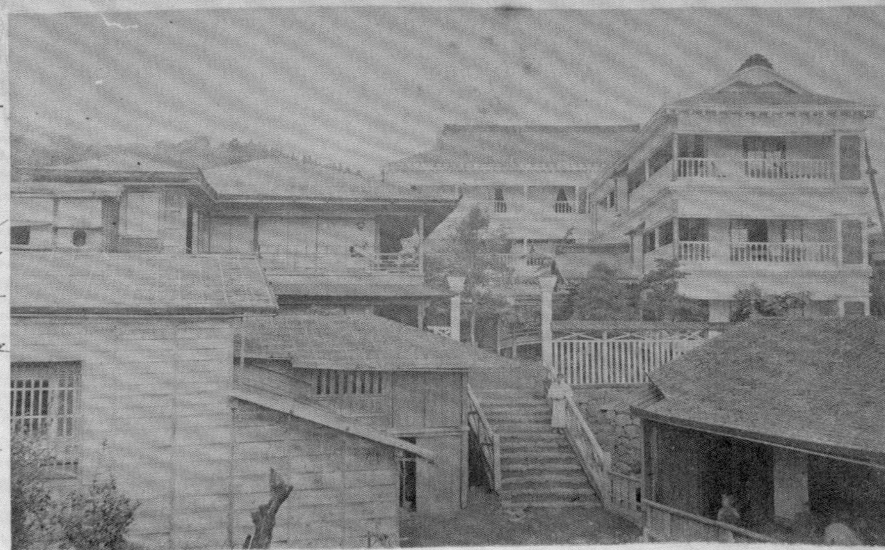

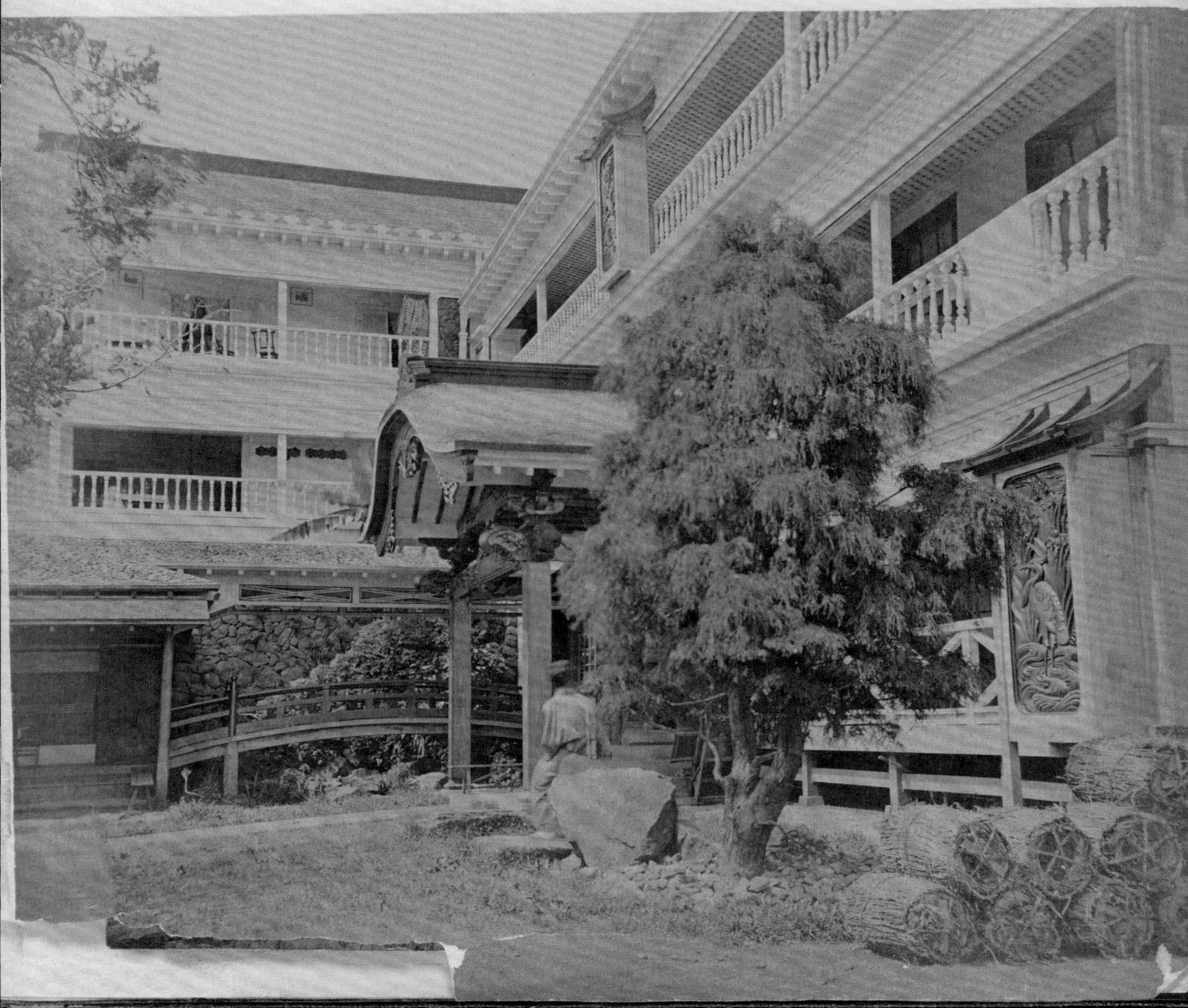

Kiōto.

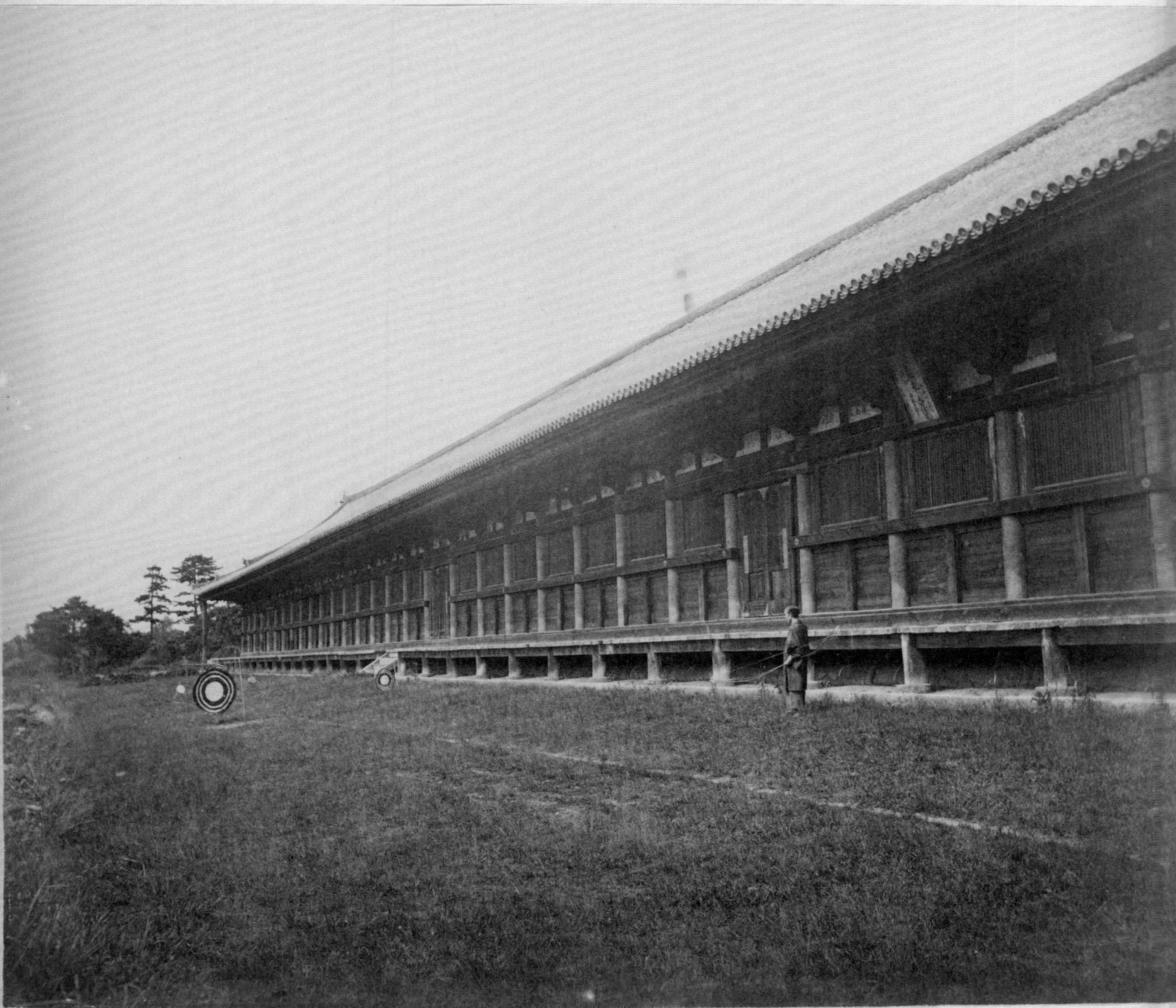

San-jiu-san-gen-Dō.
August 16.

Kioto –

The bell, nearly 14 ft. high, belonging to Dai Butsu

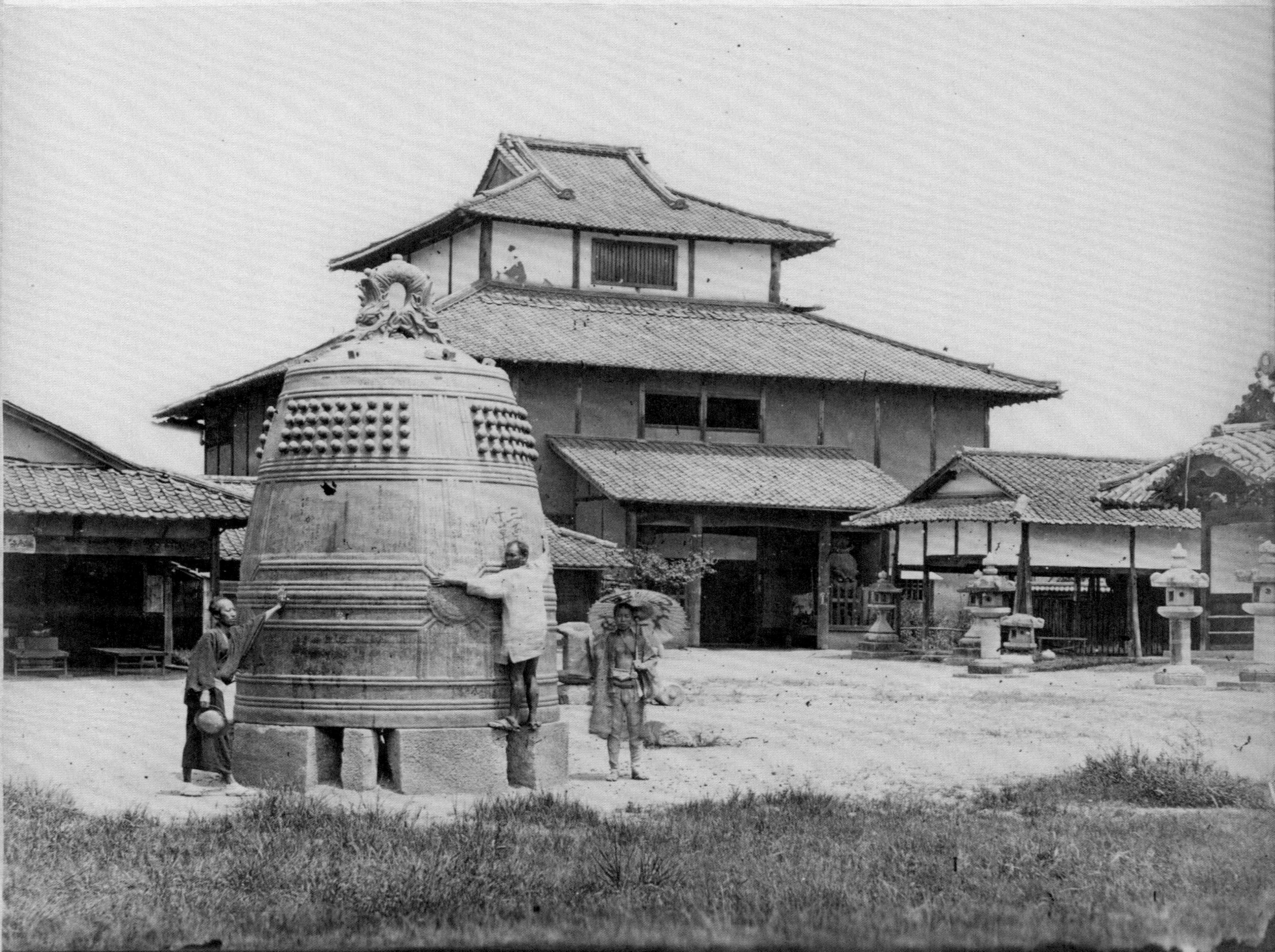

Kokka ankō – Chinese for Iye Yasu.
"May the state be peaceful & prosperous" –
"On the east it welcomes the bright moon, & on the west bids farewell to the setting sun" –

41

Went to Tea House in evg. Aug. 23 & saw 10 girls had dance & sing –

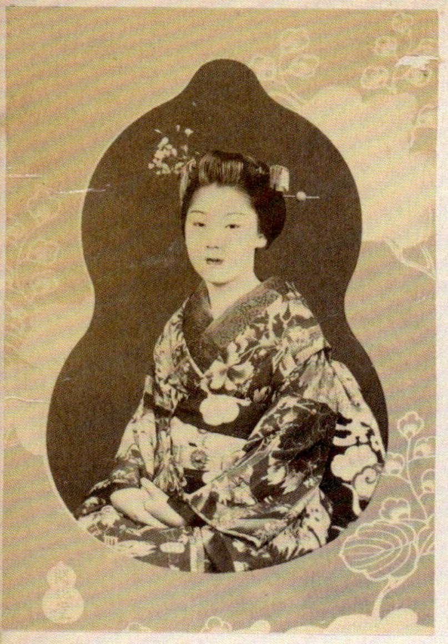

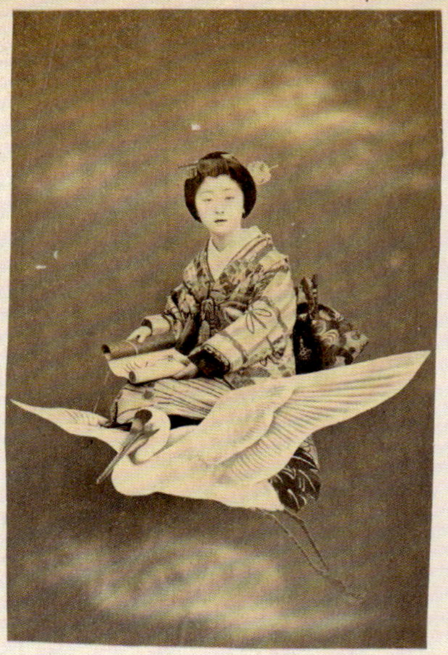

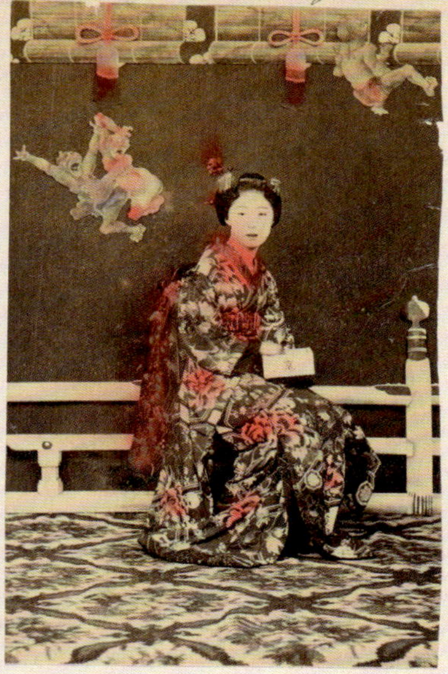

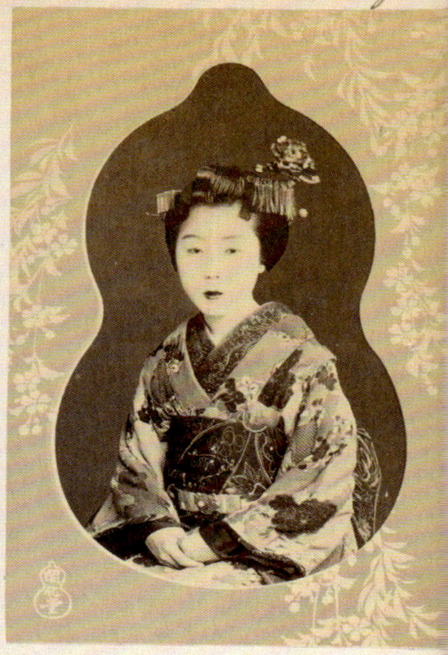

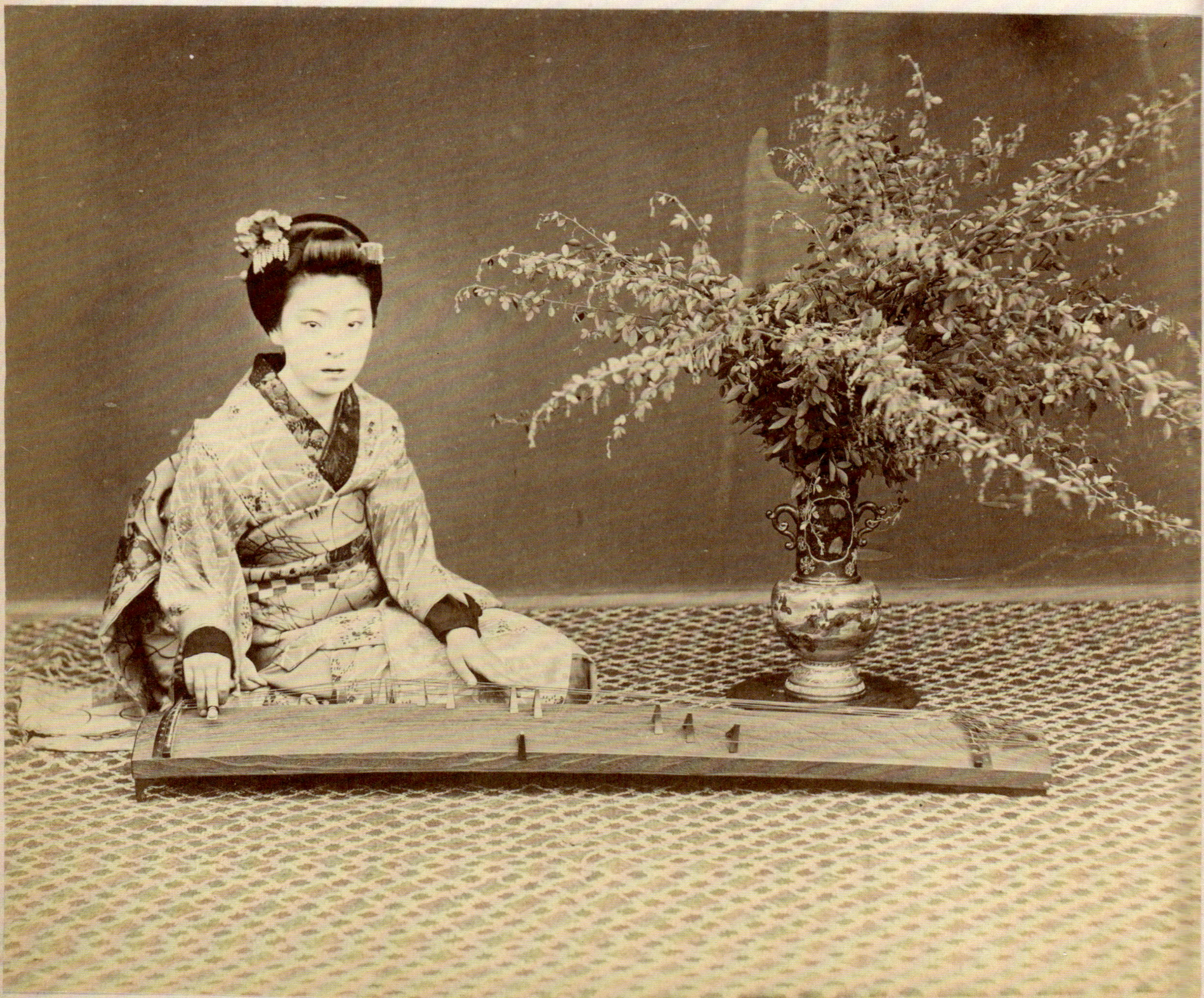

Kioto.

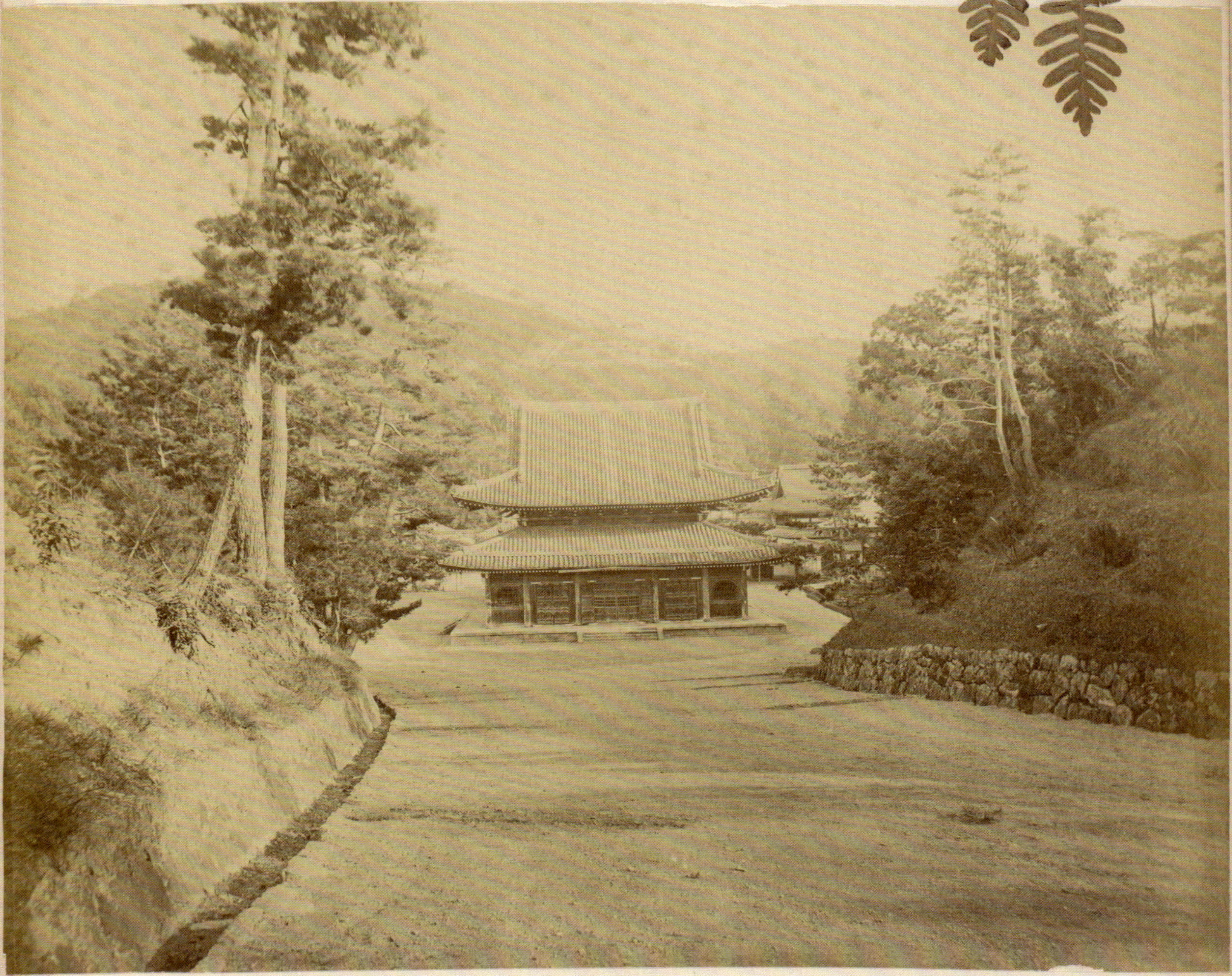

Friday August 24 –
Mikado's Tomb.

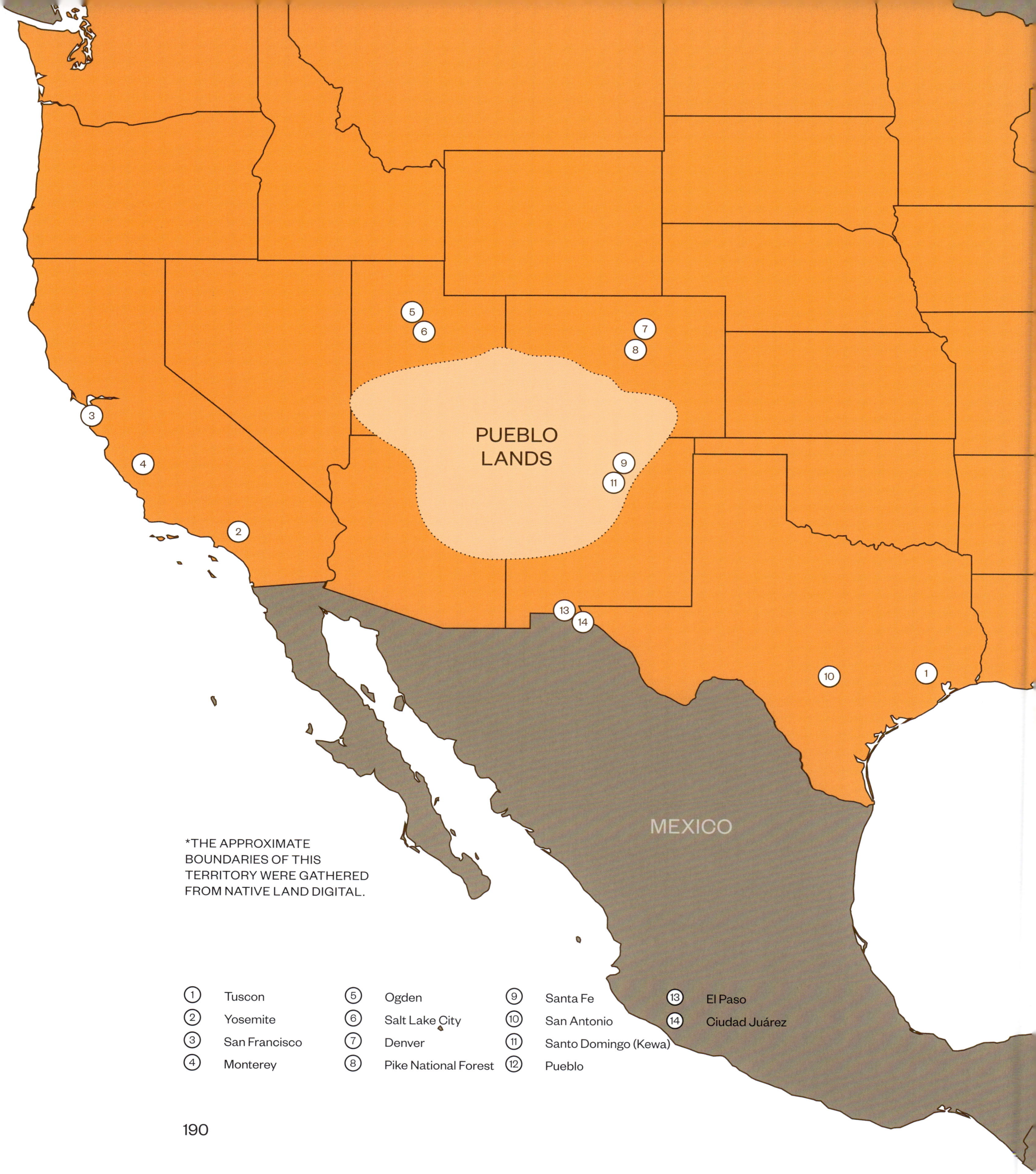
PUEBLO LANDS
MEXICO
*THE APPROXIMATE BOUNDARIES OF THIS TERRITORY WERE GATHERED FROM NATIVE LAND DIGITAL.
1 Tuscon
2 Yosemite
3 San Francisco
4 Monterey
5 Ogden
6 Salt Lake City
7 Denver
8 Pike National Forest
9 Santa Fe
10 San Antonio
11 Santo Domingo (Kewa)
12 Pueblo
13 El Paso
14 Ciudad Juárez

Pueblo

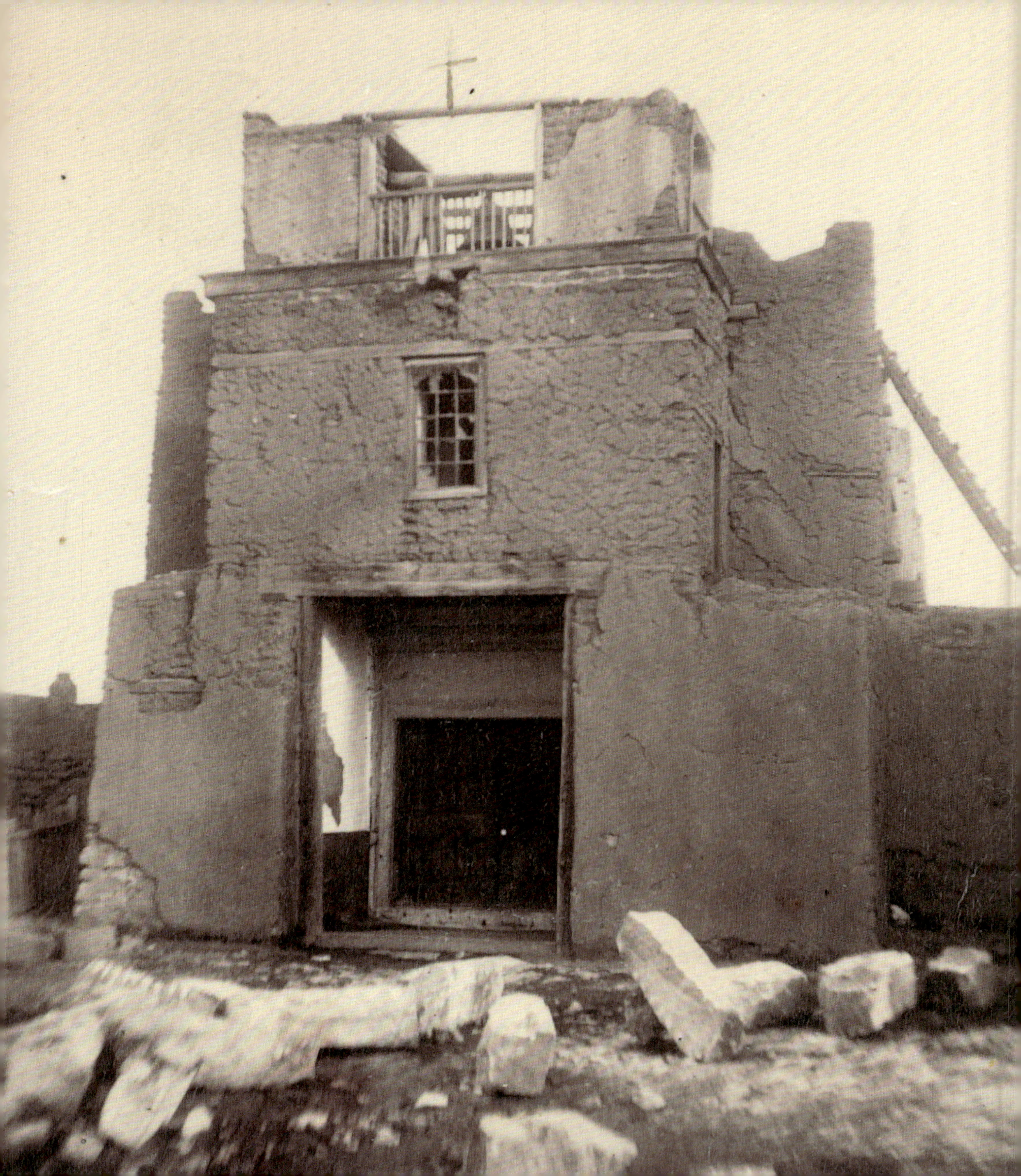

Southwest Invention

Jaclyn M. Roessel

The red sandstones, towering rock formations, rocky mountain ranges are immortal characters in the story of the Southwest. They live in the Wild West films starring John Wayne and even in more recent films like *Forrest Gump.* The backdrop of Monument Valley stars as a glowing red beacon of what is American—(or was perhaps) wild and free. The daughter of an educator, I grew up enjoying day trips to nearby national and state parks. My parents loved spending quality time in nature as a family. On one trip to Chaco Canyon, the ancestral home of the Puebloan people, a white tourist snapped a picture of my brother and me as we were observing one of the many kivas the ancestral architects built. Immediately, we went to stand alongside my parents. I confided in my dad what had happened, mentioning that they hadn't asked. My dad, a photojournalist himself, matter of factly said, "Next time, charge them money." Whether he was joking or not, I don't know but my brother and I quickly started charging for every picture taken of us. We left the park with over ten dollars in hand, a dollar for each picture taken of the two of us.

When the Santa Fe Railroad Company expanded into what is known today as New Mexico, it brought with it the promise of taming and developing the Wild West. The Fred Harvey Company built an empire on the backs of Indigenous, Black, and other people of color and created the commodification of our Indigenous culture and art. The company orchestrated jewelry vendors and entertainment at railroad stations featuring Native people in the Southwest. With these decisions the railroad not only facilitated transcontinental travel but also created a palatable Native art market that white settlers would pay for. It paved the way for tourists to consume adventures. The advent of the railroad made it possible for Isabella Stewart Gardner to travel into these homelands in 1881 (fig. 1). The completion of the railroad system ignited the invention of the Southwest with the support of white photographers like Edward Curtis (1868–1952) (fig. 2) and later Ansel Adams (1902–1984) (fig. 3). While the travel album that features Gardner's visit is limited in the breadth of photos featuring Indigenous people, what is shown is alarming yet not unique. We see images from Kewa Pueblo (formerly known as Santo Domingo Pueblo) on one page, juxtaposed by the image of an individual hanging from gallows, black hood frozen in time, on another (Plate 44).

Thinking critically about this travel album raises two questions that connect me to the subjects, questions I connected to even as a young girl in Chaco Canyon: (1) What does it mean to travel in someone else's homeland/community—that is,

Santa Fe, New Mexico, March 1881 from Isabella Stewart Gardner's
Travel Album: Northern and Eastern Europe, Russia, the United States, Mexico, and Cuba, page 19

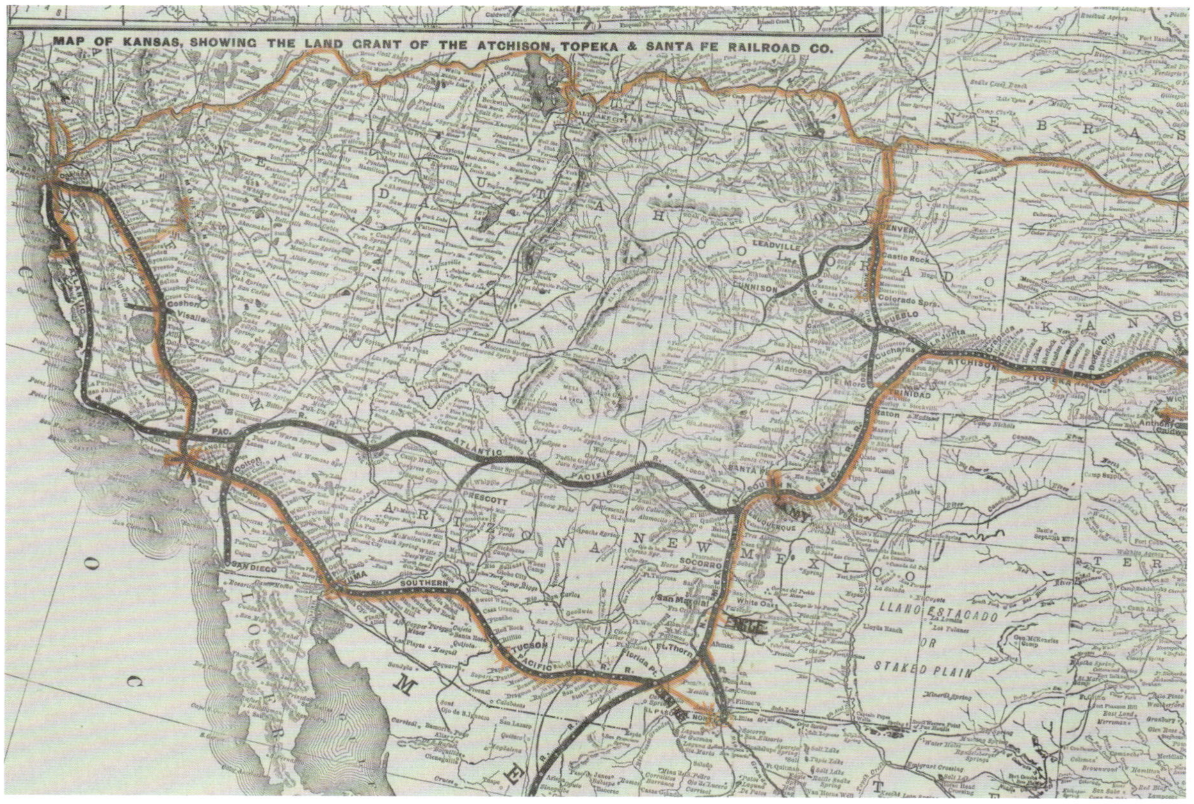

Fig. 1 Map of the Atchison Topeka Santa Fe Railroad showing the Gardners' route through the western United States and Mexico, enclosed in Isabella Stewart Gardner's *Travel Album: Northern and Eastern Europe, Russia, the United States, Mexico, and Cuba*, 3 March–6 May 1881

to be a visitor? (2) What does consent mean in the American West? Both give rise to answers that can be uncomfortable for many non-Indigenous constituencies. The creation of the railroad and the "taming" of the wild through its construction is woven in relationship to the understanding of the settler colonial project.

In the words of Cherokee scholar Adrienne Keene, "Everything in American society—the government, national language, laws, economy and 'American culture'—is all colonization."[1] The agenda of settler colonialism is to destroy the existing cultures and societies present in a place for the goal of creating a new society. Building on this understanding, it is important to note that settler colonialism is, in the words of Australian scholar Patrick Wolfe, a "structure not an event."[2] As long as the dispossession of Native land from Native stewardship continues, settler colonialism continues. What we can see clearly in Isabella Stewart Gardner's travel albums is a lack of inquiry. We see the arrangement of the pages and some script identifying place but nothing regarding interactions, names, or personal thoughts. The reality is that the photographic moments frozen in time illustrate community members who we can guess are grappling with another wave of settler disruption in their homes. There is no indication, as far as we can tell, that care was extended when these images were taken.

In Gardner's albums, we see photos of Pueblo people on top of their adobe homes, a postcard of a Pueblo woman carrying a large pottery jar atop her head (fig. 4)—images that created a monolithic view of the Southwest that carries

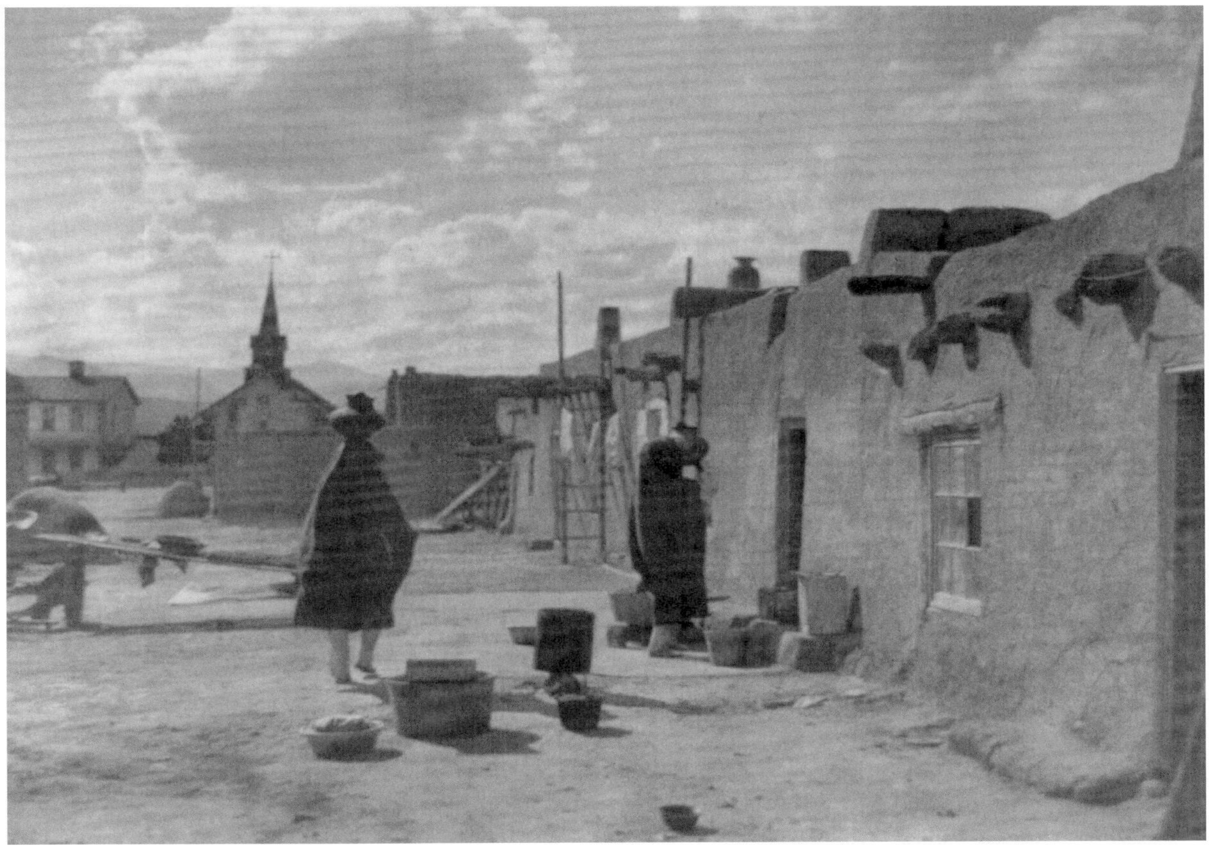

Fig. 2 Edward S. Curtis, (American, 1868–1952) *Street Scene*, San Juan, about 1927

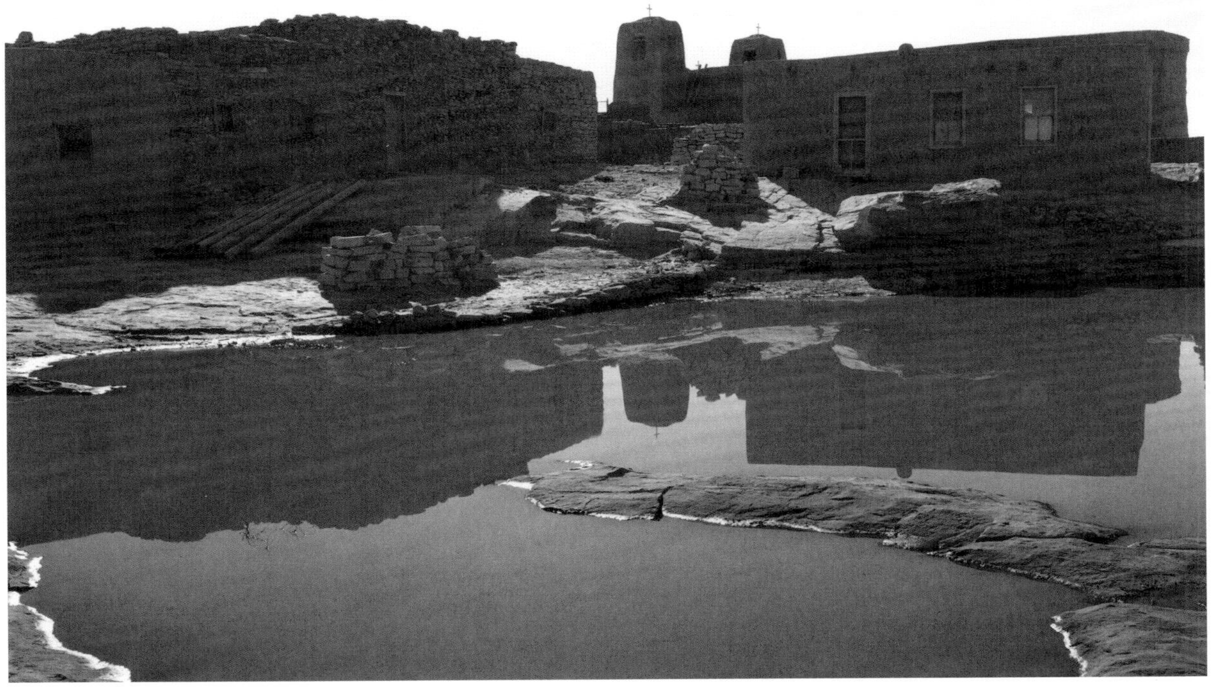

Fig. 3 Ansel Adams (American, 1902–1984), *Pool, Acoma Pueblo, New Mexico*, about 1942. Gelatin silver print, 1981. 36.8 × 25.9 cm. Gift of Virginia B. Adams National Gallery of Art, Washington

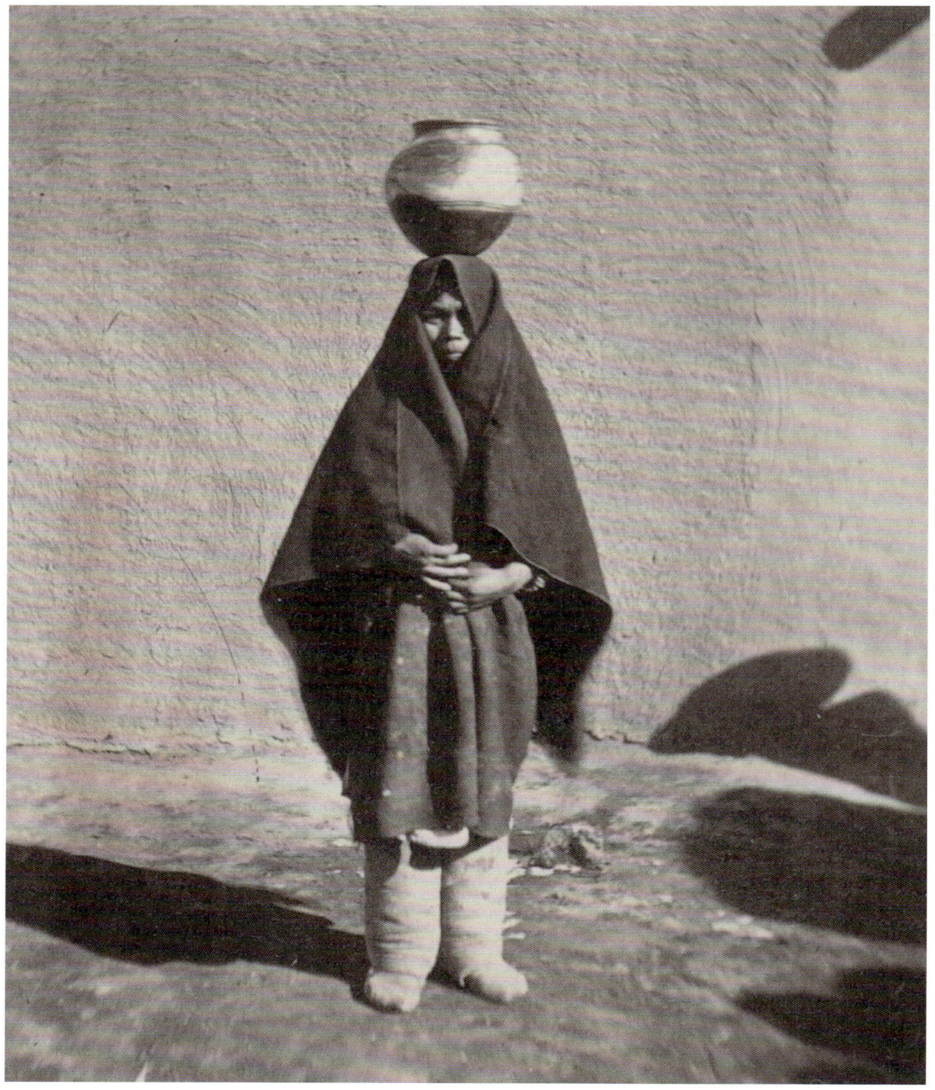

Fig. 4 Pueblo Woman Carrying Water, March 1881 from Isabella Stewart Gardner's *Travel Album: Northern and Eastern Europe, Russia, the United States, Mexico, and Cuba*, page 19

forward today in depictions of the Southwest as desolate ranges of desert promising inspiration, freedom, and the purveyal of unchanged cultures (Plate 43). In truth, the damage the colonial project had on Indigenous culture and ways of life was life-altering. Not only did the railroad allow individuals like Gardner to "explore" other cultures; it allowed for a fabricated perspective of Indigenous cultures to be seen. While one might argue that the railroad created economic growth, in general every track that was laid occupied a cubic foot taken from Indigenous stewardship and meant a challenge to the continuing access to culture in a place.

Looking back at the photographic legacy of the Southwest and more broadly of Indigenous people, we can note with great certainty that consent to photograph was not always explicitly obtained in these exchanges. In some instances, bribery, coercion, and support from federal Indian agents were used to help photographers find Indigenous sitters. There is also a legacy of photographs that share sensitive, ceremonial attire. While the images Gardner immortalized in her albums construct a selective view of the Indigenous communities, photography's power to hold the Indigenous person in a time capsule is the great paradox of documenting the Southwest. The power and responsibility of photography have been misused and misguided. It is an art form that in the case of Native people has created a dimension of distinct othering in settler colonial hands. According to the founders of the Haas Institute for a Fair and Inclusive Society at the University of California, "othering is a set of processes, structures, and dynamics that engender marginality and persistent inequality across any of the full range of human differences."[3] Longstanding impacts of this othering result in the pervasive invisibility of Indigenous peoples in present-day society.

Today Native American tourism is a $14 billion industry, according to the 2021 Economic Impact of Indigenous Tourism Businesses Report.[4] The engagement we have with a place in America impacts the construction of that place. We see the legacy of photography create images of the settler colonial dream, a vision that is romantic and free. But when I think about the layer of consent, I am led to believe that it is not solely about permission. Consent from an Indigenous lens is the relationality we agree to have with one another. Consent means we are creating a sense of belonging to each other even if for the duration of a snap of a camera lens.

What I think the images collected by Gardner lack is an awareness of what it means to be in right relationship with a place, a people, and a culture. To affix images of death next to images of cultural continuation and cultural struggle signals a lack of understanding about place.

As a young girl traveling to ancestral sites of neighboring relatives, I understood that the dwellings were homes. To make home in a place is an example of being in a relationship with the medicines, songs, prayers, and vitality of a place. When I zoom out and look at the settler colonial experiment and the way tourists like Isabella Stewart Gardner and many others experience my homelands, I want to call-in their actions. I want to remind them that these places, whether etched in silver gelatin and black and white or chromatic depictions, are and will always be Indigenous lands.

1. Keene 2021, 29.
2. Wolfe 2006, 388.
3. Powell and Menendian 2016.
4. SMS Hawai'i 2021.

A Note on Harmful Content in Gardner's Travel Album with Images of the Western United States

Diana Seave Greenwald

As this volume documents, Isabella Stewart Gardner traveled the world, collected images of the places she visited along the way, and arranged them in travel albums. Our perspectives differ from Isabella's. Viewed today, some of the images she collected are clearly outdated—and offensive, racist, colonialist, or violent. This publication, and the exhibition to which it is linked, seeks to address all of the albums' content in all of its complexity. Nonetheless, we feel this particular section of one of Isabella's albums requires a specific note about its content.

This travel album contains photographic images of Indigenous people as well as an image of two hanged men. While common in Gardner's time, buying and collecting photographs of the Indigenous people was and is an exploitative practice. It remains unknown whether these photographs were taken with their consent or under dubious circumstances. Jaclyn Roessel, whose essay appears on pages 193–197 of this volume, discusses these photographs, the question of consent in this photography, and the lasting impact that the expansion of Euro-American enterprises like the Santa Fe Railroad Company had on the Pueblo people. To complement Roessel's perspective and with her input, this note provides some further context about the album and the photographs it contains. The juxtaposition of images of Indigenous people—notably a Pueblo child—with a photograph of a hanging is particularly disturbing.

The Museum conducted additional research into the image of the hanging. Isabella inscribed the photograph of the two hanged men with the words "San Antonio Justice." San Antonio, New Mexico (not Texas), was located along the route she and Jack followed during their 1881 trip in the American West and was a stop on their rail journey. While it remains impossible to identify the victims, we researched hangings in New Mexico—both state-sanctioned capital punishment and extrajudicial lynchings—in order to better understand the context of this photograph. In cases of state-sanctioned hangings, most of those executed were white, Black, or Hispanic men who worked in mining or related fields and were convicted of murder by state authorities. Although executions took place in Socorro County in 1851, this is likely too early for this photograph. Four executions by hanging in Grant County in 1880 roughly in the area of Socorro County were the last of their kind prior to Isabella's travel through the region. The image could, possibly, be related to one of them.

That this photograph depicts a vigilante killing perpetrated against people of Mexican origin and descent in and around Socorro, New Mexico, during the 1880s remains another possibility. According to recent scholarship, lynchings of people of Mexican origin or descent in the Southwest occurred largely between 1848 and 1928. At least 49 people were victims of this heinous vigilante practice in New Mexico, in addition to at least 282 more in Texas. This period after the Mexican-American War saw a substantial increase in mob violence against people of Mexican origin or descent, where lynchings increased at a rate not seen before or since.[1]

The Isabella Stewart Gardner Museum does not endorse imagery of this kind. Nonetheless, we are committed to making our collection accessible and to exploring the complexity of our history in all of its facets. Visitors may draw their own conclusions from this object, and we commit to making this possible responsibly and respectfully by offering relevant historical context.

1. Carrigan and Webb 2003.

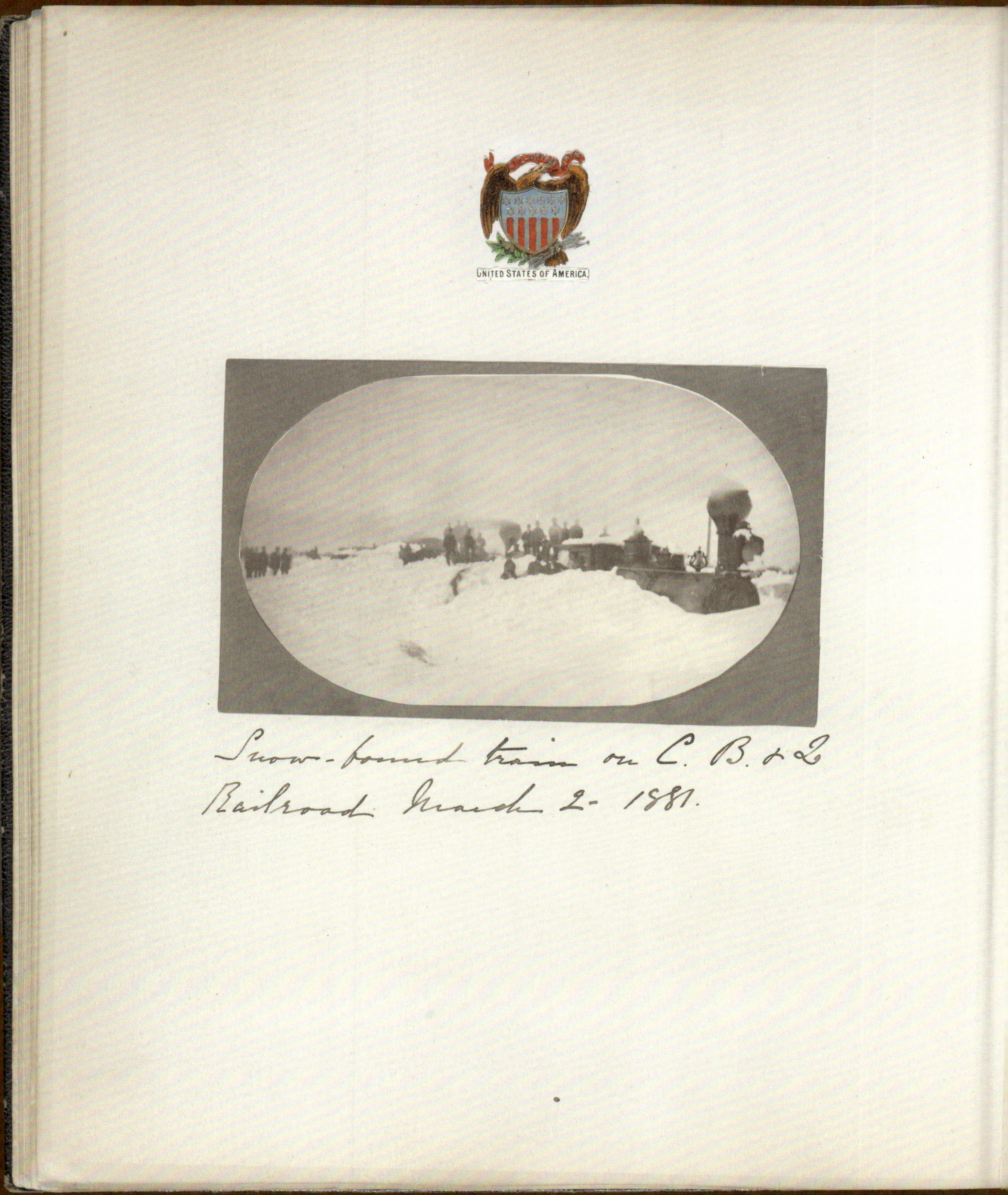
UNITED STATES OF AMERICA
Snow-bound train on C. B. & Q
Railroad March 2- 1881.

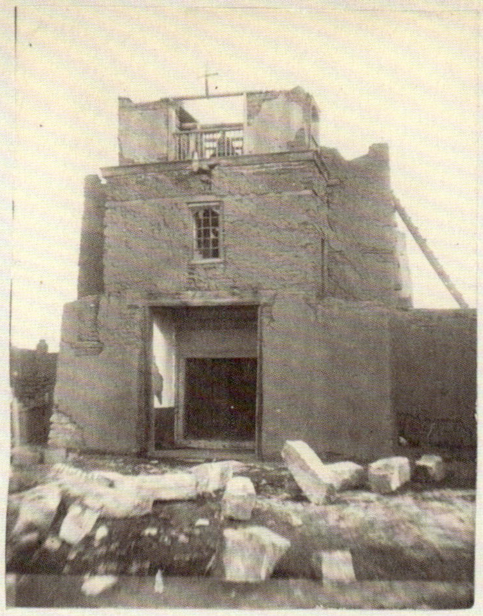

San Miguel. Santa Fe.
N. M.

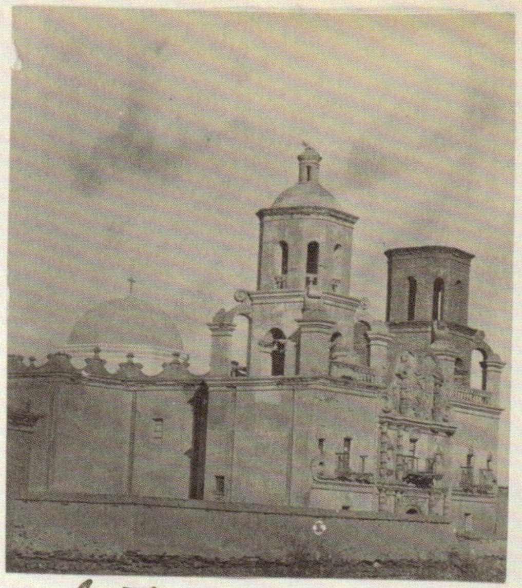

St. Xavier. Tucson.

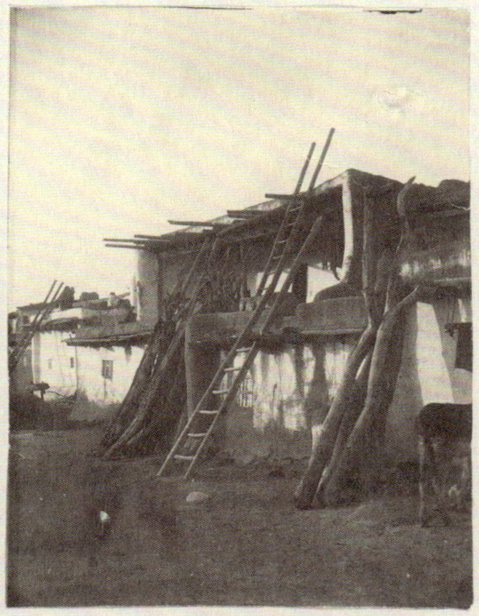

Pueblo Village
San Domingo.

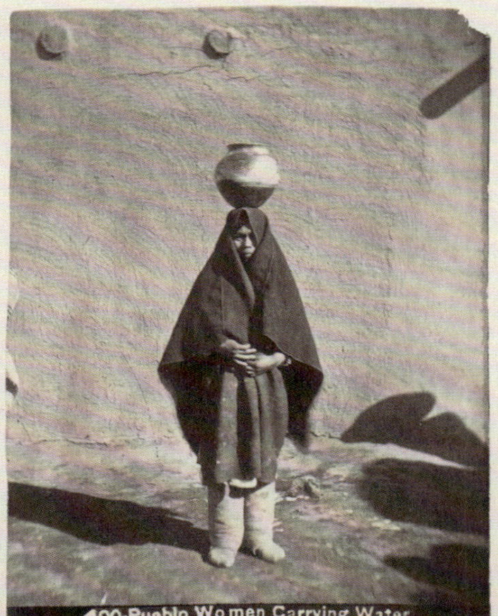

400-Pueblo Women Carrying Water.

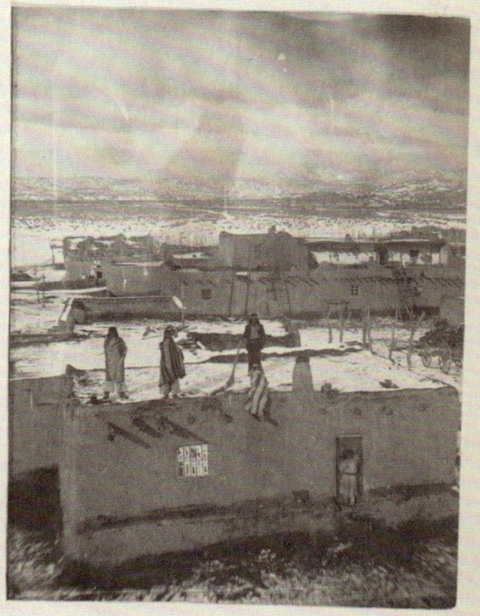

Pueblo Village.
San Domingo.

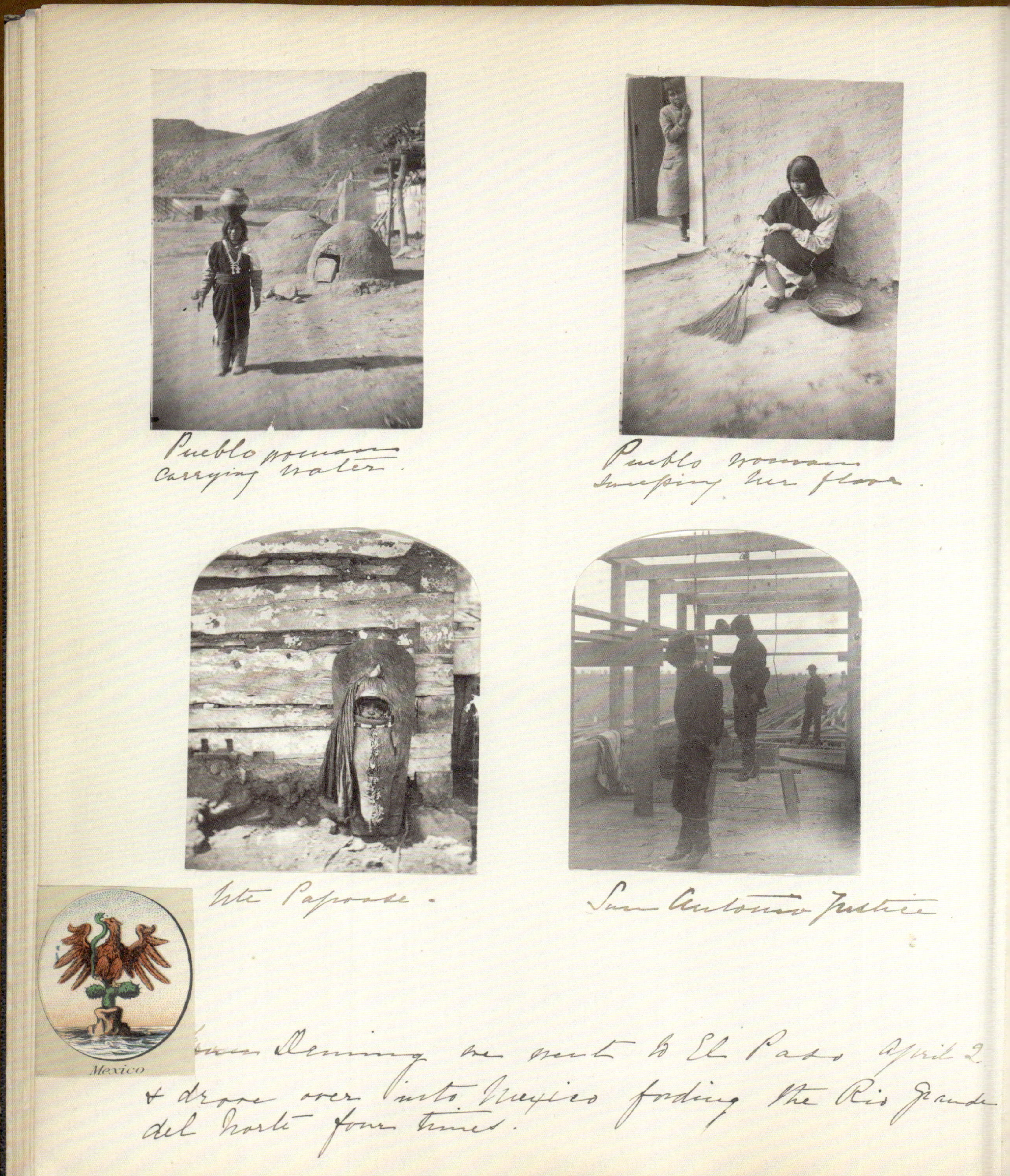
Pueblo woman carrying water.
Pueblo woman sweeping her floor.
Ute Papoose.
San Antonio Justice
Mexico
Deming we went to El Paso April 2
& drove over into Mexico fording the Rio Grande
del Norte four times.

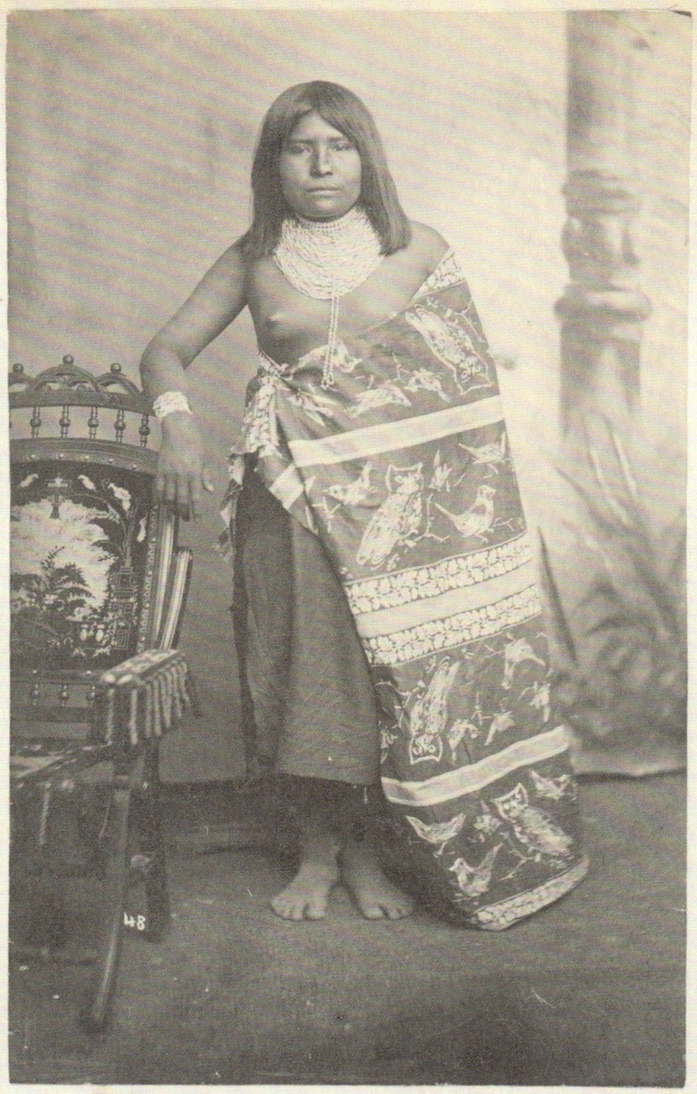

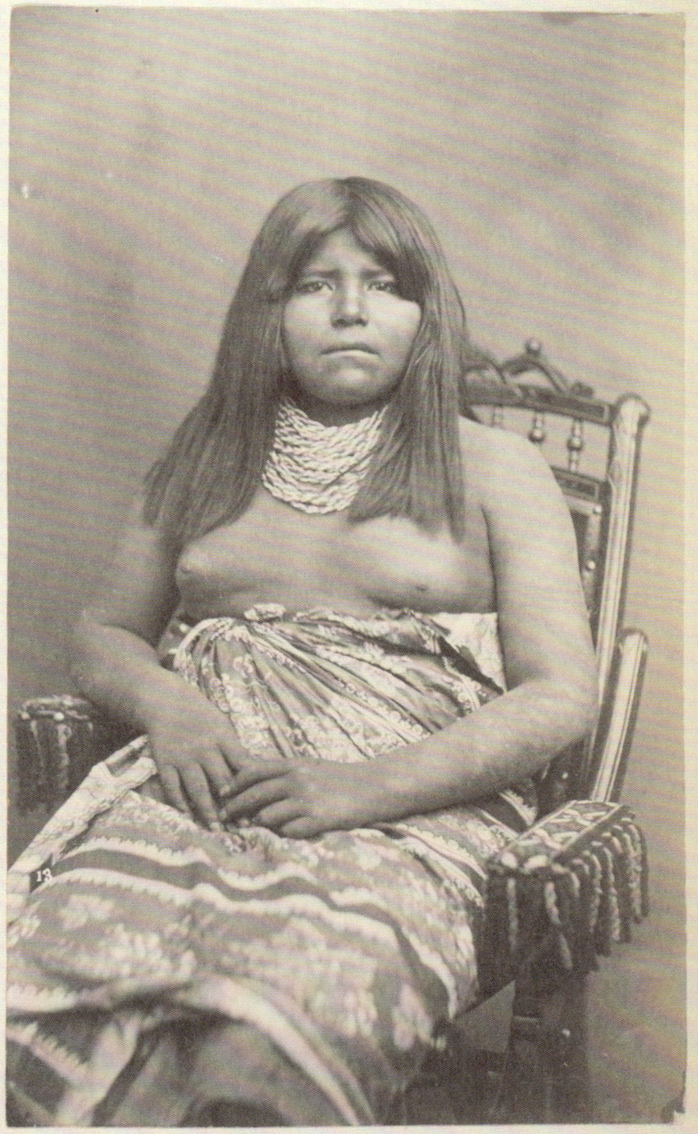

NORTHERN AND EASTERN EUROPE, RUSSIA, THE UNITED STATES, MEXICO, AND CUBA
Pages 22–23, March 1881

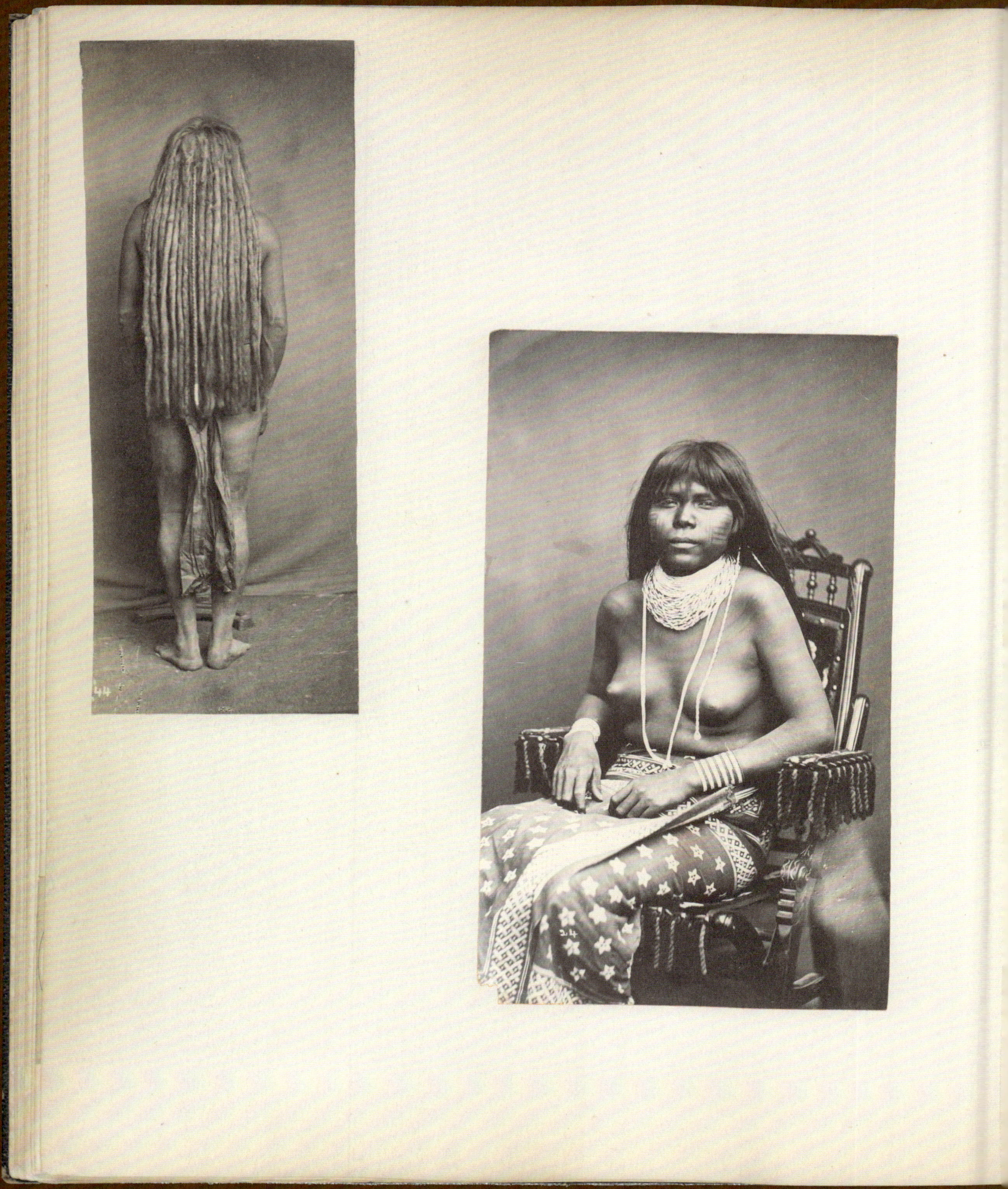

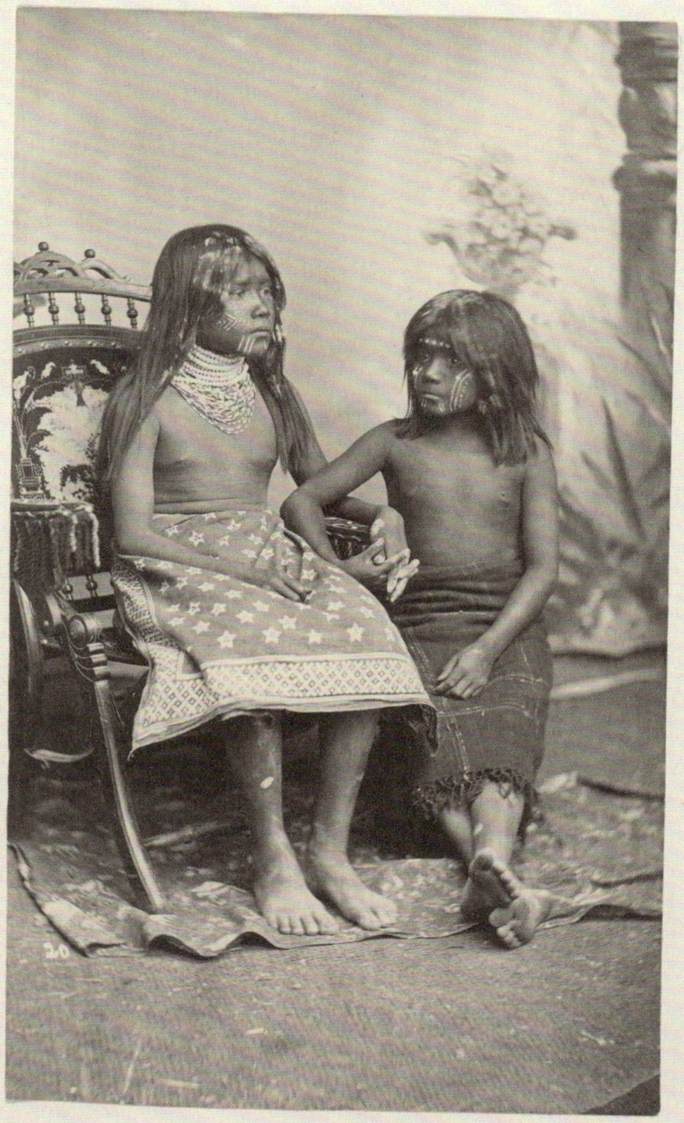

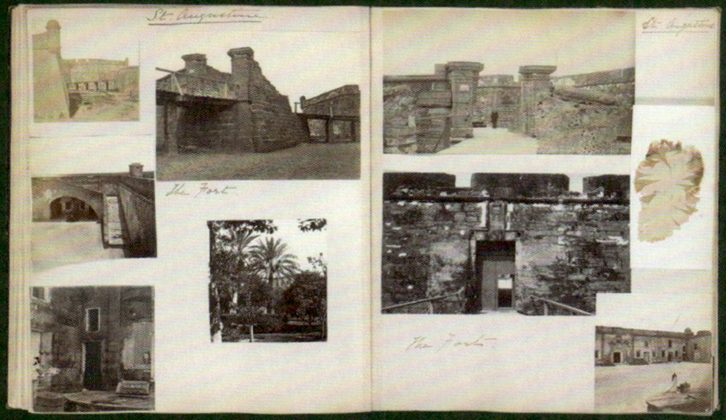

1. *Travel Album: Northern and Eastern Europe, Russia, the United States, Mexico, and Cuba,*
1867–1886
Bound album including collected photographs, found papers, pressed botanicals, and pen and ink annotations
32.1 x 27 x 3.5 cm
v.1.a.4.1

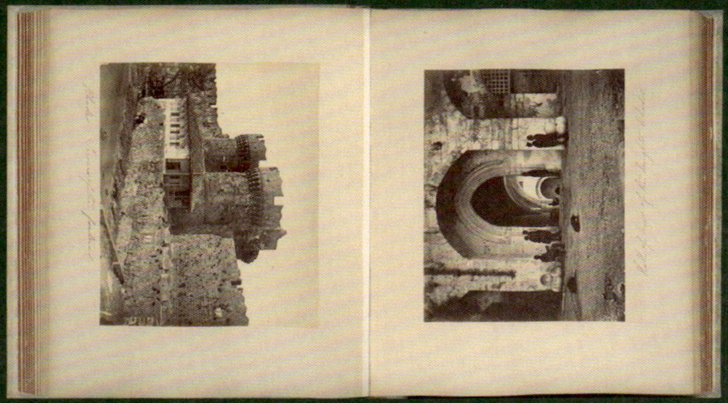

2. *Travel Album: Egypt, the Holy Land, Greece, and Turkey*
1875
Bound album including collected photographs and pen and ink annotations
39.7 x 34.9 x 4.3 cm
v.1.a.3.8

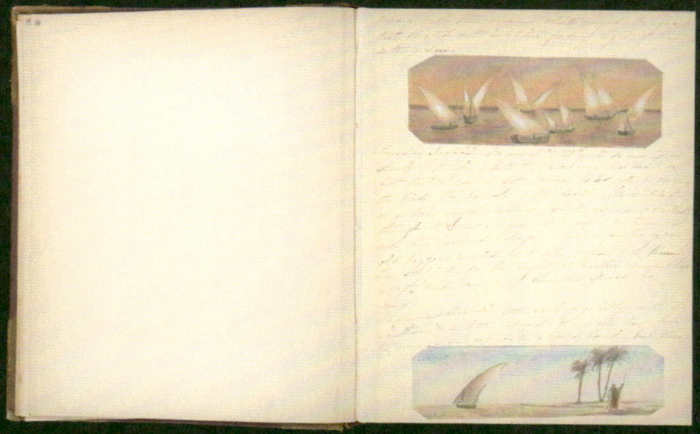

3. *Travel Album: Egypt and Sudan*
1874–1875
Bound album including collected photographs, found papers, original watercolors, and pen and ink writings
30.2 x 24 x 4.5 cm
v.1.a.4.2

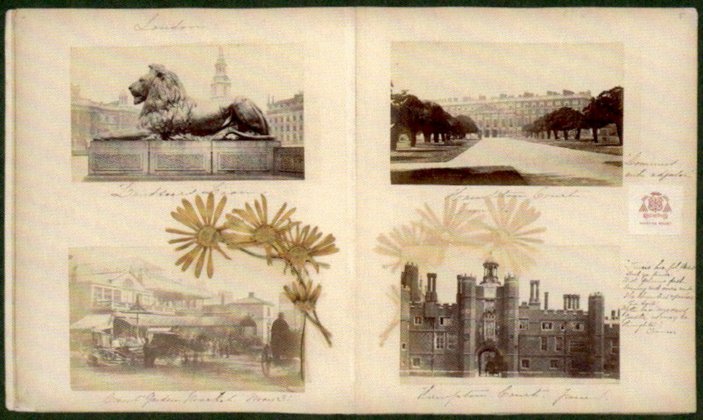

4. *Travel Album: England, Volume I*
1879
Bound album including collected photographs, found papers, pressed botanicals, and pen and ink annotations
32 x 20 x 4.5 cm
v.1.a.4.3

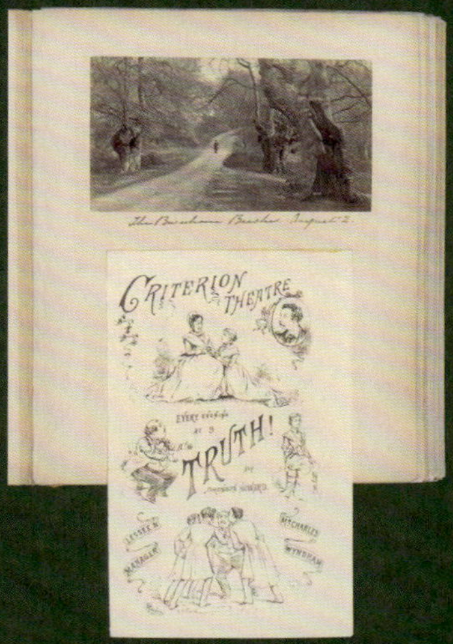

5. *Travel Album: England, Volume II*
1879
Bound album including collected photographs, found papers, pressed botanicals, and pen and ink annotations
32 x 20 x 4.5 cm
v.1.a.4.4

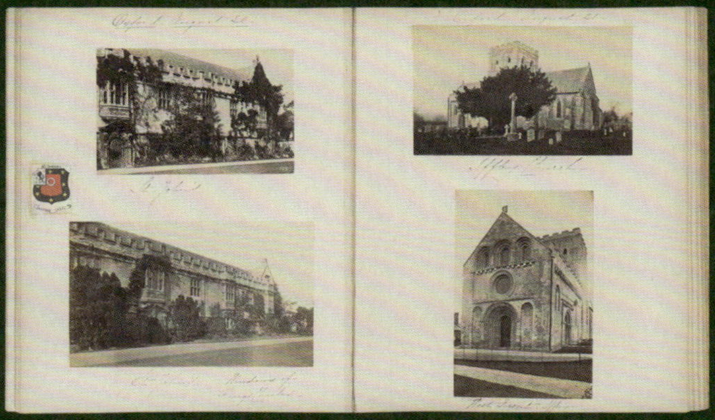

6. *Travel Album: Great Britain and France*
1879
Bound album including collected photographs, found papers, pressed botanicals, and pen and ink annotations
35.6 x 30.5 x 3.8 cm
v.1.a.4.5

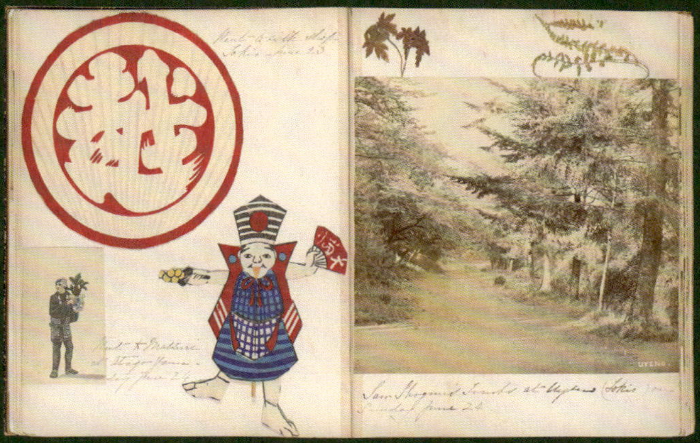

7. *Travel Album: Japan, Volume I*
1883
Bound album including collected photographs, found papers, pressed botanicals, and pen and ink annotations
31.5 x 26.5 cm x 4.8 cm
v.1.a.4.6

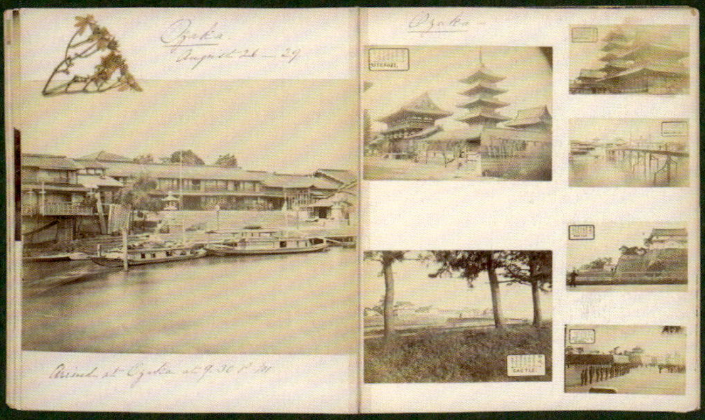

8. *Travel Album: China and Japan, Volume II*
1883
Bound album including collected photographs, found papers, pressed botanicals, and pen and ink annotations
31.8 x 26.2 x 4.6 cm
v.1.a.4.7

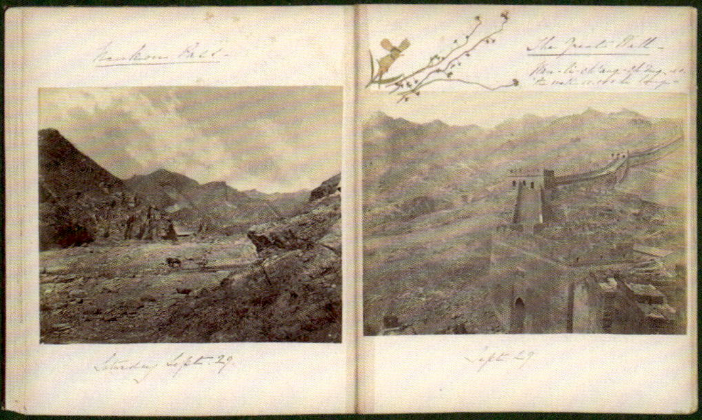

9. *Travel Album: China, Volume III*
1883
Bound album including collected photographs, found papers, pressed botanicals, and pen and ink annotations
32 x 26 x 4.6 cm
v.1.a.4.8

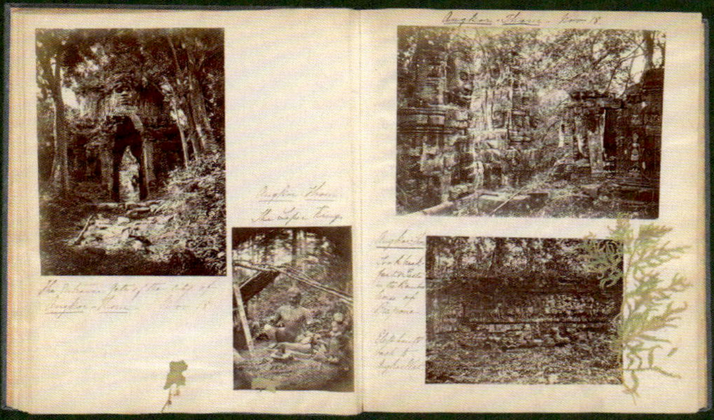

10. *Travel Album: Vietnam, Cambodia, Indonesia, Singapore, Malaysia, and Myanmar, Volume IV*
1883–1884
Bound album including collected photographs, found papers, pressed botanicals, and pen and ink annotations
36.3 x 26 x 4.6 cm
v.1.a.4.9

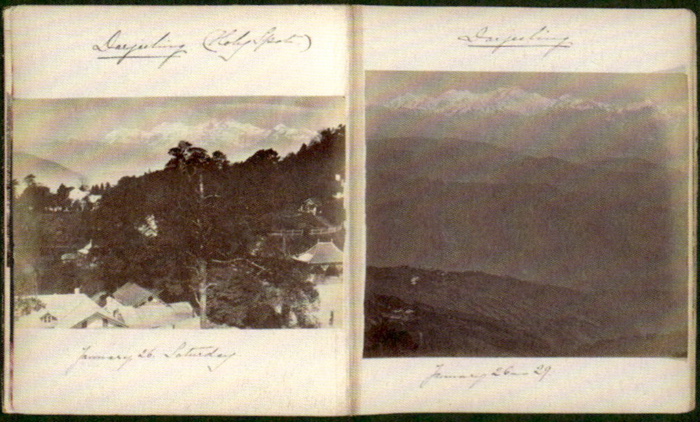

11. *Travel Album: India, Volume V*
1884
Bound album including collected photographs, found papers, pressed botanicals, and pen and ink annotations
32 x 26 x 4.5 cm
v.1.a.4.10

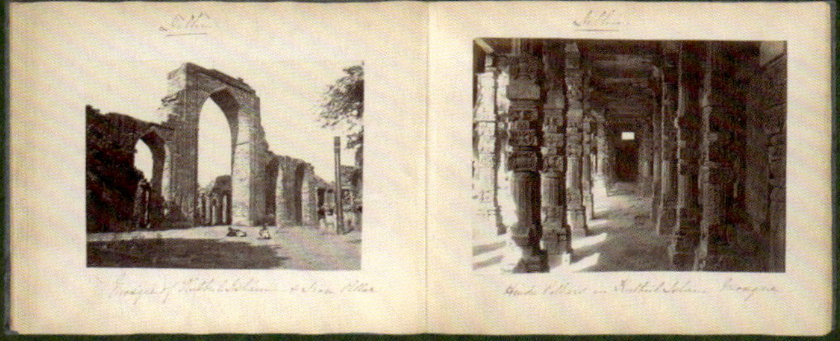

12. *Travel Album: India, Pakistan, Yemen, and Egypt, Volume VI*
1884
Bound album including collected photographs, found papers, pressed botanicals, and pen and ink annotations
30.5 x 36 x 4 cm
v.1.a.4.11

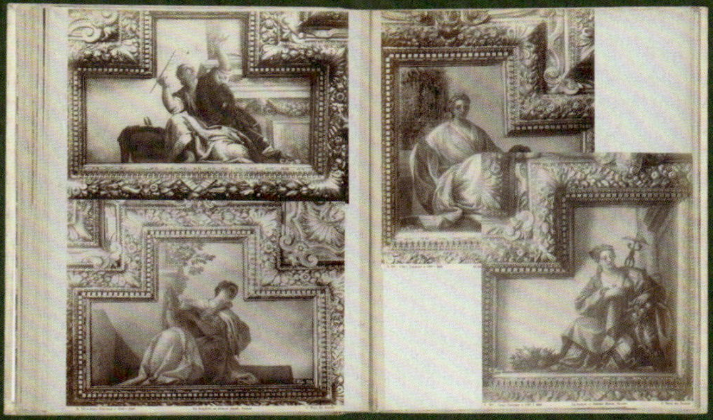

13. *Travel Album: Italy*
1884
Bound album including collected photographs, found papers, and pen and ink annotations
36.2 x 30 x 4.1 cm
v.1.a.4.12

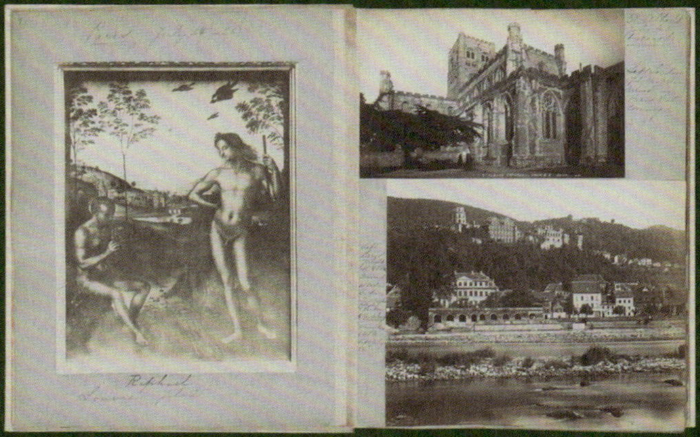

14. *Travel Album: Great Britain, France, Germany, Austria, and Italy, Volume I*
1886
Bound album including collected photographs, found papers, pressed botanicals, and pen and ink annotations
35.6 x 30.5 x 3.8 cm
v.1.a.4.13

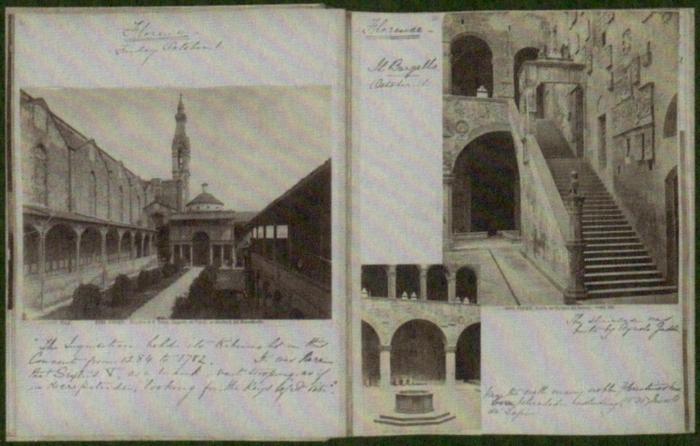

15. *Travel Album: Italy, France, and Great Britain, Volume II*
1886
Bound album including collected photographs, found papers, and pen and ink annotations
35.6 x 30.5 x 3.8 cm
v.1.a.4.14

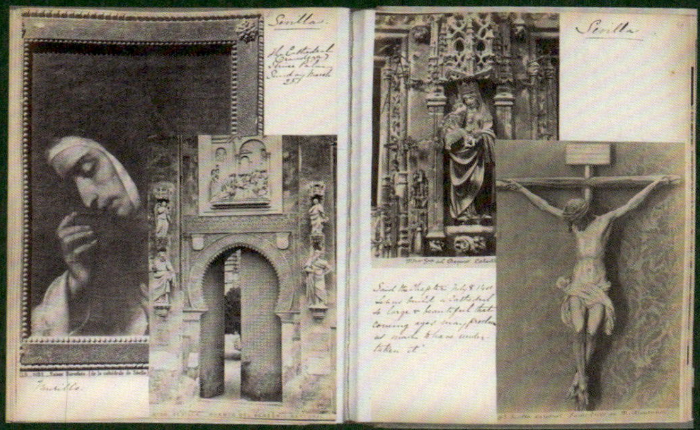

16. *Travel Album: Spain, Volume I*
1888
Bound album including collected photographs, found papers, and pen and ink annotations
35.6 x 30.5 x 3.8 cm
v.1.a.4.15

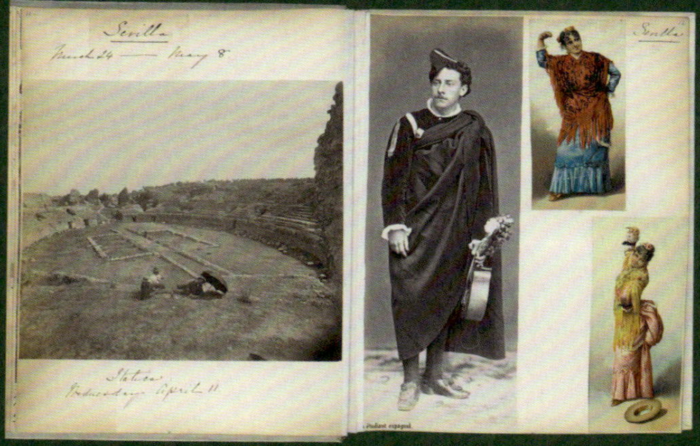

17. *Travel Album: Spain and Portugal, Volume II*
1888
Bound album including collected photographs, found papers, and pen and ink annotations
35.9 x 28.9 x 5.2 cm
v.1.a.4.16

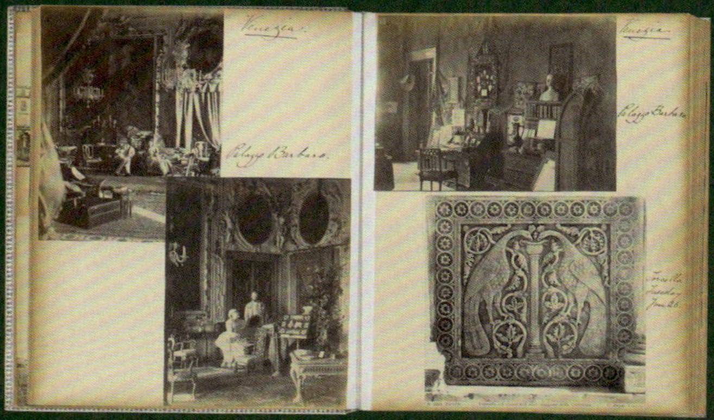

18. *Travel Album: Spain, France, Monaco, Italy, Germany, Belgium, and Great Britain, Volume III and I*
1888–1890
Bound album including collected photographs, found papers, and pen and ink annotations
35.7 x 30.2 x 6 cm
v.1.a.4.17

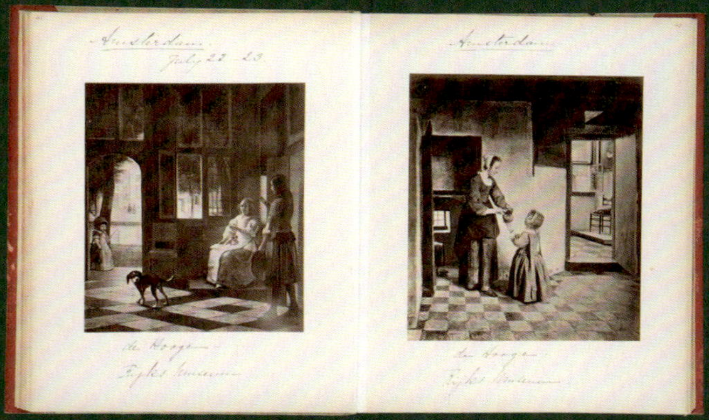

19. *Travel Album: Netherlands, Germany, and Italy, Volume II*
1890
Bound album including collected photographs, found papers, and pen and ink annotations
35.7 x 30.5 x 4.6 cm
v.1.a.4.18

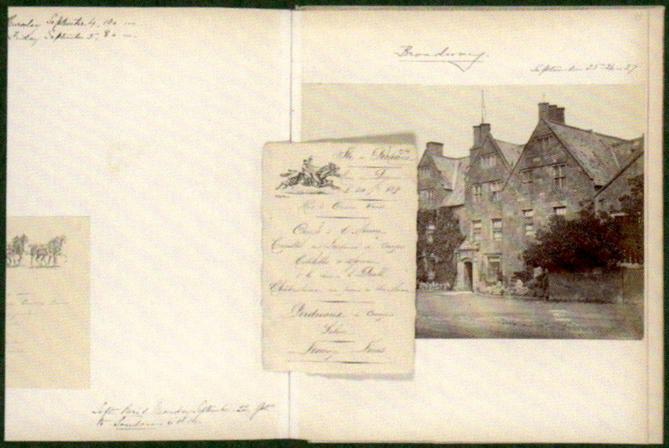

20. *Travel Album: Italy, France, and Great Britain, Volume III*
1890
Bound album including collected photographs, found papers, and pen and ink annotations
35.7 x 30 x 5.2 cm
v.1.a.4.19

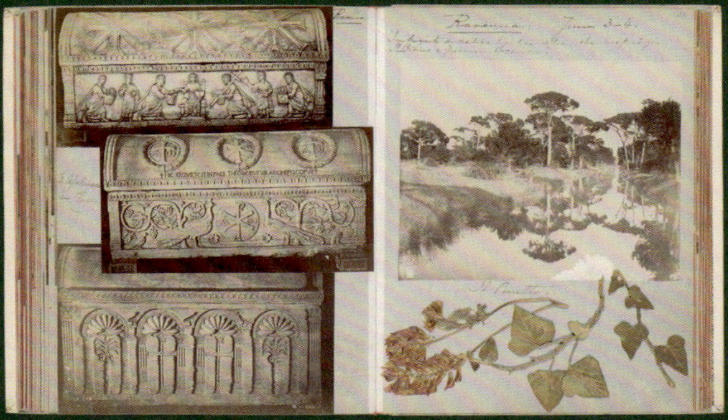

21. *Travel Album: France, Great Britain, Italy, Volume I*
1892
Bound album including collected photographs, found papers, pressed botanicals, and pen and ink annotations
34.6 x 28.6 x 4.9 cm
v.1.b.4.1

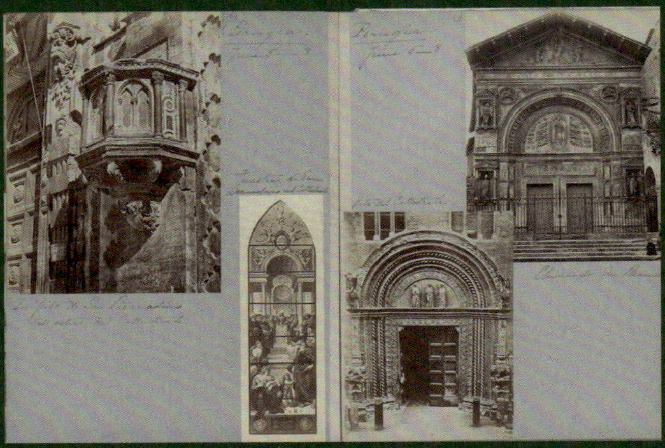

22. *Travel Album: Italy, Volume II*
1892
Bound album including collected photographs, found papers, original watercolors, and pen and ink annotations
35.6 x 30.5 x 3.8 cm
v.1.b.4.2

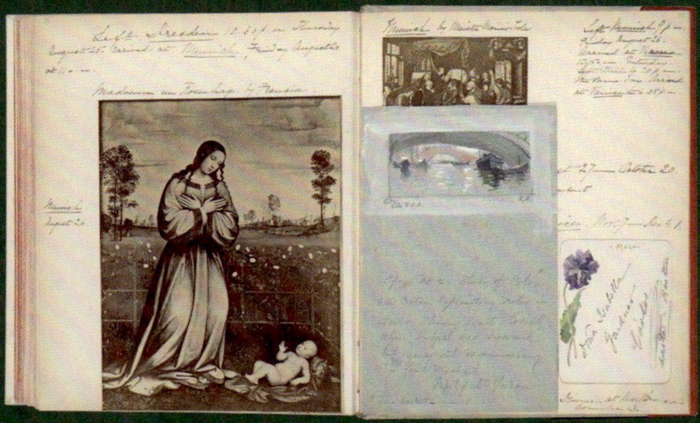

23. *Travel Album: Italy, Switzerland, Germany, and France, Volume III*
1892
Bound album including collected photographs, found papers, original watercolors, and pen and ink annotations
32 x 20 x 4.5 cm
v.1.b.4.3

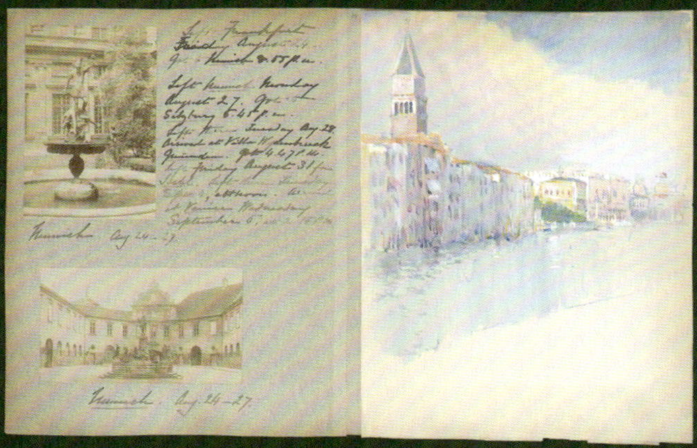

24. *Travel Album: France, Great Britain, Germany, Italy, Austria, and Belgium, Volume I*
1894
Bound album including collected photographs, found papers, original watercolors, and pen and ink annotations
32 x 20 x 4.5 cm
v.1.b.4.4

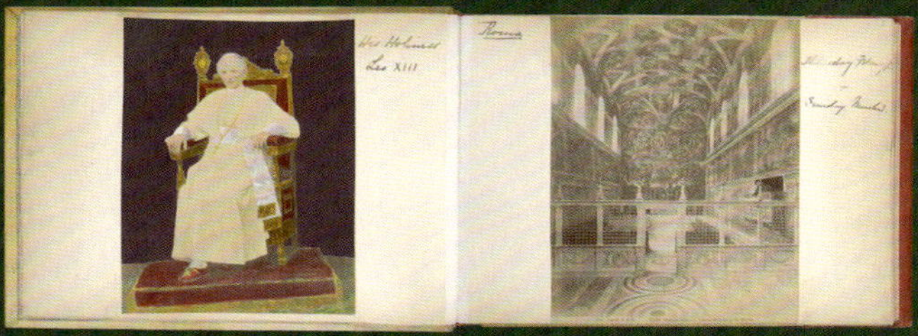

25. *Travel Album: Italy, Volume II*
1895
Bound album including collected photographs, found papers, pressed botanicals, and pen and ink annotations
32 x 20 x 4.5 cm
v.1.b.4.5

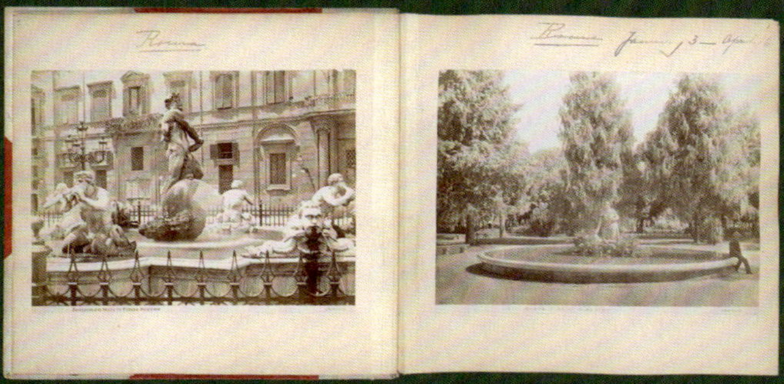

26. *Travel Album: Italy, Volume III*
1895
Bound album including collected photographs, found papers, original watercolors, and pen and ink annotations
32 x 20 x 4.5 cm
v.1.b.4.6

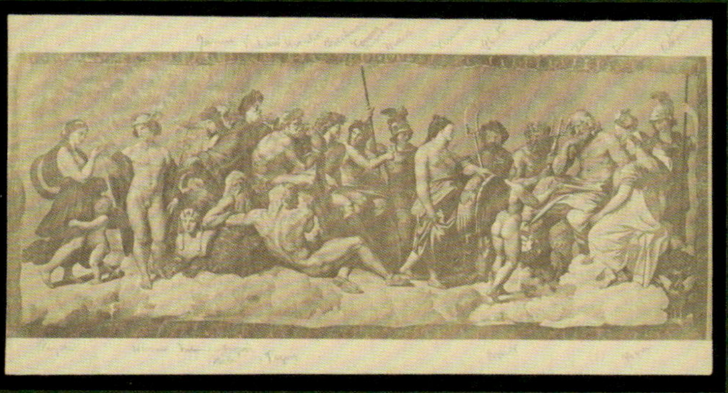

27. *Travel Album: Italy, Great Britain, and France*
about 1896–1897
Bound album including collected photographs, found papers, and pen and ink annotations
32 x 20 x 4.5 cm
v.1.b.4.27

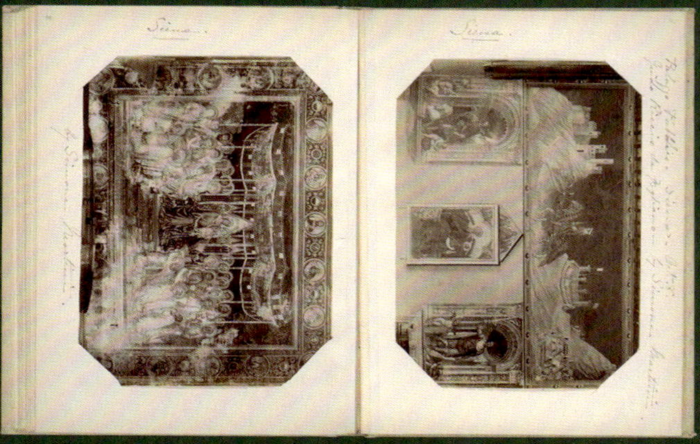

28. *Travel Album: Germany and Italy*
about 1896–1897
Bound album including collected photographs, found papers, and pen and ink annotations
32 x 20 x 4.5 cm
v.1.b.4.28

Bibliography

PRIMARY SOURCES

The Art Journal Illustrated Catalogue: The Industry of All Nations, 1851. London, 1851. http://id.lib.harvard.edu/alma/990013377850203941/catalog.

Ballou, Maturin Murray. *Due West; or, Round the World in Ten Months.* Boston, 1884.

Cartwright, Julia. "America in Europe." *The Magazine of Art* 6 (January 1883): 1–7.

Curtis, Ralph Wormeley. 9 February 1888. Letter to Isabella Stewart Gardner. Isabella Stewart Gardner Museum Archives, Boston.

Ford, Richard. *A Hand-Book for Travellers in Spain.* London, 1845, 1882.

Ford, Richard. *The Spaniards and Their Country.* New York, 1847.

Gardner, Isabella Stewart. 29 June 1906. Letter to Henry Adams. Isabella Stewart Gardner Museum Archives, Boston.

Gardner, Isabella Stewart. *Travel Diary: China, Vietnam, Cambodia, Indonesia, India, Singapore, Myanmar, Egypt, and Italy, 1883–1884.* Isabella Stewart Gardner Museum Archives, Boston.

Gardner, John L., Jr. 24 May 1888. Letter to George Peabody Gardner. Isabella Stewart Gardner Museum Archives, Boston.

Gardner, John L., Sr. 17 March 1884. Letter to John L. Gardner Jr. Gardner Family Papers. Massachusetts Historical Society, Boston.

Hadley, Rollin van N., ed. *The Letters of Bernard Berenson and Isabella Stewart Gardner 1887–1924.* Boston, 2016.

Irving, Washington. *The Alhambra: A Series of Tales and Sketches of the Moors and Spaniards.* Philadelphia, 1837.

"Mrs. 'Jack' Gardner's Palace Mystifies Back Bay Bostonians." *Philadelphia Press,* March 17, 1901.

Murray, John. *Handbook for Travellers in Egypt, Including Descriptions of The Course of the Nile Through Egypt and Nubia, Alexandria, Cairo, The Pyramids, and Thebes, The Suez Canal, The Peninsula of Mount Sinai, The Oases, The Fyoom.* Fourth Edition. London, 1873.

Royal Commission for Victoria. *Calcutta International Exhibition, 1883–84.* Report of the Royal Commission for Victoria at the Calcutta International Exhibition, 1883–84. https://www.parliament.vic.gov.au/papers/govpub/VPARL1884No50.pdf.

Saar, Betye. "The Art of Assemblage: Two Views." The Eye of the Beholder Lecture Series, Isabella Stewart Gardner Museum, Boston, MA, March 29, 1994. Transcript in the Isabella Stewart Gardner Museum Archives, Boston.

Turner, A. W. *Photographs of the Calcutta International Exhibition, 1883–84.* Photographic Office, Calcutta, 1884. https://www.indianculture.gov.in/rarebooks/photographs-calcutta-international-exhibition-1883-84.

SECONDARY SOURCES

Appolonia, Giovanni de. "A Venetian Courtyard in Boston." In *Gondola Days: Isabella Stewart Gardner and the Palazzo Barbaro Circle,* by Elizabeth Anne McCauley et al., 177–89. Boston, 2004. Published in conjunction with the exhibition *Gondola Days: Isabella Stewart Gardner and the Palazzo Barbaro Circle* shown at the Isabella Stewart Gardner Museum, Boston.

Balakrishnan, Usha. *Alamkara: The Beauty of Ornament.* New Delhi, 2014.

Boone, Mary Elizabeth. *España: American Artists and the Spanish Experience.* New York, 1998.

Boone, Mary Elizabeth. *Vistas de España: American Views of Art and Life in Spain, 1860–1914.* New Haven, 2007.

Boone, Mary Elizabeth. *"The Spanish Element in Our Nationality": Spain and America at the World's Fairs and Centennial Celebrations, 1876–1915.* University Park, 2019.

Carrier, David. "Isabella Stewart Gardner's Museum." In *Museum Skepticism: A History of the Display of Art in Public Galleries,* 110–25. Durham, 2006.

Carrigan, William D., and Clive Webb. "The Lynching of Persons of Mexican Origin or Descent in the United States, 1848 to 1928." *Journal of Social History* 37, no. 2 (2003): 411–38. http://www.jstor.org/stable/3790404.

Carter, Morris. *Isabella Stewart Gardner and Fenway Court.* Boston, 1925.

Carvalho, Pedro Mourna. "Jewelry and Objects from India." In *Journeys East: Isabella Stewart Gardner and Asia,* edited by Alan Chong and Noriko Murai, 458–64. Boston, 2009. Published in conjunction with the exhibition *Journeys East: Isabella Stewart Gardner and Asia* shown at the Isabella Stewart Gardner Museum, Boston.

Chamberlain, Basil Hall. *The Classical Poetry of the Japanese.* London, 1880.

Chong, Alan. "Mrs. Gardner's Museum of Myth." *RES: Anthropology and Aesthetics,* no. 52 (Autumn 2007): 212–20. http://www.jstor.org/stable/20167756?origin=JSTOR-pdf.

Chong, Alan. "Introduction: Journeys East." In *Journeys East: Isabella Stewart Gardner and Asia,* edited by Alan Chong and Noriko Murai, 14–61. Boston, 2009. Published in conjunction with the exhibition *Journeys East: Isabella Stewart Gardner and Asia* shown at the Isabella Stewart Gardner Museum, Boston.

Chong, Alan. "Around the World with Isabella and Jack Gardner, 1883–1884." In *Journeys East: Isabella Stewart Gardner and Asia*, edited by Alan Chong and Noriko Murai, 93–383. Boston, 2009. Published in conjunction with the exhibition *Journeys East: Isabella Stewart Gardner and Asia* shown at the Isabella Stewart Gardner Museum, Boston.

Chong, Alan, and Noriko Murai, eds. *Journeys East: Isabella Stewart Gardner and Asia.* Boston, 2009. Published in conjunction with the exhibition *Journeys East: Isabella Stewart Gardner and Asia* shown at the Isabella Stewart Gardner Museum, Boston.

Dalrymple, William. *The Anarchy: The Relentless Rise of the East India Company.* London, 2019.

Dalrymple, William, ed. "Painting for the East India Company." In *Forgotten Masters: Indian Painting for the East India Company,* edited by William Dalrymple, 9–25. London, 2020. Published in conjunction with the exhibition *Forgotten Masters: Indian Painting for the East India Company* shown at The Wallace Collection.

Dearing, Stewart. "Painting the Other Within: Gypsies According to the Bohemian Artist in the Nineteenth and Early Twentieth Centuries." *Romani Studies* 20, no. 2 (2010): 161–201. https://doi.org/10.3828/rs.2010.7.

DeGuzman, María. *Spain's Long Shadow: The Black Legend, Off-Whiteness, and Anglo-American Empire.* Minneapolis, 2005.

Dehejia, Vidya. "Fixing a Shadow." In *India through the Lens: Photography 1840–1911,* edited by Vidya Dehejia, 11–33. Washington, DC, and Ahmedabad, 2000. Published in conjunction with the exhibition *India through the Lens: Photography 1840–1911* shown at the Freer Gallery of Art and Arthur M. Sackler Gallery, Smithsonian Institution.

Dehejia, Vidya, ed. *Delight in Design: Indian Silver for the Raj.* New Delhi, 2008. Published in conjunction with the exhibition *Delight in Design: Indian Silver for the Raj* shown at the Miriam & Ira D. Wallach Art Gallery, Columbia University.

Di Bello, Patrizia. *Women's Albums and Photography in Victorian England: Ladies, Mothers, and Flirts.* Aldershot, 2007.

Dobson, Sebastian. "Guiding the Sitter: Matsuzaki Shinji's Dos and Don'ts for the Photographic Customer." In *Portraiture and Early Studio Photography in China and Japan,* edited by Luke Gartlan and Roberta Wue, 211–17. London and New York, 2017.

Dower, John W. "Ways of Seeing, Ways of Remembering: The Photography of Prewar Japan." In *A Century of Japanese Photography,* edited by Japan Photographic Association, 3–20. New York, 1980.

Downs, Jacques M. "American Merchants and the China Opium Trade, 1800–1840." *Business History Review* 42, no. 4 (Winter 1968): 418–42. https://doi.org/10.2307/3112527.

Emery, Elizabeth. *Reframing Japonisme: Women and the Asian Art Market in Nineteenth-Century France, 1853–1914.* London, 2020.

Gartlan, Luke. *A Career of Japan: Baron Raimund von Stillfried and Early Yokohama Photography.* Leiden and Boston, 2016.

Graburn, Nelson H. H., and Université de droit, d'économie et des sciences d'Aix-Marseille. *To Pray, Pay and Play: The Cultural Structure of Japanese Domestic Tourism.* Aix-en-Provence, 1983.

Guha-Thakurta, Tapati. "The Compulsions of Visual Representation in Colonial India." In *Traces of India: Photography, Architecture, and the Politics of Representation, 1850–1900,* edited by Maria Antonella Pelizzari, 108–40. New Haven, 2003. Published in conjunction with the exhibition *Traces of India: Photography, Architecture, and the Politics of Representation, 1850–1900* shown at the Canadian Centre for Architecture, Montreal.

Guth, Christine M. E. *Longfellow's Tattoos: Tourism, Collecting, and Japan.* Seattle, 2004.

Haddon, Madeleine. "Local Color: Race, Color, and Spanishness in French and Spanish Painting, 1855–1927." PhD diss., Princeton University, 2021.

Hawley, Anne. "Director's Foreword." In *Journeys East: Isabella Stewart Gardner and Asia,* edited by Alan Chong and Noriko Murai, 10–11. Boston, 2009. Published in conjunction with the exhibition *Journeys East: Isabella Stewart Gardner and Asia* shown at the Isabella Stewart Gardner Museum, Boston.

Heath, Elizabeth. "Albums of Empire: Photography, Collective Memory, and the British Raj." *The Journal of Decorative and Propaganda Arts* 27 (2015): 74–103. http://www.jstor.org/stable/24739833.

Higonnet, Anne. "Secluded Vision: Images of Feminine Experience in Nineteenth Century Europe." In *The Expanding Discourse: Feminism and Art History,* edited by Norma Broude and Mary D. Garrard, 170–85. New York, 1992.

Higonnet, Anne. "Museum Sight." In *Art and Its Publics: Museum Studies at the Millennium,* edited by Andrew McClellan, 133–48. Malden, 2003. https://doi.org/10.1002/9780470775936.ch6.

Higonnet, Anne. *A Museum of One's Own: Private Collecting, Public Gift.* New York, 2009.

Hockley, Allen. "The Business of Photography." *Felice Beato's Japan: Places: An Album by the Pioneer Foreign Photographer in Yokohama.* MIT Visualizing Cultures, 2010. https://visualizingcultures.mit.edu/beato_places/fb1_essay01.html.

Hockley, Allen. "Globetrotter Tourism." *Globetrotter's Japan: Places: Foreigners on the Tourist Circuit in Meiji Japan.* MIT Visualizing Cultures, 2010. https://visualizingcultures.mit.edu/gt_japan_places/ga2_essay01.html.

Hoffenberg, Peter. "Photography and Architecture at the Calcutta International Exhibition." In *Traces of India: Photography, Architecture, and the Politics of Representation, 1850–1900,* edited by Maria Antonella Pelizzari, 174–96. New Haven, 2003. Published in conjunction with the exhibition *Traces of India: Photography, Architecture, and the Politics of Representation, 1850–1900* shown at the Canadian Centre for Architecture, Montreal.

Ishimori, Shuzo. "Popularization and Commercialization of Tourism in Early Modern Japan." *Senri Ethnological Studies,* no. 26 (1989): 179–94.

Ivy, Marilyn. *Discourses of the Vanishing: Modernity, Phantasm, Japan.* Chicago, 1995.

Iwasaki, Haruko. "Western Images, Japanese Identities: Cultural Dialogue between East and West in Yokohama Photography." In *A Timely Encounter: Nineteenth-Century Photographs of Japan,* edited by Melissa Banta and Susan B Taylor, 23–37. Cambridge, 1988.

Jiménez Díaz, Pablo, Óscar Muñoz, and Carlos Teixidor Cadenas. *La España de Laurent (1856–1886): Un paseo fotográfico por la historia.* Madrid, 2019.

Kagan, Richard. *The Spanish Craze: America's Fascination with the Hispanic World, 1779–1939.* Lincoln, 2019.

Keene, Adrienne. *Notable Native People: 50 Indigenous Leaders, Dreamers, and Changemakers from Past and Present.* Berkeley, 2021.

Kinoshita, Naoyuki. *Tamoto Kenzō to Meiji no shashinka-tachi.* Tokyo, 1999.

Llewellyn-Jones, Rosie. *The Last King of India: Wajid Ali Shah (1822–87).* New Delhi, 2014.

Llorens Serra, Tomás, and Nerea Sagredo Ezquerra. *Sargent/Sorolla.* Madrid, 2006.

Macartney, Hilary, and Jose Manuel Matilla, eds. *Copied by the Sun: Talbotype Illustrations to the Annals of the Artists of Spain by Sir William Stirling Maxwell.* Madrid, 2016. Published in conjunction with the exhibition *Copied by the Sun: The Talbotypes from 'The Annals of the*

Artists of Spain' by William Stirling Maxwell (1847) shown at the Museo Nacional del Prado, Madrid.

McCauley, Elizabeth Anne. "A Sentimental Traveler: Isabella Stewart Gardner in Venice." In *Gondola Days: Isabella Stewart Gardner and the Palazzo Barbaro Circle,* by Elizabeth Anne McCauley et al., 3–51. Boston, 2004. Published in conjunction with the exhibition *Gondola Days: Isabella Stewart Gardner and the Palazzo Barbaro Circle* shown at the Isabella Stewart Gardner Museum, Boston.

McCauley, Elizabeth Anne, et al. *Gondola Days: Isabella Stewart Gardner and the Palazzo Barbaro Circle.* Boston, 2004. Published in conjunction with the exhibition *Gondola Days: Isabella Stewart Gardner and the Palazzo Barbaro Circle* shown at the Isabella Stewart Gardner Museum, Boston.

McKenna, Shana. "Green Hill and Its Gardens." *Inside the Collection Blog.* Isabella Stewart Gardner Museum. 28 July 2020. https://www.gardnermuseum.org/blog/green-hill-gardens.

Motz, Marilyn. "Visual Autobiography: Photograph Albums of Turn-of-the-Century Midwestern Women Author(s)." *American Quarterly* 41, no. 1 (March 1989): 63–92.

Muller, Gregor. *Colonial Cambodia's "Bad Frenchmen": The Rise of French Rule and the Life of Thomas Caraman, 1840–87.* London, 2006.

Nielsen, Christina, Casey Riley, and Nathaniel Silver. *Isabella Stewart Gardner Museum: A Guide.* New Haven and London, 2017.

Odo, David. *The Journey of "A Good Type": From Artistry to Ethnography in Early Japanese Photographs.* Cambridge, 2015.

Pelizzari, Maria Antonella. "From Stone to Paper: Photographs of Architecture and the Traces of History." In *Traces of India: Photography, Architecture, and the Politics of Representation, 1850–1900,* edited by Maria Antonella Pelizzari, 20–60. New Haven, 2003. Published in conjunction with the exhibition *Traces of India: Photography, Architecture, and the Politics of Representation, 1850–1900* shown at the Canadian Centre for Architecture, Montreal.

powell, john a., and Stephen Menendian. "The Problem of Othering: Towards Inclusiveness and Belonging." *Othering & Belonging,* no. 1 (Summer 2016). http://www.otheringandbelonging.org/the-problem-of-othering.

Prokop, Ellen. "'Here One Feels Existence': Isabella Stewart Gardner's Spanish Cloister." In *Collecting Spanish Art: Spain's Golden Age and America's Gilded Age,* edited by Inge Jackson Reist and José Luis Colomer, 97–123. New York, 2012.

Riley, Casey. "From Page to Stage: Isabella Stewart Gardner's Photograph Albums and the Development of Her Museum, 1874–1924." PhD diss., Boston University, 2015.

Riley, Casey. "Self Assembled: Isabella Stewart Gardner's Photographic Albums and the Development of her Museum, 1902–24." In *Photographs, Museums, Collections,* edited by Elizabeth Edwards and Christopher Morton, 47–63. London, 2015.

Riley, Casey. "Commerce and Connoisseurship: Isabella Stewart Gardner's Catalogue MCM." *Mitteilungen des Kunsthistorischen Institutes in Florenz* 62, no. 1 (2020): 93–107.

Riley, Casey. "To Make a Case: Isabella Stewart Gardner's Archival Installations at Fenway Court." In *Boston's Apollo: Thomas McKeller and John Singer Sargent,* edited by Nathaniel Silver, 154–73. Boston, 2020. Published in conjunction with the exhibition *Boston's Apollo: Thomas McKeller and John Singer Sargent* shown at the Isabella Stewart Gardner Museum, Boston.

Sampson, Gary. "Photographer of the Picturesque: Samuel Bourne." In *India through the Lens: Photography 1840–1911,* edited by Vidya Dehejia, 163–75. Washington, DC, and Ahmedabad, 2000. Published in conjunction with the exhibition *India through the Lens: Photography 1840–1911* shown at the Freer Gallery of Art and Arthur M. Sackler Gallery, Smithsonian Institution.

Siegel, Elizabeth, "Society Cutups." In *Playing with Pictures: The Art of Victorian Photocollage,* by Elizabeth Siegel et al., 13–36. New Haven, 2009. Published in conjunction with the exhibition *Playing with Pictures: The Art of Victorian Photocollage* shown at the Art Institute of Chicago, the Metropolitan Museum of Art, New York, and the Art Gallery of Ontario.

Silver, Nathaniel. "Isabella and the Hispanic Society." *Inside the Collection Blog.* Isabella Stewart Gardner Museum. 13 October 2020. https://www.gardnermuseum.org/blog/isabella-and-hispanic-society.

Silver, Nathaniel, and Diana Seave Greenwald. *Isabella Stewart Gardner: A Life.* Boston, 2022.

Sladen, Douglas. *Club Hotel Guide to Japan.* Yokohama, 1891.

SMS Hawai'i. *Economic Impact of Indigenous Tourism Businesses Report.* Prepared for the American Indian Alaska Native Tourism Association (AIANTA), October 2021. https://www.aianta.org/wp-content/uploads/2021/10/20211018-Native-Owned-Firms-in-the-Tourism-Sector-National-FINAL-v2.pdf.

Stelle, Charles C. "American Trade in Opium to China, 1821–1839." *Pacific Historical Review* 10, no. 1 (March 1941): 57–74. https://doi.org/10.2307/3633184.

Tharp, Louise Hall. *Mrs. Jack: A Biography of Isabella Stewart Gardner.* Boston, 1965.

Topsfield, Andrew. "The Master Artists of the Impey Album." In *Forgotten Masters: Indian Painting for the East India Company,* edited by William Dalrymple, 40–77. London, 2020. Published in conjunction with the exhibition *Forgotten Masters: Indian Painting for the East India Company* shown at The Wallace Collection.

Tseng, Alice Y. "Imperial Portraiture and Popular Print Media in Early Twentieth-Century Japan." *The Journal of Japanese Studies* 46, no. 2 (2020): 305–44. https://doi.org/10.1353/jjs.2020.0043.

Tweed, Thomas. "Smells and Bells: Buddhism, Catholicism, and the Therapeutic Aestheticism of William Sturgis Bigelow and Isabella Stewart Gardner." In *Inventing Asia: American Perspectives around 1900,* edited by Alan Chong and Noriko Murai, 51–61. Boston, 2014.

Volk, Mary Crawford, Nicolai Cikovsky, Warren Adelson, and Elizabeth Oustinoff, eds. *John Singer Sargent's El Jaleo.* Washington, DC, 1992. Published in conjunction with the exhibition *John Singer Sargent's El Jaleo* shown at the National Gallery of Art, Washington, DC, and the Isabella Stewart Gardner Museum, Boston.

Wolfe, Patrick. "Settler Colonialism and the Elimination of the Native." *Journal of Genocide Research* 8, no. 4 (December 2006): 387–409. https://doi.org/10.1080/14623520601056240.

Photographic Credits

Unless otherwise noted, works are in the collection of the Isabella Stewart Gardner Museum.

Amanda Guerra: Haddon fig. 7, Odo figs. 1, 2

(c) The Ansel Adams Publishing Rights Trust: Roessel fig. 3

Canadian Centre for Architecture: Gandhi figs. 1, 3

David Joyall: Checklist plate 15

David Mathews: Introduction figs. 1,2, 14, Haddon fig. 2, Gandhi fig. 8

The Digital Ark, Corp.: Haddon fig. 5

The Getty Research Institute: Tung fig. 5

Historic Collection / Alamy Stock Photo: Introduction fig. 9

Isabella Stewart Gardner Museum, Boston: Gandhi fig. 11, Odo fig. 3

Jean S. and Frederic A. Sharf Collection, The Wolfsonian-Florida International University: Gandhi figs. 6, 7

Jean–Pol GRANDMONT: Introduction fig. 8

John Adams Whipple (American, 1822-1891): Introduction fig. 3

Library of Congress, https://www.loc.gov/item/2003652738/: Roessel fig. 2

Metropolitan Museum of Art, New York: Haddon fig. 4

Minneapolis Institute of Art: Gandhi fig. 10

The Miriam and Ira D. Wallach Division of Art, Prints and Photographs: Photography Collection, The New York Public Library: Gandhi fig. 5

Photograph © 2023 Museum of Fine Arts, Boston: Odo fig. 4

National Parks Service, Longfellow House, Cambridge, Massachusetts: Odo fig. 8

NEDCC: p. 6, 10, 54, 95, 134, 160, 192, Introduction figs. 4, 7, 11, plates 1-8, Haddon fig. 6, plates 9-19, Gandhi figs. 4, 12, plates 20-30, Tung figs. 1, 3, 4, plates 31-37, Odo figs. 5, 6, 7, 9, 10, plates 38-42, Roessel figs. 1, 4, plates 43-45, Checklist plates 1-14, 16-28

Courtesy of the Peabody Essex Museum: Tung fig. 2

Photographic Office, Calcutta: Gandhi fig. 2

Qatar National Library, World Digital Library: Haddon fig. 3

Royal Collection Trust / © Her Majesty Queen Elizabeth II 2022: Gandhi fig. 9

Sean Dungan: Cover, Introduction figs. 5, 6, 10, 12, 13, Haddon fig. 1

Selected Index

ISABELLA
STEWART GARDNER
MUSEUM

Copyright © 2023 by the Isabella Stewart Gardner Museum, Boston. All rights reserved.

This book may not be reproduced, in whole or in part, including illustrations in any form (beyond that copying permitted by Sections 107 and 108 of the U. S. Copyright Law and except by reviewers for the public press), without written permission from the publishers.

Requests for permission to reproduce material from this work should be sent to permissions@press.princeton.edu

Published by the Isabella Stewart Gardner Museum
25 Evans Way, Boston, MA 02115

Distributed by Princeton University Press
41 William Street, Princeton, NJ 08540
99 Banbury Road, Oxford OX2 6JX
press.princeton.edu

ISBN 978-0-691-97386-9

Library of Congress Control Number: 2022945376
British Library Cataloging-in-Publication Data is available

Designed by world
Ria Roberts, partner
www.publications.world

Typeset in Founders Grotesk and Century Old Style
Copy edited by Kim Hastings
Printed in Bruges by die Keure, Belgium

Isabella Stewart Gardner Museum Publications
Elizabeth Reluga

Cover: Courtyard, Isabella Stewart Gardner Museum, Boston. Photo: Sean Dungan

Endpapers from Isabella Stewart Gardner's *Travel Album: Italy,* 1884.
Isabella Stewart Gardner Museum, Boston (v.1.a.4.12)

The Isabella Stewart Gardner Museum resides on the homelands of the Massachusett Tribe, who belong to this place and continue to regard these lands and waters as sacred.

10 9 8 7 6 5 4 3 2